östberg™

VALUE REDESIGNED

9/05
Best Wishes,
Kyle V. D[illegible]

VALUE REDESIGNED

New Models for Professional Practice

Kyle V. Davy, AIA, and Susan L. Harris, PhD

Greenway Communications

östberg™

Östberg Library of Design Management

ISBN: 0-9755654-1-9

Cover design: Austin Cramer
Layout: Karen Berube, K.Designs

Published by:
Greenway Communications, LLC
A division of The Greenway Group
30 Technology Parkway South, Suite 200
Atlanta, GA 30092
800.726.8603
www.greenway.us

Contents

Illustrations

Figures

Tables

Acknowledgments

As we finish the last edits of this book, we've come to the end of a long, fulfilling journey. The journey began almost five years ago when, supported and encouraged by friends and colleagues at the Advanced Management Institute for Architecture and Engineering (AMI), we decided to launch our "Discovery" research effort to explore new models of practice for architecture and engineering firms.

The journey actually began almost a decade before that when we joined founder Lou Marines to help him establish AMI, an institution that teaches architects and engineers the leadership and management skills they need to build thriving, successful practices and to lead satisfying professional and personal lives. We would like to express our gratitude to the faculty and staff who have made our years of teaching and consulting with AMI so rich and rewarding. In particular, we would like to thank Lou Marines and David Aitken for their support, counsel, and input into our thinking over the years. We would also like to acknowledge the AMI students and clients who have played powerful roles in our learning and development. In addition, we would like to highlight the vital role in our learning played by participants in the American Council of Engineering Companies' Senior Executive Institute (SEI). Each year since this program was launched in 1995, a class of 25 leaders from architecture and engineering firms has joined us in an extended learning experience. We have been continually blessed by the support and encouragement of the participants in these classes as well as the active learning partnership that has flourished among all of us.

We were joined on the first leg of our Discovery journey by a remarkable group of people. Leaders from a dozen organizations made a special commitment to an open-ended process of research and learning, one in which even the guides were unsure of the ultimate destination. The contributions of these visionary and adventurous members of our industry to our effort were extraordinary, and we thank them for their energy, enthusiasm, imagination, and creativity.

At what we imagined would be the final stop for Discovery, we paused to pull together the ideas and insights we had gained along the way. To our surprise our trip was extended. A significant number of the participants were not ready to stop; both the learning and the relationships that were growing out of the process were too valuable to give up. So, it was agreed that this emerging community would continue the journey together, stopping periodically to share our latest insights and best practices and to engage in dialogue and discussion as we explored new territory together. Along the way a few new people joined the march and enriched our efforts.

At the urging of participants, we agreed to write down what we had learned on the journey. We had no idea how much hard work would be involved or how long it would take. Three years later we have completed this second leg of the journey. Countless other authors in the opening pages of their books have spoken eloquently about how difficult it is to write a book. Our experience was no different, and we echo their sentiments. One of the greatest difficulties was the constant pull into rich detours and uncharted byways as we tried to sum up our ideas. With each step it seemed that the path grew longer, there were always more avenues to explore, and we often felt the need to return to paths already traveled in order to examine them more deeply. As we took time to reflect on what we were learning, we realized that there is still much rich territory to be examined. Nevertheless, we have chosen this time to call a halt. The paths not yet traveled offer rewarding prospects for the future.

We were also fortunate to find extraordinarily helpful books to guide us along our way. Books by Arie de Geus, Ronald A. Heifetz, Clayton M. Christensen, James F. Moore, and Peter Senge were central to the evolution of the ideas that form the heart of this book. The wisdom of Dee Hock, Ray C. Anderson, and Janine M. Benyus also inspired and informed our work.

We want to express particularly heartfelt appreciation for the respectful, thorough, and artful editing of the entire manuscript by Janet Rumnarger. We are also very grateful for the support we received from Jim Cramer and Jennifer Evans Yankopolus at Greenway Communications.

Finally, we also relied heavily on and are grateful for the support provided by family, friends, and colleagues outside the AMI circle. Kyle's wife, Karen Yencich, played a special role, serving both as a sounding board for him throughout the process and helping him with detailed editing for his sections of the book. She also patiently endured with grace and good humor the times when the effort took him away from home, and his inattentiveness as he wrapped himself up in the writing process.

Susan would like to express special appreciation to the members of her longstanding writing/discussion group—David Bradford, Alan Briskin, and Peggy Umanzio—who endured the messiness of early drafts and offered much-needed encouragement at difficult moments of the journey. Bruce Beverly, Larry Smith, Donna Garske, and Jenifer Hedrick offered crucial moral support. Her brother, Frank Harris, brought the perspective of his long career in music composition and production to discussions about the rigors of the creative process. Frank and Susan's mother, Lillian Harris, patiently accepted prolonged periods of unavailability for family events. Thanks to Don Benson, Helen Spector, Cate Gable, Michael Stone, Jo Mortensen, and Valena Williams for maintaining steady, caring communications. And thanks to all the friends who have borne with Susan as she felt compelled to opt out of the delights of time together. It is now time to reconnect!

Discovery Participants

Pamela Brule, AIA*
Anderson Brule Architects, Inc.

Sam McBane*
Anderson Brule Architects, Inc.

C. Ronald Capps
The Beck Group

Betsy del Monte, AIA*
The Beck Group

Laurence S. Courtney
Independent

Donald Simpson, AIA
KPS Group, Inc.

Gray Plosser, FAIA
KPS Group, Inc.

John Haddad
Lucchesi, Galati Architects, Inc.

Raymond Lucchesi, AIA*
Lucchesi, Galati Architects Inc.

Denise Cook, AIA*
Lucchesi, Galati Architects Inc.

Christopher P. Ratcliff, AIA*
Ratcliff

Carolyn Silk, AIA*
Ratcliff

Phil Hodgin, AIA*
RDG Planning & Design

John R. Birge, FAIA
RDG Planning &Design

Nora R. Klebow, AIA*
RMW architecture & interiors

Bart J. McClelland, AIA*
RMW architecture & interiors

Frederick G. Aufiero, PE
SEA Consultants Inc.

Mira A. Furth*
SEA Consultants, Inc.

Mark D. Heaney
SEA Consultants Inc.

Arthur Spruch, PE*
SEA Consultants, Inc.

Tom Olesak
Setter Leach & Lindstrom, Inc.

George Theodore
Setter Leach & Lindstrom, Inc.

Robert W. Swanagon, AICP
Washington Infrastructure Services

Larry R. Thomas, PE
Washington Infrastructure Services

* *Individuals who have been part of the extended Discovery process.*

Preface

This book represents a profound meeting of two very different minds on a subject that we are both passionate about. We did not plan to write it. The book grew organically in response to a request from client participants in a collaborative research project that we co-facilitated. That effort was one of a long series of consulting and teaching efforts undertaken together over the last 10 years. Developing our synergy over that period was a gradual process—because we are so very different.

Kyle is a licensed architect with a Stanford MBA. After practicing as an architect for 16 years, he began teaching and consulting to design firms, focusing on the way these professionals do their work and manage their firms, operations, and projects. Susan is an organization development consultant with a PhD in English literature from the University of California, Berkeley. Following many years of consulting to public schools and a variety of industries, she began offering leadership development and strategic planning services to the architecture and engineering professions.

In the tradition of professionals and consultants, we gave each other plenty of room to do our own thing. As we did this more and more in shared contexts with shared purposes, we became increasingly interested in what the other was reading and thinking. We were also increasingly intrigued by the insights and discoveries that sprang from our shared efforts. Kyle's interests expanded into strategy and leadership, and Susan tackled the operational and financial aspects of the professional service business model.

As this convergence increased, so did the quality of what we were able to deliver together. We were addressing a wider range of shared concerns from an emerging set of shared insights, while preserving our very different styles and approaches. Professionals participating in efforts we co-facilitated were learning and growing more by engaging with both of us than they would have by working with either of us alone.

As we have reflected on this result, it makes perfect sense. Even in the relatively homogeneous groups of professionals we work with, a significant range of learning styles and worldviews are present. Individually, we always struggle to find the best way to pitch what we have to offer to a particular audience. But trying to reach everybody with a single approach becomes less effective as diversity increases in a group. Why wouldn't this be equally true for a book?

This book is an experiment in presenting a single, cohesive argument through two very different voices. With this experiment, we are living by two important convictions: first, the need for relentless pursuit of innovation, and second, the need for diversity in disciplines and ways of thinking as the best means of achieving innovation. We believe that like our teaching, our writing becomes more powerful if we don't homogenize our differences, but leverage them for more generative outcomes. In leaving you to make sense across these differences, we believe you will be more deeply and creatively engaged in shaping your own version of the insights we present.

As you read, you will find sections that present ideas illustrated with stories and examples of what is possible—what people are already doing to advance business model innovation in architecture and engineering firms. You will also find other sections that present ideas more theoretically. These sections may challenge you to think in new ways and to deal with more complexity than you may be accustomed. While we recognize that many readers will be more comfortable with one style than the other, we hope that in each case, that comfort will be a bridge to taking on and benefiting from the other as well.

We believe that innovation thrives on learning from other fields. As a consequence you may find much that is new and unfamiliar in this book. For example, we will challenge you with new knowledge from the fields underlying our understanding of social systems because they offer important insights about exercising effective leadership. Don't expect that you will be doing all freeway driving on this trip. Prepare yourself to take some back roads too, and to shift speeds and gears as needed. We hope that you will not turn away from challenging sections, but will slow down to take in and comprehend different terrain.

We both have deep love for and profound faith in the architects, engineers, scientists, and other built environment professionals with whom we work. In the hope that our efforts will help spark and support transformation, we have endeavored to offer the very best thinking that is available today related to business model innovation and evolving the core of professional practice to achieve that innovation. We hope you will be provoked, challenged, and inspired to join in this creative endeavor—for the sake of a vibrant future for professional firms involved in the built environment. And for the sake of a better world.

Introduction

It should have been the best of times. At the dawn of a new millennium, architecture and engineering firms were riding the crest of a long economic expansion. Demand for professional services in support of building new facilities and infrastructure was strong and firms were busy.

But as we listened to leaders of firms across the country, it sounded more like the worst of times. Clients were choosing firms primarily on the basis of price, not expertise or experience, treating architecture and engineering services—including, for the purposes of this book, landscape architecture, interior design, planning, survey, geoscience, and other built environment professions—like they were commodities. Competitive pressures were rising. Traditional relationships between client and professional were being co-opted by construction managers, management consultants, and design-build companies. Fees were not keeping up with the escalating costs of practice. An onslaught of litigation was fraying bonds of trust between firms and clients. Architecture and engineering schools no longer attracted the best and brightest into their programs; many of their graduates were choosing employment in other industries with more exciting work and brighter prospects. Would they advise a son or daughter to follow them into their chosen profession? No, these leaders shook their heads; they hoped for better for their children.

It was clear that there were no simple answers to these problems. Conventional management prescriptions weren't helping; in fact, they often made matters worse. This situation demanded systemic analysis and change, not treatment of isolated symptoms. In the spring of 1999, we decided to lead a research effort, sponsored by the Advanced

Management Institute for Architecture and Engineering (AMI), to better understand how this had happened and to identify alternative paths that firms could use to move toward a more fulfilling future.

Our intuition told us that the place to begin was with the prevailing business model used by most architecture and engineering firms. A business model is a holistic expression of how an enterprise works—how it delivers value to its customers and what it receives in return; it describes an organization's recipe for success. A successful business model combines a powerful narrative with numbers that add up. The history of business is filled with examples of companies that prospered with sound business models (American Express, Nucor, Kroger, and Amazon). Other examples describe organizations that fell on hard times as their business models lagged behind shifting markets (Sears, Bethlehem Steel, A&P), or models that were conceived with numbers that ultimately didn't add up (many of the dot.com failures of recent years). There are also examples of enterprises that innovated their business models as a key step toward organizational rebirth and renewal, transforming old into new (Kroger, Walgreens, and Abbott Laboratories). With this is mind we wondered, is the business model used by most architecture and engineering firms healthy or failing?

It seemed to us that the technical work that has been the bread and butter of architecture and engineering firms for the past 50 years was losing its value in the eyes of clients. Although 25 years earlier, it had represented a recipe for success, the business model used by architecture and engineering firms was now on its last legs; its value proposition was less compelling and the numbers no longer added up. Firms were being pressured to reduce fees, while their costs of doing business continued to escalate. It appeared that many of the difficulties being experienced by firms could, at least in part, be traced back to a failing business model. Given this perspective, our inquiry began with a focus on the following questions:

- What are the underlying dynamics and mental models of the existing business model? What historical forces shaped it, and how does it influence the thinking and behavior of both members of firms and their clients?

- Are alternative models of practice available that offer a brighter future? What might they look like?
- Are there examples of firms moving toward new models of practice? What could we learn from their efforts?
- How should firms proceed with business model innovation efforts? What challenges will they face as they move forward?
- What values shifts and new social and leadership capacities will be required, both to create new forms of value for clients and to support transformational efforts within firms?
- Could business model innovation help architects and engineers establish a new leadership position in society?

Our thinking about how to approach this research was significantly influenced by Ronald Heifetz's groundbreaking body of work on leadership and adaptive work. Heifetz distinguishes between two types of "work" that organizations or communities do to solve problems or resolve difficulties. The first type, *technical work*, is appropriate for situations in which the problem is well understood and amenable to authoritative expertise or standard operating procedures. The second type, *adaptive work*, is required in situations where the solution is not known (or in some cases, not knowable), and significant learning is required to even understand the nature of the problem. Adaptive work involves changing values, beliefs, and behaviors and requires learning, new discoveries, and experiments across the organization.

Business model innovation clearly involves adaptive work. Heifetz advises that leaders guiding adaptive work must "give the work back to the people" rather than attempt authoritative solutions. With that in mind, we decided to involve architects and engineers who had a stake in the situation with us. Rather than observing the situation from outside, we needed to engage with these stakeholders in a collaborative learning process directed at achieving a deeper understanding of the problem. Fortunately, leaders of a dozen firms from across the country (architects, engineers, and a design-build organization) were eager to join us in a search for new models of practice. We named the effort AMI's Discovery process.

Beginning with an initial one-year research effort and then extending into a commitment to ongoing dialogue and learning, these leaders generously offered their thoughts and perspectives. They shared experiments and best practices, and acted as sounding boards and coaches as we generated and refined our ideas. Most of them were simultaneously engaged in their own adaptive work, leading business model innovations within their own firms. It was at their urging that we decided to write this book, documenting the ideas that grew out of the Discovery process and codifying our perspective and advice into a guide for architecture and engineering firms. We remain deeply grateful for their energy and enthusiasm, assistance and wise counsel, and for the push to get our ideas down on paper.

The major theme of our book is business model innovation, creating a new, highly diverse population of architecture, engineering, and design organizations thriving in an increasingly challenging world.

In part I, we describe how new models of practice, built around the inherently creative capacities of living systems, will replace the static economic model that dominates firms today. These new "living firms" will continuously learn and adapt to their rapidly changing environments. We also outline the ways these new models of practice can respond to the transformation imperative that comes when technologies mature. Both the existing business model and many of the practices and technologies of architecture and engineering firms are relatively mature, and ripe for overthrow by new, disruptive technologies wielded by outside competitors. Architecture and engineering firms need to reinvent themselves and the way they work before they inevitably fall victim to this threat. These new practices and technologies will be shaped into bold new value propositions—innovative value creation accompanied by new pricing strategies.

For firms to successfully transform their business models, a parallel wave of innovation needs to take place within the architecture and engineering professions. These professions remain largely unchanged from the way they were defined and shaped during the first 75 years of the last century. In the past quarter century, rapid social and technological change has widened the gap between societal needs and desires and professional values, ethics, practices, and capacities. Professionals have remained in place while clients and society have moved forward. In part II, we examine these changes more closely and describe shifts in professional mind-set that can help architects and engineers not only catch up and close this gap but also get far enough ahead to be able to exercise significant leadership in society. These new mind-sets will serve as starting points for building the new social and leadership capacities firms will need in order to move toward an innovative business model.

At the center of a business model is the following value proposition: how a company creates value for its customers and what it is paid in return. Part III focuses on the first half of this equation—how architecture and engineering firms can create new forms of value for their clients and society and how they can use those strategies to form robust new models of practice. It details emerging trends and provides new ideas that firms can choose from to develop innovative new offerings.

To implement these value-creating strategies, firms need to develop new social and leadership capacities. As they ascend from doing technical work to collaborative work to transformative work for clients, opportunities for value creation increase. But so do demands for social and leadership skills. Part IV lays out a comprehensive framework identifying the core capacities needed to do each of these three types of work. It also describes two overarching capacities that firms must develop to be successful, regardless of the level at which they choose to work. The first is the capacity to lead innovation. The second is developmental awareness, an understanding of how human systems learn and grow.

In addition to creating value for clients, these leadership capacities are essential to internal change efforts directed at business model innovation. Part IV also includes a discussion of how firms can use the advanced leadership capacities for transformative work along with developmental awareness to successfully address the adaptive challenges involved in transforming themselves.

Finally, in part V we examine the second half of the new value proposition: how do firms earn a fair return on the value that they create? We offer a menu of innovative pricing strategies and share examples from architecture and engineering firms that are already using these strategies. Although the menu is not definitive, it provides leaders a place to start rethinking this part of their business models.

The next decades can be the best of times for architects and engineers. Architects and engineers can be the preeminent value creators in this new century, offering creative leadership and technical innovation to help communities, the nation, and the world confront and conquer critical problems. However, that future is only possible if leaders of architecture and engineering firms are willing to actively commit themselves to the difficult adaptive work that comes with business model innovation. Venturesome innovators are already attempting transformation, launching bold experiments and initiatives. Early adopters are now learning from the innovators' efforts and are beginning to take their first steps into adaptive work. Our hope is that this book will fuel these efforts, generate new thinking and ideas, and encourage others to join these innovators and early adopters in reinventing firms, transforming our industry, and designing a better future for society.

PART I

THE CASE FOR CHANGE

I
Sailing into White Water

If somebody wants to learn from history,

there may be some lessons here.

—William W. Moore, cofounder of Dames & Moore[1]

In the summer of 1938, two young civil engineers, Trent Dames and Bill Moore, set up a consulting practice in Los Angeles to do foundation engineering and soil mechanics work. They scrambled through the war years, doing work across the country to support the war effort. By the early 1950s, the firm had multiple offices, its staff had grown to almost 100 people, and it had established a reputation as one of the leading consulting engineering practices in the country. By the late 1970s, the firm was several thousand strong, and was operating around the world. In 1992 Dames & Moore made financial news with a $115 million initial public offering and became a publicly traded company. In 1998 management projected the company would achieve a milestone $1 billion of revenue in the next year. Unfortunately, Dames & Moore never made it through that year. In the spring of 1999, Dames & Moore ceased to exist.

In the early years, Dames & Moore pioneered the emerging discipline of soil and foundation engineering, establishing its role as "interpreters of soil mechanics findings for design engineers." The firm convinced skeptical clients that soil mechanics provided information vital for designing and building new facilities and infrastructure. They invented new processes for taking and analyzing data from soil samples and formed new working relationships with structural engineers and other designers who relied on their results and recommendations.

The decades of abundance for the U.S. economy following World War II provided Dames and Moore an ideal environment for building their consulting engineering firm. Led by Trent Dames, the firm developed structures and systems to facilitate the growth of its professional practice, addressing technical performance, quality control, staff development, and client relationship and business development issues. They invested in ongoing research and development efforts, inventing analytical techniques, sampling equipment, and work processes to do engineering work more effectively. At one time, the firm was having difficulties obtaining useful soil samples from beds of sand and gravel. So, according to Moore, "over a weekend we invented a new kind of sampler, and soldered it together on the floor of my kitchen."

After almost 40 years of steady growth, the good times came to an end. During the 1960s, Dames & Moore had gotten into the nuclear power plant business and by the mid-70s nuclear work constituted roughly 40 percent of their revenue. Then, in the late 1970s, the nuclear business came to a crashing halt. The near catastrophic meltdown of the Three Mile Island nuclear reactor combined with a dramatic shift in public opinion to shut down a once burgeoning nuclear power construction market. Overnight, construction of new nuclear power plants stopped, and Dames & Moore was left scrambling to find sources of work that could replace this lost revenue.

This revenue loss was followed almost immediately by an equally severe cash-flow crisis. Over the years, the firm had developed a significant presence doing geotechnical engineering in support of energy projects in the Middle East, particularly in Iran. With the overthrow of the Shah in 1979, all of the firm's projects in the country ceased and left it holding unpaid invoices for over $5 million of completed work. It was only years later that the firm was able to collect a fraction of these funds.

Dames & Moore's ability to respond to this unfolding crisis was further complicated by a change in the firm's leadership. In 1975 the two founders had decided to retire from active management of the firm, replacing Trent Dames with a new CEO who was, says Moore, "one of our senior partners who had been involved in the operations of the company for quite some years. He was smart, and a good engineer, although he did not have any special training for fiscal retrenchment." Nor had his experience prepared him to lead the firm through the severe storm that it was now facing.

The firm struggled under the new CEO's leadership for several years, sinking ever deeper into financial difficulty. Finally, to prevent the firm from going broke, Moore reassumed his leadership position on the firm's board and replaced that CEO with George Leal, another long-time Dames & Moore employee. Leal appeared to possess the skills and savvy needed to cope with this crisis. Although an engineer, Leal had expressed a deep interest in business throughout his career and had earned an MBA from the University of Chicago. Moore noted, "we asked Leal if he could do something to stop the hemorrhaging. He said, 'I don't know, but I would like to try.' Leal realized that we had to stop spending money at the rate we had been, so he instituted a lot of financial constraints." These actions signaled a new business orientation and cost-consciousness for the firm. Management began to concentrate on utilization (billability), weekly or monthly profit-and-loss accounting, and on the latest bottom-line figures.

Leal's prescription worked. By the mid-1980s, the firm seemed to have gotten back on track. It had also established itself as a dominant player in the newly emerging hazardous waste market. Revenues were up and cash-flow problems had eased. Firm management retained a tight grip on the organization with its new business ethic.

Buoyed by their renewed success, management embarked on new strategies to grow the firm's revenues and profitability, mainly through an aggressive program of expansion through acquisition. By the early 1990s, this strategy had proven unsustainable for the partnership, the organizational form it had retained since its founding. In a bold move to acquire the capital it needed to continue growing, the firm made a $115 million initial public offering (IPO) of its shares to the financial markets.

From then on, the demands of public shareholders for an attractive return on their investment had to be satisfied. Quarterly results, projections for future earnings, and stock price movements became the benchmarks for management. By the mid-1990s, however, the business environment had shifted once again. The hazardous waste market had flattened. Project fees and profits were squeezed as services became more commodity-like in the minds of clients and competition heated up. Debt soared as the firm continued to pursue revenue growth through costly acquisitions. Although revenues did increase, earnings were insufficient to satisfy the demands of analysts and investors. The company's share price fell from the $20 at the time of the IPO to $12 in 1998.[2]

The board and shareholders wanted management to get the stock price up. The firm's management believed they could accomplish this by recapitalizing the firm through a leveraged buyout with a financial investor. With the right deal, they could return a premium to current shareholders and access additional funds that could be used for future expansion plans. However, at the last moment negotiations with the potential investor failed. Within days, another consolidator in the environmental market, URS (URS/Greiner/Woodward Clyde), stepped in with a bear hug offer and swept Dames & Moore off the table.[3]

Approaching the Rapids

Why did this once successful firm run out of gas? Was the experience of Dames & Moore an isolated example, or is it representative of a broader trend among professional architecture and engineering firms?

Dames & Moore's growth and development during the postwar boom economy of the 1950s and 1960s was typical of many architecture and engineering firms. The shift away from a wartime production footing, coupled with pent-up demand from the Depression years, set off an explosion of spending on capital construction projects. Young architects, engineers, scientists, and designers of all types set up shops to practice their professions and satisfy the needs of the expanding economy. As Moore noted, "After the war, a great rebuilding was going on in this country and overseas. There were industrial facilities, transportation facilities, highway programs, school building. There was a lot of engineering and construction work, and it seemed like the market was unlimited."

Clients and society needed the technical solutions to building, transportation, and infrastructure problems that architecture and engineering firms offered. Demand was strong enough and fees high enough that practitioners could concentrate on doing the work, developing innovative new processes, practices, and designs, and contributing to the growth and evolution of their professional disciplines. Many aspired to solve pressing societal problems and create a brave new world through the application of their professional expertise to their client's projects. Nathaniel A. Owings, co-founder of the architecture firm Skidmore, Owings, and Merrill, noted, "We were not after jobs as such. We were after leverage to influence social and environmental conditions."[4]

Design firms worked independently, with little need for collaboration with clients, contractors, or other stakeholders in the design and construction industry. Project delivery processes were standardized to handle the volume of work and codified in standard agreement forms produced by professional societies. The standard fee curves built into these contracts, while not generous, were set high enough that they allowed professionals to not only do the work, but also spend time on experimentation, learning, coaching and mentoring, exploration, and innovation. Office structures, roles and responsibilities, and work processes were set in place, designed around the delivery of the work. In a time of abundance, business needs were secondary to the primary technical and professional work of the firms.

In the 1970s, conditions changed dramatically. The era of postwar abundance had passed. A series of shocks, including recessions, hyperinflation, antitrust rulings, societal turmoil, global conflicts, and an explosion of litigation, forced firms out of a comfortable status quo and eliminated the surplus that had allowed them to operate from privileged professional positions.

Hyperinflation and rising global competition drove clients to demand reduced fees and faster schedules. Projects became more complex and uncertain as new stakeholders (community groups, environmentalists, and government regulators) forced their way into seats at decision-making tables. People began to question the status and leadership position society had granted to professionals. Clients and other project stakeholders attempted to hold professionals accountable through the use of lawsuits. Competition among professionals heated up. Federal antitrust rulings eliminated protections against competition that professional societies had established. New competitors emerged in the form of construction managers, global design and construction firms, and design-build contractors. For professional architecture and engineering firms, fees were falling, costs were rising, risks were skyrocketing, and pressures were intensifying. Dames & Moore's financial crisis and brush with bankruptcy were all-too-common experiences for firms, both large and small, throughout the country.

Clearing the Decks

Facing these challenges, leaders of firms woke up to the necessity of running their practice like a business. Revenue growth, profitability, and cash flow could not be taken for granted. Financial and accounting acumen was needed to chart a path out of current predicaments. Organizational structures and controls needed to be put in place to ensure compliance with management direction and accountability for financial performance. Business development and marketing systems had to be developed to help firms find attractive market niches and win new work.

Individual engineers, architects, and designers with business skills were given increasing power and responsibility for guiding firms through these difficult times. George Leal, at Dames & Moore, was typical of a wave of management-oriented leaders who took the helms of firms throughout the industry. "Becoming a business" and "acting like businessmen" became rallying cries for leaders determined to mitigate the stress their firms were experiencing.

An expanding group of management consultants provided guidance for these business newcomers. These consultants preached a gospel of management practices, providing advice in the form of books, newsletters, training, and hands-on coaching to management teams who were determined to rebuild their firms in this new image. Like Dames & Moore, management in many firms came to believe that its primary responsibility was monitoring and controlling the firm's financials.

The most elegant version of this economic business model for professional practice came from outside the design and construction industry. David Maister, a Harvard Business School professor and professional practice guru, offered a comprehensive framework drawn from his work with legal, accounting, and management consulting firms. It combined a business imperative—maximizing partner (shareholder) profit—with the professional service firm's traditional commitment to serving clients and developing staff. Maister's model demonstrated how management could set and control utilization, billing rates, multipliers, staff composition, and other factors to achieve the firm's financial and practice goals.[5]

Maister's business model, and the many similar versions of it promoted by other business consultants and financial managers inside firms, offered architecture and engineering practices exactly what they were looking for. It provided a language, roadmap, and set of tools that professionals could easily grasp and use to manage their finances and achieve renewed success. It was powerful because it was simple and relatively easy to apply.

Frozen in the Headlights

In the short term, this new business discipline worked. Most firms, just like Dames & Moore, made it through this difficult period. By the mid-1980s, architecture and engineering firms appeared to have regained some level of stability and work was starting to accelerate again.

Firm leaders were lulled into a false sense of security by the success of these management strategies. This temporary sense of well being allowed them to ignore, or minimize, the threat posed by the underlying brutal facts present in their situation. Not only were many of the competitive dynamics and economic and social trends that had emerged in the 1970s still present in the market, but new, unanticipated threats continued to emerge.

Their new business practices allowed firm managers to avoid making more farsighted changes to the way their organizations practiced. Firms continued to retain their deeply entrenched views about whom they worked for, how they did their work, and who was in their firms. Project delivery methods and other deeply held ways of working remained largely unquestioned. Traditional professional values and beliefs could continue to be acknowledged and celebrated, though they often played second fiddle to this new economic imperative. Firms learned to survive in an increasingly competitive market by becoming more efficient in their use of resources. Utilization targets were raised and funds for quality management, staff development, research and development, and service to the professions were cut.

Bill Moore commented on the state of design professionals during this period of time from his vantage point at Dames & Moore: "Neville [Donovan, a longtime Dames & Moore employee] talked to me about the loss of the sense of fun in the Dames & Moore work. In a nutshell, he said that for maybe 45 years or so, working at Dames & Moore had been fun and was exciting. Things were going on, and people were doing things. But then there was a change, at about the time when some major management changes occurred. The fun seemed to go away, and instead work became mostly a matter of struggling to watch the financial details and the computer printouts. I think what Neville says is true."[6]

Moore went on to underscore the downside of this mind-set for his firm: "[In the 1980s] we were in a really very fortunate position to take a lead role in the other environmental and hazardous waste work. Unfortunately, however, some of the cost-cutting and belt-tightening attitudes and management techniques have persisted long past the original crisis. We continued pinching pennies and expecting employees to work 100 percent of the time for billable accounts, and provided no money for client and business development. And no money for professional development . . . They did this to reduce expenses, but they took out the wrong thing. Others have done the same. The result nationally has been a drastic reduction—practically a cessation—of support for professional activities."[7]

Large-scale changes continued throughout the remainder of the 1980s and 1990s, including a massive economic expansion, globalization, and a technological revolution highlighted by the rise of the Internet. Organizations reengineered, outsourced, and became virtual. Clients began demanding new project delivery processes capable of delivering higher quality projects in shorter times at lower cost. Management consultants, design-build companies, and other types of organizations began offering clients new services that were not part of a traditional architecture and engineering firm's repertoire.

Despite these changes, many architecture and engineering firms remained—and remain—frozen in the headlights. Firms continue to cling to an increasingly outdated model of practice that is poorly adapted to the needs of their clients and society. Managers make incremental improvements rather than experiment with new models and radical change. The way most firms are organized and managed has not materially changed from when business practices were introduced in the early 1980s. The primary goals of that business model, profitability and revenue growth, remain central in the minds of many firm leaders.

Charting a New Course

"Looking back now," Moore concluded, "what Dames & Moore did at the end of the nuclear boom was probably the right action at the time, but it should not have persisted . . . If we had been wise enough, we might have avoided this permanent or long-term overcorrection. I use the analogy of a ship that is running aground. First the mariners saved it from running on the rocks on the right side, but then the retrenchment-type philosophy has the ship in danger of running on the rocks on the other side. The basic point is that when you try to get out of trouble you may overcorrect. Then you may find that the overcorrection itself gets you into some other kind of trouble . . . I emphasize this because it is easy to say that Dames & Moore made a mistake when the nuclear boom was over. On the other hand, this same kind of illness has widely affected other parts of U.S. business and industry."[8]

For many architecture and engineering firms the situation is even more serious than the one Moore describes. Firms not only need to weather gale force winds and steer a course clear of obstacles, but also to completely refit their ships while still moving forward, transforming their organizations into a new form better suited to the needs of the emerging knowledge economy. One misstep in an increasingly turbulent world and firms may falter and fail.

Equally problematic, the legitimacy of the professions themselves is at risk if clients lock in attitudes that see architecture and engineering firms as businesses first and professionals second. Firms cannot afford to waste another decade. A continued failure to adapt will lock many into a future that offers, at best, marginal returns and, at worst, loss of independence and identity as professionals.

2 Becoming a Living Firm

What if we thought about a company as a living being . . .

not as a machine for making money?

—Peter Senge

During the second half of the last century, knowledge began replacing capital as the primary driving force behind economic progress and corporate success. Organizations are increasingly being shaped to optimize people and knowledge first and reward capital second.

A breakthrough study on organizational success, conducted by Royal Dutch/Shell in the early 1980s, highlighted the significance of this shift. This effort, led by Arie de Geus, former director of planning for Royal Dutch/Shell and author of the concept of the learning organization, uncovered the genetic makeup of companies that "survive for very long periods in an ever-changing world" and "attain extraordinary long-term performance."[9] The study examined such companies as DuPont, the Hudson Bay Company, W.R. Grace, Kodak, Mitsui, Sumitomo, and the Swedish company, Stora.

According to de Geus, these "living companies" differ profoundly from the "economic companies" that grew up with and dominated the Industrial Revolution. Economic companies evolved to attract the capital needed to fuel a growing industrial economy. They accomplish this by offering the owners of capital assets quick, lucrative returns on their investments. To generate these returns, economic companies strive to maximize short-term profitability. For this type of company, profitability is both a dominant goal and primary indicator of corporate health. Revenue growth and increased market share, two means of increasing short-term profitability, are assumed to be inherently good.

In contrast, a living company exists "to fulfill its own potential and become as great as it can be . . . and to perpetuate itself as an ongoing community."[10] A living company is goal oriented, has its own consciousness, and is capable of learning and adapting. Its dominant purpose is to "realize the development of its potential from its talents and its aptitudes." Profitability is seen as a "symptom of corporate health, not a predictor," and growth is balanced with other attributes important for organizational longevity.

Living companies can weather significant changes in the world around them without losing their corporate identity. Economic companies are much less resilient and die off at an alarming rate. De Geus notes that "the average life expectancy of a multinational corporation is between 40 and 50 years." In addition, "a full one-third of the companies listed in the 1970 Fortune 500 had vanished by 1983—acquired, merged, or broken to pieces." The life expectancy of smaller companies is probably even shorter. De Geus's figures showed that in some countries only 40 percent of all newly created companies survive more than a decade. Despite (or perhaps as a result of) maximizing short-term returns on investment, relatively few economic companies live and flourish more than a half century. In contrast, living companies survive centuries, not just decades, and consistently exhibit an ability to transform themselves and adapt to changing conditions. More than just survive, these living companies also thrive, achieving long-term financial success by cultivating their inherent potential as living systems.

The Royal Dutch/Shell findings are corroborated by research that James Collins and Jerry Porras presented in their book *Built to Last.*[11] Collins and Porras identified a set of "visionary companies" that combined premier status in their respective industries with extraordinary performance and a long life, including 3M, American Express, Boeing, Citicorp, General Electric, Merck, Nordstrom, Sony, Wal-Mart, and Walt Disney.

These were compared to a matching set of unvisionary companies, such as Norton, Wells Fargo, McDonnell Douglas, Chase Manhattan, Westinghouse, Pfizer, Melville, Kenwood, Ames, and Columbia.

Just like de Geus's living companies, Collins and Porras found that for visionary companies, "the company itself is the ultimate creation." They quote Richard Schickel, author of *The Disney Version*, talking about Walt Disney: "Above all, there was the ability to build and build and build—never stopping, never looking back, never finishing-the institution . . . In the last analysis, Walt Disney's greatest creation was Walt Disney (the company)."

For visionary companies, "profitability is seen as a necessary condition for existence and a means to more important ends, but it is not the end in itself." Nevertheless, in the Collins and Porras research, the portfolio of visionary companies outperformed the comparison set of companies by a factor of six over a 65-year period. A dollar invested in a visionary company portfolio on January 1, 1926, would have been worth $6,356 by December 31, 1990. A dollar invested in the comparison portfolio would have grown to only $955.[12]

The shift from capital-based economic companies to knowledge-based living companies is accelerating. This trend can be seen in the rise of asset-poor, brain-rich companies such as management consultants, software developers, and advertising and media relations firms. Other companies such as IBM are transforming themselves into organizations that emphasize consulting services and intellectual property. Companies are being redesigned to support the growth of communities of knowledge workers, to produce new forms of value for customers and society by using their knowledge and wisdom, and to earn a fair return on the value they create.

Joining the Economic-Company Camp a Day Late

For a hundred years, from the 1870s to the 1970s, architects and engineers were able to focus on the practice of their chosen professions rather than spending much time thinking about the business aspects of their firms. Leaders of firms were able to simultaneously build their firm's practice and contribute to the evolution of their professional domains by designing buildings, structures, and infrastructure that satisfied the needs of a growing economy and maturing society. However, by the 1980s, the widespread adoption of business strategies, to cope with the turbulence they were experiencing, brought many architecture and engineering firms solidly into the economic-company camp. For many, making a living became more important than practicing their profession as a way of life.

These architecture and engineering firms wrapped themselves in business practices designed to restore and maintain profitability. Top management focused on selling hours, increasing billability (utilization), decreasing expenses, avoiding risks, and collecting receivables. Firms adopted strategies built around market segmentation, finding and dominating ever more precise niches in order to appropriate as much value as possible from clients and other niche stakeholders. Managers built organizational structures to direct increasingly complex operations, and hard-wired control systems around new financial accounting tools.

These measures allowed firms to survive the turbulence they were facing at that time. But they also left firms less able to recognize or respond to the unfolding demands of the growing knowledge-based economy. Many firms yielded to a tyranny of utilization, where time for learning, innovation, and creativity was sacrificed to satisfy ever-escalating demands for higher billability. New, more nimble competitors positioned themselves to solve client problems and provide services that architecture and engineering professionals had long considered their exclusive territory. Fear crept in and self-esteem and confidence flagged as firms faced rising dissatisfaction, and ultimately litigation, from clients and other stakeholders in the design and construction process.

Ironically, architecture and engineering firms began their movement toward the economic company camp just as the knowledge economy was starting to boom and increasing numbers of client organizations began shifting toward living company principles. This only ended up exacerbating a growing gap between client expectations and firm performance.

It also cut these firms off from much of the creative energy being generated by the knowledge economy, and left them exposed to significant new risks as the pace of change in society quickened.

Becoming a Living Firm

Although the pace of change in the knowledge economy is accelerating, there is still plenty of opportunity for architecture and engineering firms that are willing to build, or rebuild, their professional practices on the four cornerstones of the living company that de Geus identified and become a living firm:[13]

1. *Persona*: A living company is a cohesive community with a strong sense of identity. This persona is manifest in a deeply held vision, shared values, and common culture.
2. *Learning*: A living company is a learning organization. It is sensitive to its environment, exhibiting a capacity to learn, create new ideas, and adapt to changes in its world.
3. *Tolerance*: Living companies are diverse and particularly tolerant of activities on the margin, outliers, experiments, and eccentricities within their boundaries.
4. *Conservative Financing*: Return on investment is important, but optimization of capital is balanced by optimization of people. The living company minimizes debt to provide the financial flexibility it needs to adjust to a changing environment. Profitability is important, but not as an end itself. Profitability is a means to achieving the organization's goals of growth and development.

Beyond the opportunity to align the firm with the evolutionary path of the knowledge economy, becoming a living firm also offers professionals an organizational form that closely fits their particular needs. Society grants professionals certain rights and privileges in return for their commitment to do work of high social value in proscribed domains. In accepting this mission, professionals agree to use their specialized knowledge and skills to do "good work," or "work of expert quality that benefits the broader society."[14] The four cornerstones of a living company provide an effective foundation that professionals can build on to satisfy this basic covenant with society.

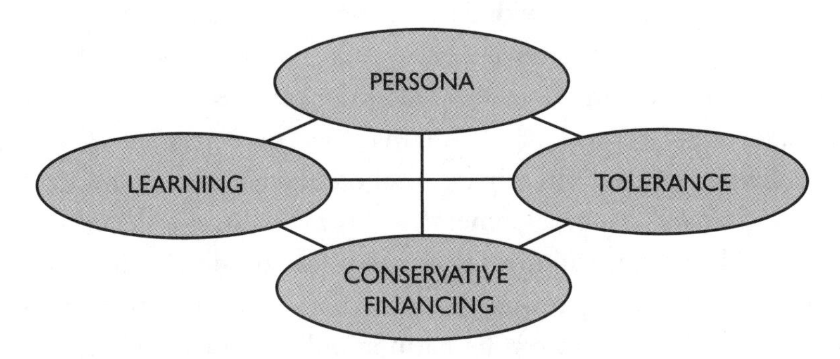

FIGURE 2.1. Cornerstones of a Living Company

Cornerstone 1: Persona

A living company is a work community that shares a common purpose, set of values, and culture. It is more than the simple aggregation of individual members. It has its own unique identity—what de Geus calls a persona. This identity is pervasive across a living company, no matter how diverse the organization. It provides the glue that binds the community together into a cohesive whole. It also establishes a boundary for the organization and provides the means for members of the community to know who belongs and who doesn't. People or groups who accept this identity, and are prepared to live according to the company's basic precepts, can become members of the community and thrive. Those that are not willing to accept it are quickly identified, confronted, and forced out.

Professionals have traditionally been attracted to the organizational metaphor of community. Whether in the form of the guilds of the Middle Ages or more contemporary studios, communities have often been created by professionals as preferred settings to practice their chosen domains. Unfortunately, the adoption of an economic company mentality drove many design and professional engineering firms away from this organizing concept. For an economic company, people are assets to be managed, not valued members of a community. There is no purpose or identity for the economic company beyond the mechanistic creation of profits for current owners. Becoming a living firm offers an opportunity to reconnect professionals with their heritage and to rebuild practices into thriving communities. It can also help restore relationships between individual practitioners and firms that have been damaged by the tyranny of utilization that has ruled many firms in the last twenty years.

A living company's persona is "goal-oriented . . . driven toward self-preservation and self-development" and is manifest in a shared vision, values, and common culture. Goals and values of individuals and subunits within the company coexist in harmony with this persona. The persona of a living firm can be harmonized with many of the deeply held, long-term commitments that individual practitioners make to their chosen professional domains. Those professional commitments are often oriented toward helping society—the larger living system surrounding the living firm—reach its full potential. These commitments are, in their essence, complementary to the living company's goals of self-preservation and self-development. These goal structures, the professional and the living firm, can support and reinforce each other. Movement toward one facilitates movement toward the other, and vice versa. These goal structures also provide a means by which diverse professionals can bind themselves together within a single living firm.

Cornerstone 2: Learning Organizations

Living companies are skilled at rapidly identifying, creating, and refining capabilities needed for future success. They are, in de Geus's words, "learning organizations." A living company is highly sensitive and quickly adapts to the changing conditions of its surrounding environments. Its feelers are constantly extended, gathering intelligence about the world, how it works, and how it is changing. Information is absorbed, processed, and acted upon by people throughout the organization, not just by top management. This ability to learn and adapt helps a living company maintain harmony with the world around it and allows it to thrive over a long period of time.

Knowledge management is a key success factor for learning organizations. The knowledge, wisdom, and experience embodied in people and shared by the community are primary assets of a living company. Acquiring, creating, interpreting, retaining, and transferring knowledge must be primary pursuits of a living company. Learning is given a high priority by both members of the community and top management.

By contrast, the economic company's focus on short-term profitability places severe restrictions on learning processes. When a learning activity cannot be justified by an immediate payback to the company's bottom line, it is likely to be deferred or cut. Investments in human assets are held to a minimum in order to produce the greatest possible return on investment in the near-term, ignoring potential longer term benefits. As professional architecture and engineering firms became economic companies, time for learning was significantly reduced or cut out entirely. The rich coaching environment present in many firms during the 1950s and 1960s disappeared as tight project budgets required staff to remain task-focused. With the exception of training for CAD and other computer software needed to do project work, training budgets were minimized. Support for professional activities dwindled. Overall, little management time was taken up thinking about either individual or organizational learning needs.

Becoming a living firm corrects this situation. This change places an emphasis on learning squarely where it must be for design professionals and consulting engineers. The ability of professionals to do good work is directly correlated to their knowledge and wisdom, particularly in a rapidly changing environment. Pervasive change in professional domains creates the need for practitioners to continually relearn their craft. Knowledge and skills must be updated and kept in step with the emerging needs of both clients and society.

A living firm supports the personal growth and development of its individual members by helping them stay abreast of changes in and around their disciplines. This support can include coaching and mentoring, formal training activities, on-the-job learning through challenging project assignments, and other learning opportunities. A living firm also invests in knowledge management systems and practices that allow members of the community to share their most current learning across groups and offices and contribute to knowledge repositories that endure beyond a single person's tenure with the organization. Relationships outside the firm are nurtured as sources for new knowledge and insight and to provide early warning signs of significant shifts in the firm's environment.

Cornerstone 3: Tolerance

The living company's emphasis on learning requires a high degree of openness. This openness allows the company to absorb both the new people and the new ideas it needs to learn and adapt. Living companies value diversity. Knowledge generated by diverse individuals and groups can be shared and used to leverage growth and development throughout the organization.

Living companies also exhibit both tolerance and patience for the disruptive dynamics that inevitably accompany diverse people, knowledge, and ideas. They are particularly tolerant of activities on their boundaries. They empower staff to experiment, try new things, and build new relationships outside the company. They support entrepreneurial actions that stretch the realm of possibilities and open up new opportunities for the company.

By contrast, economic companies maintain a high degree of centralized control in hierarchical authority structures. Top management takes primary responsibility for scanning the organization's environment, building relationships outside the firm, and for crafting new strategies to be executed by troops deployed down the organizational hierarchy. Control is favored over empowerment as the means of ensuring profitability and growth.

The work of professional architecture and engineering firms is predominantly project work. Unfortunately, a shift by firms toward an economic company philosophy results in significant gaps between these business and management principles and the way that projects need to operate, gaps that are seldom bridged. Hierarchical control often conflicts with the need to field project teams that are empowered to work directly with clients and other project stakeholders. Professionals, project managers, and project teams can grow frustrated at the loss of their professional autonomy. Given this frustration, they often refuse, either explicitly or tacitly, to bow to centralized control of top management and the firm's control systems. Managers are frustrated by the lack of accountability exhibited by project teams to the firm's profitability goals and accounting control systems. Hoping to correct these problems and bring teams into compliance, management layers on new operating constraints and reporting requirements. The net result is a loss of initiative, energy, and productivity, all of which have contributed to a decline in the ability of firms to do good work.

The tolerance inherent in a living company fits professional design and consulting firms much better. Project teams are empowered to react quickly to meet shifting client expectations in changing project environments. Entrepreneurial teams and professionals can identify and take advantage of opportunities that come with these changes, offering new services and trying out new ways of working. As teams interpret and respond to project needs, they expand the firm's pool of experience and knowledge. Professionals in living firms can exercise a high degree of autonomy and freedom to practice in their professional domains. They also are able to proactively expand the limits of those domains through their work. The persona of the firm guides these empowered employees to do the right thing and make choices that not only solve immediate problems, but also contribute to the overall growth and development of the organization.

Cornerstone 4: Conservative Financing

Living companies are frugal and know the value of having cash in the bank and of minimizing debt. They know that being financially conservative offers both protection from sudden economic downturns and the flexibility to take advantage of new opportunities as they emerge. Members of the community recognize the role that profits play in building the community. Profitability is important, but it is not seen as an end itself. Profitability is viewed as a means to reach the community's more fundamental goals of ongoing development and growth.

A living company strives to achieve sustainable growth. Revenue growth is a result of successful development and evolution. It is not seen as a driver of that development or a primary goal on its own. For a living company, growth should be governed by the ability of the organization to self-fund the activities required to create that growth.

Again, in contrast, an economic company's primary purpose is to maximize the return paid to owners of its capital assets. Top management manipulates the company's production factors (land, people, and capital) to achieve high levels of profitability. The major goal is to increase revenue, with the implied promise of increasing profits. Leverage, in the form of debt, can be used to fund this growth and provide a higher rate of return to shareholders. However, these expansions can quickly tip the company's balance sheet toward liabilities that demand repayment, and significantly increase the firm's exposure to financial risk. They also soak up cash that could otherwise be directed toward the development of new skills and capabilities.

The landscape of the design and construction industry in the last 20 years is littered with the remains of architecture and engineering firms that focused on rapid revenue growth as a primary means of increasing profits. Many pursued this growth through an aggressive strategy of mergers and acquisitions. Many firms were able to expand quickly, but often the trade-off was adding significant levels of debt to their balance sheets. This debt reduced their flexibility and independence and left them at risk when they encountered unanticipated recessions or sudden shifts in client fortunes or market trends. Having encountered difficulties, many retrenched brutally, or offered themselves up for acquisition when it was clear they had no other alternative.

The short-term, bottom-line orientation of the economic company also often conflicts with values held by individual professional practitioners. They are pressured to compromise or even abandon professional commitments and aspirations in favor of doing work that is primarily focused on generating revenue and profits. Doing good work is constrained by rigid demands for profitability and high utilization from the organization.

Professionals in living firms are rewarded according to their skills and contributions. As learning organizations, living firms also offer the support practitioners need to grow those competencies. The absolute amount of compensation an individual receives is less important than the fairness of the firm's compensation system, and whether there are opportunities for further developing personal skills through training, coaching, and challenging work assignments.[15]

Investments that improve the capabilities of the living firm in the long term, foster long-term relationships with clients and other industry stakeholders, and enhance the firm's reputation for doing good work and operating with high standards are seen as essential for success. The payoff for these investments is revenue growth and higher profits. Investments can include funding geographic and market expansions, but those investments are driven by fundamental factors, not by strategic dictates simply to increase revenues and profits.

Professional architecture and engineering firms have a tremendous reservoir of creative potential embodied in their people. These people are deeply committed to the higher ideals of their professional disciplines. This commitment has remained largely untapped by firms operating as economic companies. It is waiting to be released and channeled by leaders willing to transform their organizations into living firms, built on a foundation of the four cornerstones of a living company.

3 Creating Value, Not Selling Hours

Community is not about profit. It is about benefit. We confuse them at our peril. When we attempt to monetize all value, we methodically disconnect people and destroy community.

—Dee Hock[16]

A sound business model is fundamental to organizational success. It explains how an enterprise works and presents its recipe for success. A business model provides answers to a set of critical questions: Who are the company's customers? What products or services does the company offer those customers? How does the company produce those goods or deliver those services? And, how does the company make money as a reward for its efforts and the risks it takes on?

The answer to these questions should "tell a good story."[17] This story should combine a compelling narrative with numbers that add up. American Express built its original business model around traveler's checks. That model offered "all the elements of a good story: precisely delineated characters, plausible motivations, and a plot that turns on an insight about value."[18] For customers, the checks neatly solved two problems: they were convenient to use (offering wide acceptance) and heightened the traveler's feeling of security (insurance against loss or theft) in exchange for a small fee. For merchants, the American Express traveler's checks acted like letters of credit ensuring payment. And, by accepting them, merchants could attract more customers. Finally, for American Express, traveler's checks not only provided revenue in the form of fees but also, more importantly, operated as the equivalent of interest-free loans, with the required up-front payment by the customer. Both the narrative and the numbers made sense. American Express' business model "changed the rules of the game, in this case, the economics of travel."[19]

At the center of the business model sits the company's value proposition, a statement of the value that a business creates for its customers and society and what it receives in return. For example, stockbrokers such as Merrill Lynch traditionally gave their retail customers investment advice and the ability to trade, in exchange for commission fees. Classic grocery store chains from the 1950s and 1960s, like A&P and Kroger, offered customers cheap, plentiful groceries at cost plus a small margin, in utilitarian stores located in their neighborhoods.

Business models are holistic in nature. All of the elements in the model must work together. Successful business models exhibit the dynamics of a self-reinforcing system, where resources generated by the firm's value proposition are used as fuel to strengthen the distinctive competencies that it uses to create that value. American Express not only used the float provided by its value proposition to pay out profits to investors, but also reinvested the cash flow to expand its business across wider geographies, sign up more merchants, and offer enhanced services to travelers. This reinforcing loop shaped the identity and behavior of the company.

Formally expressing a business model provides a powerful lever for organizational innovation and alignment. Business models provide a medium for communication and collaboration between an organization's leaders and stakeholders. Strategic conversations swirl around the models, as leaders engage in give-and-take explorations of ways to make them better. The articulation of a model helps force a company to confront the brutal facts of its current situation (Does it have a viable value proposition? Are its products and services valued by its customers? Is it able to make a sufficient return on the value it creates?) and begin to plot strategies for moving forward. Finally, an innovative business model can create the equivalent of a magnetic field for attracting smart people who buy into the idea and see an opportunity to enrich it with their own ideas and efforts.

Left unarticulated and unexamined, business models operate as a set of deeply held mental models that can restrict the thinking of a firm's leaders. Necessary changes or initiatives don't happen because they conflict with unspoken ideas, assumptions, and opinions about what "business" the firm is in and what value it offers its customers. For nearly 20 years, despite mounting evidence about the mismatch between its current business practices and the changing world around it, A&P executives and managers held onto a business model and set of mental models shaped by the organization's original founders almost a century earlier. New data and information were filtered through the mask of these models and summarily ignored or reinterpreted, leading to a series of operational and strategic mistakes. By the end of the century A&P had shrunk to a shadow of its former self, while competitors such as Kroger had changed their business models and learned how to thrive.

Successful business models are aligned to the forces driving their markets, industries, and society. Companies must be willing to change their business models as times change. The failure to adjust results in diminished performance, and severe misalignment can even threaten the continued existence of the business. Merrill Lynch failed to adapt to changes in trading patterns brought about by the Internet and lost ground to new online brokerage services. In the 1970s and 1980s, Kroger shifted its business model toward a superstore concept that offered a broader range of goods in bigger, nicer stores and became the number-one grocery chain in the country. A&P failed to make the shift and "dwindled to a sad remnant of a once-great American institution."[20]

New companies that grew up around the Internet triggered a burst of business model innovation. Their innovative business models offered new value propositions, explored new organizational designs, and created new wealth for customers and company stakeholders. Amazon reinvented the book-selling business. EBay and Priceline figured out how to leverage the neighborhood garage sale into a global marketplace by matching buyers and sellers online. Established businesses launched bold reinventions. Charles Schwab launched eSchwab, a new online brokerage service experiment, and not only wiped out its own existing brokerage business but also redefined the way stocks are bought and sold.

Ford reengineered the way it manufactured cars using business-to-business Web applications to integrate its supply chain, improve quality, and cut costs. IBM has shifted from a business model centered on the design and assembly of computers to a new model that combines the aggressive sale of its technology, components, and subsystems in the open market with new consulting and systems integration businesses. The new model has boosted both its performance and profitability.[21]

Selling Hours: An Economic Business Model

Architecture and engineering firms remain largely unaffected by the trend toward business model innovation. Firms continue to cling to the economic-company model adopted a quarter century ago, a model that is increasingly mismatched to the needs of the emerging knowledge-based economy. The heart of this economic business model is found in a simple statement: "We sell hours." Hours are the common denominator used to assess, calculate, and exchange value. More fully expressed, the value proposition of such a model can be characterized as, "We provide technical solutions to client problems in the form of professional services that are paid for by the hour."

The value a firm provides is equated with the number of hours clients are willing to buy and the rate (dollars per hour) clients are willing to pay for those hours. For many firms this value proposition is explicit. For others it lies unstated, a deeply held mental model that shapes the behavior of firm managers and practicing professionals.

The simplicity and tangibility of this value proposition were major factors contributing to its widespread adoption by architecture and engineering firms. It is easy to assign a value to a professional's work by equating it with the number of hours he or she spends doing a particular task or activity. Billable hours are easily quantified and measured. They can be used as the basic building blocks for cost-based compensation processes.[22] Billable hours appear to reduce the complexities of practice to a least common denominator that everyone can understand.

Projects can be planned, priced, and negotiated by the hour. Operations can be fine-tuned to sell an optimum number of hours. Staff members can be assigned and held accountable to personal targets for billing hours to projects. Financial systems can be set up to measure and monitor the number of hours sold and how many remain unsold. Managers can be taught how to read utilization reports (billable hours versus total hours)—and to read them first. Many management consultants and firm leaders have proclaimed utilization to be the most important measure of a firm's productivity and that utilization reports are the most valuable information produced by a firm's financial system.[23] Ask a group of executives from architecture and engineering firms what metric they spend the most time analyzing and discussing, and the majority will answer utilization.

However, for all its simplicity, the billable hour and its corresponding emphasis on utilization have pushed architecture and engineering firms into a vicious cycle of declining performance. During difficult times, firm revenues and profits may decline, driven down by competitive pressures and escalating client demands. In response, managers reach to adjust the firm's utilization control knob to rebalance firm profitability. Alternative actions for restoring firm profitability are much more difficult, time-consuming, and uncertain (for example, negotiating higher fees with reluctant clients, inventing and putting in place new service offerings for which clients will pay more, diagnosing and correcting quality problems, or improving staff performance through training).

The easiest way to raise utilization, given the difficulties and time lag involved in winning new work as a means of selling more billable hours, is to cut nonbillable time. Firm management can eliminate overhead positions, slash training and development hours, and stop internal change initiatives that focus on building the firm's infrastructure and preparing it for the future. Staff members who are not currently assigned to projects can be laid off, and the remaining people can be asked to work longer hours. In the short term, this strategy works. Utilization goes up and profitability appears to be restored.

However, in the long run, this tyranny of utilization saps the strength of the firm. The learning, reflection, creativity, and innovation needed to solve client problems effectively and cope with a rapidly changing economy don't happen. Coaching, which once occurred naturally, withers and dies. Firms continue to struggle along with antiquated infrastructures built of systems, processes, and procedures that are ill suited to current project delivery demands. Key staff members become burned out, and many not only quit the firm but also leave the profession. Remaining staff struggle just to survive. And managers don't seem to notice—they're too busy watching the numbers.

As the quality of service erodes, clients demand lower fees and turn to other providers to satisfy critical project needs. Driven by dissatisfied, price-conscious buyers, firm profits continue to slide, so management once again focuses its attention on utilization. After all, it seemed to work the last time. Industry newsletters recommend further cuts in overhead costs, reductions in benefits, and elimination of nonbillable hours. The vicious cycle spins on.

Equating hours with value creates a second major problem for design and consulting professionals. In the eyes of many clients, hours sold by one firm look remarkably similar to those offered by another. After all, many firms appear to practice the same way, employ the same types of people, and have the same performance problems discussed above. Consequently, it makes sense to clients that those hours should be treated and purchased like any other commodity. Price sensitivity escalates further, and fees move downward.

It was remarkable that during the last five years of the 1990s, as demand for architecture and engineering services peaked with the U.S. economic boom, the laws of supply and demand didn't seem to apply to these firms. As demand increased, classic economics would have predicted that the price firms were able to command for their services should have risen. That wasn't the case. The price that clients were willing to pay for the hours they purchased remained fixed; net multipliers for firms hardly budged.[24]

The Transformational Imperative

The commodity status many architecture and engineering firms suffer is only one sign of a larger threat inherent in their economic business model. To understand the larger threat, we need to look at the relative maturity of their technology of practice, or the way that these firms transform labor, capital, materials, and information into professional services of value to their clients.

Technologies, whether products (disk drives), project delivery processes used in the design and construction industry, or services (security brokerages), move through a predictable developmental pattern, according to Harvard Business School professor Clayton Christensen. That pattern, called a "technology S-curve," plots a trajectory showing the increase in performance of a given technology over time (Figure 3.1). Performance gains are relatively slow during the early stage of a technology's development. The rate of improvement accelerates as the technology spreads and becomes better understood. Eventually, performance gains level off, as the time and effort required for each additional increment of improvement escalates. At that point, the technology can be thought of as having matured, and it then moves toward the end of its natural life cycle.[25]

If the environment within which a mature technology exists is relatively benign, that technology may be able to maintain this flattened trajectory for some time. It will continue to generate income, although usually with decreasing returns as it becomes more and more of a commodity in the eyes of many customers.

However, in a changing environment, a mature technology is at great risk, ripe for overthrow by a successor technology. In *The Innovator's Dilemma: When New Technologies Cause Great Firms to Fail*, Christensen documents numerous cases in which mature technologies, companies, and industries were unseated by new disruptive technologies and were literally driven out of business. The owners of those mature technologies had remained complacent, choosing neither to recognize nor respond to the perils inherent in their positions. This fixed position exposed their organizations to the risks of rapidly occurring technological realignments, when customers switched to the new competing technologies faster than the established technology suppliers could react. Even if they were able to stay in business, these companies or industries faced a long process of decline and decay, living in the shadows of more successful innovations. To avoid such a fate, Christensen urges leaders to recognize when their company's technology is moving into this red zone and to innovate before it is too late.

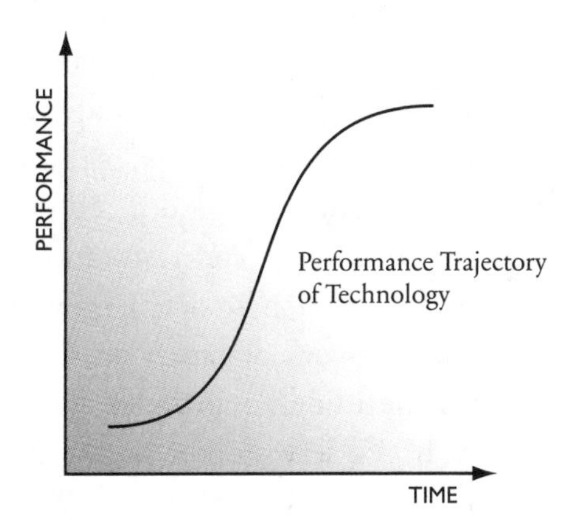

FIGURE 3.1. Technology S-Curve

Unfortunately, as we will discuss in chapter 14, many of the services and project delivery technologies—as well as the prevailing economic business model—used by architecture and engineering matured long ago. Further, most firms have not recognized the precarious nature of their current situation. They cling to their tried-and-true methods with the mistaken belief that somehow things will turn around with the next economic boom or that clients can still somehow be convinced of the wisdom of qualifications-based selection procedures.

Firms must face up to this transformational imperative and act before they either fall victim to new disruptive technologies or are sentenced to long-term commodity status. They must reinvent their business models, launch innovative new project technologies, establish new ways of working with clients, and develop new knowledge and competencies—creating the beginnings of a new generation of technology S-curves.

Creating Value: A New Value Proposition

The creation of value lies at the heart of a professional's performance. Architects and engineers create value as they shape the built environment where people live, communities thrive, companies prosper, and society progresses. Architects "design and build the multiplicity of shelters needed for man's habitat."[26] Engineers "protect people from the destructive force of water while harnessing water for the enormous good it can do; provide people with electricity, the motive force of modern life; make great cities habitable and vital; and create the pathways that connect place to place and person to person."[27] Together with planners, landscape architects, scientists, and other design professionals, they conceive and guide the creation and restoration of the built and natural environments within which society's growth and evolution unfolds. This "good work" or "work of expert quality that benefits the broader society"[28] fulfills the underlying social contract between the professionals and society. It also serves as the basis for a bold new value proposition matched to that professional status: "We create value and are fairly rewarded for our efforts by our clients and society."

This new value proposition reframes how professionals are compensated for their efforts. They aren't paid for the hours they spend working for clients. They are fairly compensated for the value they create on behalf of those clients. Create more value, even if you use fewer hours to do so, and earn greater rewards. It becomes a virtuous circle. By compensating architects and engineers fairly, clients and society provide the resources those individuals and firms need to further develop their skills, knowledge, and expertise and create even higher levels of value. As competence and creativity grow, professionals are able to be even more effective stewards of the built and natural environments, enabling the evolution of a thriving, sustainable society within which their clients can become even more successful.

It is important to recognize that value exists in both monetary and nonmonetary forms (value can be both tangible and intangible); it's not just about money. The knowledge economy places a premium on many forms of nonmonetary value, including learning, knowledge, personal and organizational development, relationships, and connections. Many companies are now adopting balanced scorecards that recognize the importance of nonmonetary value—adding customer satisfaction, operational effectiveness, and learning and growth metrics—to traditional financial measures to provide a more well-rounded assessment of how the firm is performing.[29]

Even Wall Street recognizes the significance of nonmonetary value. A study of investment decisions made by institutional investors, prepared by the Ernst & Young Center for Business Innovation, showed that over a third of the information used by these investors to justify a given investment decision, valuing a company stock, is nonfinancial.[30] Another study by PricewaterhouseCoopers reported that about 80 percent of the market value of public companies is represented by intangible assets such as brand awareness, human capital, knowledge assets, and other complementary business assets such as the leadership qualities of management.[31]

The best work of design professionals has always created nonmonetary value far in excess of the direct monetary benefits. The symbol of the Golden Gate Bridge and the experience of crossing it, particularly for the millions that have crossed it on foot, have a value exclusive of the income from bridge tolls collected every day. The enlightenment offered to students at Thomas Jefferson's new University of Virginia campus created value for a fledgling democracy far beyond the tuition paid by students. The awe that accompanies one's first sight of Hoover Dam cannot be valued on the same balance sheet as the electricity that flows from it generators. The refreshment of cool, pure water on a hot afternoon, brought to thirsty city dwellers by an aqueduct that comes hundreds of miles from mountain streams, is a value that goes beyond the utility bill paid each month. It's not that one form of value is necessarily greater or lesser, it's just that they are different, and both real and capable of appreciation.

By leveraging their insight, imagination, and creativity, architects and engineers can become preeminent creators of value in our society. But architecture and engineering firms will never be able to realize this potential from the confines of their economic business model. It will take business model innovation to recognize, celebrate, and exploit this potential.

From One to Many: A New Ecology of Business Models

The explosion of business model innovation over the last two decades has resulted in a diverse new population of business species, from evolved manufacturing and service businesses to bold new high-tech companies. Architecture and engineering firms have inverted this process, devolving over the past two decades toward a single business model. This lack of diversity has become a major source of difficulty. Most firms look alike and offer the same value proposition to clients. Consequently, clients treat them more like commodities and less like providers of valued professional services.

From a living systems perspective, this lack of diversity bodes ill. In nature, lack of diversity is often a precursor of disaster. When confronted with radical change, species that fail to cultivate enough diversity to adapt suffer extinction. Those that adapt survive, carried through by individuals whose experimental changes meet the demands of the new environment.

For example, the long isolation of the Hawaiian Islands gave rise to flora and fauna that lacked the diversity, and the defenses, of similar species found in rain forests in less isolated settings. Birds nested on the ground and plants possessed neither thorns nor toxins. When Europeans suddenly introduced new animals that preyed on the birds and browsed on the plants, many species were quickly devastated.[32] By contrast, faced with similar threats, the rain forests of Costa Rica survived relatively unscathed. Costa Rica's position on the bridge between two continents ensured constant change, forcing ongoing adaptation. When Europeans arrived, its diverse species once again adapted and survived.

The lack of business model diversity among architecture and engineering firms has left an entire population of firms at risk. The turbulent economy promises just the type of traumatic change that could force their extinction. New species of organizations better adapted to client needs, whether design-build companies, management consulting firms, or foreign competitors, are even now gaining ground in the struggle for survival.

Architecture and engineering firms still have time to act. They can begin to transform their current business model to create new living firms matched to the needs of particular clients, enlivened by different ways of working, and focused on providing varied types of value. Through a shift to business models based on diversity, firms (and, by inference, the professions) will not only survive but also thrive in an increasingly turbulent world.

This transformation presents no small challenge. Business model innovation will require that architecture and engineering firms learn new skills; change attitudes, beliefs, and behaviors; and adopt new ways of interacting with clients and the world surrounding their practices. To create an array of highly diverse business models, architects and engineers must

- *Innovate the core of professional practice* to close a widening gap between the values, ethics, and practices of the professions and societal needs and expectations. While professional mind-sets have remained anchored in place for the last quarter century, clients and society have moved forward. As a consequence, architects and engineers have lost standing as leaders in society. Part II of this book will examine the genesis of this gap and propose shifts in professional mind-sets that will allow firms to develop the social and leadership capacities they will need to create innovative business models.
- *Transform the ways in which they create value* by capitalizing on emerging trends and conditions, adopting innovative strategies for working with clients, and building new capacities for both creative

leadership and technical work. Part III offers a guide to the driving forces that firms must respond to and to new ideas that firm leaders can consider as they innovate the first half of the new value proposition, creating value for clients and society.

- *Develop the social and leadership capacities* that will allow the people in these firms to implement innovative business models based on these value creation strategies. In part IV, we present a comprehensive framework of social and leadership capacities that can help professionals not only catch up, but get far enough ahead of clients and society to be able to exercise significant leadership within the realm of the built and natural environments. Developing these capacities will also be centrally important for firms to successfully tackle the difficult work of transforming their business models.
- *Implement new pricing strategies* that allow them to earn a fair return for their efforts, a return that will support a thriving living firm and provide the resources needed for it to develop its full potential. A menu of new pricing strategies, many of which are already being tried by architecture and engineering firms, is presented in part V.

PART II

EVOLVING THE CORE OF PROFESSIONAL PRACTICE

4 Stakeholder Expectations vs. Professional Mind-Set: A Serious Gap

A professional realm is healthiest when the values of the culture are in line with those of the domain, when the expectations of stakeholders match those of the field, and when domain and field are themselves in sync.

—Howard Gardner, Mihaly Csikszentmihalyi, and William Damon[33]

A year or so into our collaborative research, we were gathered around a conference table in a modest Midwest hotel. During a reflective dialogue, one of our more seasoned participants opined in a distraught tone, "Trust . . . , trust is the problem—we don't have it from our clients anymore." Our group's thoughts had wound around this theme in various ways before. Through this participant's very personal and painful statement, it now seemed to well up from deeply shared feeling as well.

These practitioners were lamenting an absence of trust not just in a particular firm or relationship gone wrong, but in general between clients and professional firms. Credentialed professions operate in a different relationship to society than other kinds of occupations. Professionals receive the privilege to practice in exchange for their commitment to serve society above monetary reward. A kind of sacred trust resides in this exchange. The client perception our group was registering was not of a violation of business ethics or contract—though these might be involved in individual cases—but of a fundamental breach in this sacred trust. Such a breach could only have evolved as a result of a substantial gap between what clients expect and what they believe professionals are delivering.

Though the degree of this breach varies widely across firms, clients, and projects, in the aggregate, its implications for business have been deeply damaging. Most clients now view the value they receive more as a technical product than as a precious service from a trusted adviser. This change in perceived value has contributed significantly to the commoditization pressures firms have been experiencing. The implications for practice have been even more serious. Without trust, architects and engineers cannot lead, and they find their creativity and contribution reduced to narrower and narrower confines.

During the period of our research, a number of professionals we were working with who had teaching roles in schools of architecture and engineering reported declining enrollments. In their recently published *Good Work: When Excellence and Ethics Meet*, Howard Gardner, Mihaly Csikszentmihalyi, and William Damon point out that a decreasing number of aspiring new entrants, as well as diminished trust and leadership position in society, are symptoms of decline in a profession.[34] Their book explores the conditions in a profession that either promote or obstruct an individual professional's ability to do "good work," defined as work that is both highly skilled and beneficial to society. The model these authors offer for understanding a profession's current state of health consists of four components. *Individual practitioners* are "persons who elect to enter a professional realm, secure training, and pursue their own personal and professional goals." The *domain* is the "knowledge, skills, practices, rules, and values captured in various codes." It expresses the mind-set and ethics of a profession. The *field* is the realm of practice, including supporting "institutions" and the "roles" played by individuals who are working with codes of the domain, such as elite gatekeepers, expert practitioners, apprentices, and students. *Other stakeholders* include "corporate shareholders and executives" as well as the "general public," including "consumers and citizens."[35]

These authors see a variety of ways things can go wrong inside a profession, including failure to respond to rapid technological or social changes and the exhaustion of the profession's knowledge base. They believe that the underlying causes driving such developments, and therefore a profession's health, involve the degree to which its four key elements are aligned with one another and with their larger societal context. This model can help us understand what has been happening in architecture, engineering, and related professions, and why individual practitioners have been facing more and more obstacles to doing good work.

Misalignment 1: Domain vs. Field

Three interrelated misalignments are in play. The first is a disconnect between the domain and the field. The professional entering an architecture or engineering firm fresh out of school experiences a shock that makes this disconnect immediately visible. Table 4.1 attempts to capture what young practitioners believe are their professions' core values contrasted with the conditions they perceive to drive their firms' business environments.

The values in the left column are firmly rooted in the early to mid-20th-century construction heyday—though they may have longer histories. These values at the heart of the domain reflect the heroic, independent, and well-funded view of architecture and engineering that successfully drove the very significant achievements of that period. In contrast, the right column combines new client constraints and expectations with the rigors of the economic business model that was adopted to cope with the stresses of the 1970s. This column reflects the very different conditions of practice today.

This dissonance parallels the universal experience of moving from the ideality of an academic setting to the reality of a business or organizational environment. However, the professions have been suffering an intensifying form of this dissonance as their businesses have come under assault by commercializing economic forces. And engineers and architects have experienced this in a different way from doctors and lawyers, for example. Unlike practitioners in these professions, engineers and architects have never developed a comfort level with wealth. They are averse to talking and thinking about money—to the point where many have difficulty accepting the idea of being well paid for their services. This aversion to economic concerns has made it particularly difficult for firms to meet the economic challenges they have faced and continue to face. Not surprisingly, architects' and engineers' version of the economic business model has been especially unable to support professional practice in the highest sense. In reducing the activity of the professional to production, it has effectively eliminated the learning and experimentation required for optimal creativity and contribution.

Young professionals' mind-set based on their education (the domain)	**Conditions perceived in their firm's business environment (the field)**
Autonomy/expert authority	Participatory politics across complex web of stakeholders
Power to shape the built environment	Marginalized role and diminished influence
Access to and ability to preserve integrity of whole project process	Access to small part of project process
Contribution to society/legacy (long-term view)	Current economic and functional realities (short-term view)
Aesthetic or technological ideals	Pragmatism of client decisions
Requirements of design, problem-solving, or analytical process	Tight budgets and schedules
Interesting problems to solve	Relatively predictable needs that are the firm's bread and butter
Opportunity to be creative	Production based on existing procedures and solutions
Large-scale ambition	Small scale of actual work
Desire to keep learning	Requirement to be billable

TABLE 4.1. Gap Between the Domain and the Field

As the authors of *Good Work* define it, the field technically includes professional schools and societies—where values are transmitted, licensing tests administered, continuing education provided, and awards conferred—as well as professional firms. The disconnect depicted in table 4.1 thus also describes the range of values operating in the field itself. Producing those young professionals, professional schools are aligned with the left column. Forming the dominant field of active practice, professional firms have had no choice but to respond, at least to some extent, to new realities operating in their clients' world. Professional societies struggle between the two columns, attempting to adjudicate the split. As a field institution, firms are structured to support the values expressed in the right column, but the practitioners within them express the whole range defined by these poles.

In general, the tension between designers and project managers on the one hand and the firm's managers and management systems on the other has produced an ongoing conflict. This conflict has frozen practitioners in defensive allegiance to the domains they assimilated in school, while their firms' structure and roles have been shaped by current client expectations and the economic business model to a very significant degree.

Misalignment 2: Practitioners vs. Other Stakeholders

In the context of the tension between domain and field that fractures the field, practitioners tend to experience clients' constraints as consumer-like concerns that bring out the worst in the economic business model. Clients' needs are thus often perceived as forces opposing good work rather than as elements of the design or problem-solving challenge. This perception has in turn brought about the second interrelated misalignment: a serious gap between practitioners' professional mind-set and the expectations of their key stakeholders—their clients.

Professionals who have linked clients' concerns to the ills of the economic business model tend to place a low value on clients' ability to contribute meaningfully to design or technical outcomes, at least relative to their own ability to do so. This value drops even further when they must deal with multiple stakeholders who want to be involved in the project process. Professionals often view this cumbersome social process as obstructing their ability to do the good work they were trained to do. Often somewhat unconsciously, they set up an adversarial relationship with the client and other stakeholder groups. In the best cases, productive conflict leads to more creative outcomes. However, in far too many cases, consultants attempt to satisfy the client at one level but behind the scenes follow their own convictions—in the interests of being true to the domain, of course! In these situations the client ends up feeling ill-served or even betrayed by the professional. The breach of trust caused by this failure to truly embrace the client's values erodes the ability of the client to authorize the leadership position that the consultant so desires.

Some firms have attempted to meet these challenges by becoming primarily service firms, oriented more to taking good care of their clients than to their own aspirations for the best possible design or technical solution. Interestingly, however, clients of these firms often end up feeling less well served than they would have liked because the solution they received was not as creative or powerful as it could have been. In fact, this way of addressing the social and business challenges of current practice has also contributed to the loss of leadership position that these professions have suffered. An exclusively client-driven perspective ironically provides little more basis for conferring leadership authority upon a professional than does a perspective driven by lack of sufficient respect for the client.

Misalignment 3: Domain vs. Values of Society

To the extent that they have held their clients' values in an adversarial relationship to their domain, architecture and engineering practitioners have failed to understand the extent to which clients and other stakeholders represent changes in the larger society to which they must attend and adapt. Locked into a defensive posture, these practitioners have been unable to distinguish between situations where their professional opinion is truly needed to challenge the client's view and situations where the client's view needs to challenge their own. They have also been unable to develop the collaborative capacity to deal with this complexity.

The decision to hold fast to their domains as defined in the first 70 years of the 20th century has prevented these professionals from being open to creative adaptation at the very time that social change—driven by technological change—has been accelerating at an unprecedented rate. This defensive posture has resulted in the loss of the most fundamental alignment required for a profession to be healthy: alignment of the domain with the values of the larger society in which its professionals operate.

Like any living system, a profession must continue to develop and adapt to changes in its environment. Over time, these professions' lack of creative response to the extraordinary forces that have transformed our society has wrought a very fundamental form of damage. Even where firms and individuals were otherwise equipped to lead, these professions' domains and, thus, their fields, have not provided a viable base from which to exercise leadership. This loss of leadership capacity is a problem not only for the professionals themselves but for society as a whole. In the absence of the commitment to higher values that effectively adapting professionals can provide, the built environment has been left primarily to market and bureaucratic forces—endangering both its quality and the quality of social life that depends on it.

In another recent study, the sociologist Eliot Freidson has argued that during the late 20th and early 21st centuries, all of the traditional professions have undergone a parallel loss of leadership position and capacity for creative contribution, primarily at the hands of market forces.[36] As architects, engineers, and other professionals involved in the built environment take stock of their situation, it is important to acknowledge the scope and scale of social, political, and economic forces in play. A defensive posture has no doubt been encouraged and exacerbated by the macro level on which these disempowering forces are operating, and creative response inhibited by their sheer magnitude.

Each profession has experienced this larger dynamic in its own way and must find its own way out of it. Educational institutions must help.

Unfortunately, schools of architecture and engineering—experiencing blowback from the adversarial dynamics unfolding in firms—have tended even more insistently to hold to the domains as currently defined. Thus, at the very time young practitioners might have brought greater openness to a broad adaptation process, most of these schools have continued to turn out professionals woefully at odds with the expectations of contemporary clients and society as a whole.

A Pressing Need to Update the Domains

The current business model itself poses a significant challenge. It has reached the mature point of its life cycle in which commoditization will force a wave of innovation accompanied by a wave of extinction. In this process, the economic business model has become even less able than it was originally to support the high level of professional practice that both professionals and society need. It is thus certainly responsible for some of the damage that professional practice has sustained. However, the business model has been midwife rather than mother to the deeper crisis in the perception of professional value. That more primary role has been played by professionals themselves. Responding defensively rather than creatively to the market and social forces transforming their world, they have caused their services to be seen as less important and less relevant to that world's emerging issues and challenges.

Given these dynamics and relationships, we are convinced that innovation in the prevailing business model is a requirement for the long-term health of these professions. This innovation will be the subject of sections 3 through 5 of this book. We are equally convinced, however, that such innovation is impossible without simultaneously building a new foundation for these professions' leadership position in society. In other words, the basis for creating new value through business model innovation must be vital and healthy professions aligned with stakeholders and society as a whole. Achieving this healthy state requires a thorough update of these professions' domains.

Freidson believes that the power for renewal lies in what he calls "the soul of the profession"—namely, its ability to hold and exercise moral responsibility for the well-being of society.[37] We agree, and will conclude this section on "Evolving the Core of Professional Practice" with a new approach to the domains' overarching definitions and an updated set of values and ethics that we believe can enable renewal by this means. A profound evolution of this nature will necessarily entail profound shifts of professional mind-set. In chapter 7, we will explore one set of shifts required simply to catch up with where clients and other stakeholders have been for some time. In chapter 8, we'll take up another, more challenging set of shifts needed to get out ahead enough to be able to exercise significant leadership for society.

Practitioners can only get to these new professional mind-sets—and the values and ethics that will buttress them—by being willing to adapt to the stunning changes that have rapidly been restructuring and transforming our society. Effective adaptation to those changes must begin with understanding them in a more comprehensive way. Let's turn now to the new social and technological complexities that entwine our 21st-century world. We offer two powerful models that can help professionals understand and work with these new forms of complexity in creative and, yes, exciting ways.

5 Understanding New Social Complexity: Organization Ecosystems

The world is becoming smaller. Faster innovation, global development, and environmental and social decline are forcing us to attend to problems as whole cloth. Business is just a subset, a distinctly human activity within the social and natural environments. The[se] so-called externalities . . . are in fact internal to the world system as a whole, and business leaders will increasingly have to address them.

—James F. Moore[38]

At a recent workshop for a group of more than 100 consulting engineers, Susan invited participants to reflect on what is different about practicing today than 20 years ago. Two comments were particularly telling. One gentleman observed, "now we have to do everything by consensus." A few minutes later, another stated, "we used to just be able to build things." Both of these participants were senior practitioners, and both spoke in tones that conveyed the feeling that something precious had been lost. While they were thus expressing their attachment to the domain defined by former practice, they were at the same time alluding, respectively, to new social complexity and new technological complexity that now challenge engineering and architecture professionals at every turn.

A New Social Fabric

Today, any built environment project larger than a single residence involves not just a single client group or organization, but other stakeholder groups as well. Even an expansion of a small, private elementary school, for example, requires an architect or engineer to work with donors, parent groups, neighborhood groups, city inspectors, and fire marshals, as well as school administrators and faculty. Projects with a larger geographic footprint, such as a highway interchange, a water treatment plant, or a hospital expansion, significantly multiply the stakeholder and regulatory groups involved. For an airport expansion or a big city transit terminal, the stakeholders multiply still further and the number of professional disciplines required increases by an order of magnitude. A recent potential airport expansion in the San Francisco Bay Area involved more than 20 environmental groups, a number of mayors and city councils, as well as regional organizations, the port authority, and untold business and neighborhood groups who believed they would be affected by the project outcomes.

When the interests of stakeholder groups are in fierce opposition to those of other groups, as is often the case, finding common ground can be an enormous challenge. Successfully navigating this new social complexity has become a requirement for professionals who would contribute to the built environment. What are the factors that have so quickly transformed our social landscape into this dense web of diverse and empowered participation? An examination of the key trends that have been reshaping the world will help provide an answer.

Long before the advent of the Internet, communication and transportation technologies were already changing the relationships between countries and societies. Television made it possible to broadcast events live across the globe and to see people in other parts of the world in present time for the first time. Many of us could hop on planes to see those other parts of the world. And many of us began regularly to pick up the telephone to talk to people not just in a neighboring state but on the other side of the globe. Diplomats, soldiers, and business executives were no longer the only people who got connected to other parts of the world. For good or ill, products from developed countries advertised on TV began to be sold and then manufactured in the poorest countries. And for good or ill, people in the poorest countries began to see how differently people in the richest countries lived.

As the first pictures of the earth from space captured for many people, we were now beginning to experience a state of connection without boundaries that had no precedent in human history. As economies became more and more intertwined through the inexorable process of globalization, people began to see how political and economic events in a distant part of the world affected their own economy. The global appeal of British and American popular music combined with that of American and European products to catalyze the emergence of a global culture, especially among young people. The seeds of a major backlash from traditional cultures in less developed nations were planted,[39] whose full growth perhaps did not become visible until the fateful events of 9/11.

At the same time that globalization was evolving, the world's population was increasing and becoming concentrated in and around large urban areas. As part of this process, large numbers of people from developing countries, exposed to the attractions of more developed countries through media and other communications, began to move to developed countries. Significant pressure from crowding and diminishing resources began to be felt in the United States and other developed countries, along with its intensification in developing countries, where it had been a longstanding way of life. With this new wave of immigration, a significantly higher degree of cultural diversity began to characterize America's public processes, especially those involving education and health care.

In the midst of this increased social interdependence on a global scale, the 1970s oil crisis gave us our first collective glimpse of the limits to growth that might be inherent on our finite planet as population continued to explode. Around the same time, early environmentalists began to raise our awareness of the toxic effect of human activity on our biosphere. An initial wave of regulation moved across the developed world to begin to address the problems of limited resources and pollution. And as a response to greater awareness of social and economic inequity within and across societies, social and political norms began to shift toward greater sensitivity to cultural differences and inclusiveness across these differences.

The unprecedented experience of local and global connectedness, driven by multinational corporate commerce, television, travel, and migration, was dramatically accelerated by leaps in information and communications technology. By the late 90s, Internet-enabled e-mail provided people with almost free access to this connectedness. For those on the have side of the digital divide, the experience of connection thus became even more personal than before. With the wireless phone and now the wireless Web, this connectedness can be achieved with even less physical infrastructure.

The rapid development of information technology has had two, perhaps even more, stunning effects on our social processes. First, it has set off a wave of accelerated research and innovation that has driven an exponential rise in the rate at which new knowledge is created. This unprecedented acceleration has in turn driven an exponential rise in the amount of knowledge that has begun to inform and to be required for most human activity. Second, in addition to connecting people at a whole new level, information technology has taken information and knowledge out of the heads and hands of the privileged few and made it available to any reasonably literate person with a computer or other user interface device and an Internet connection.

In a world driven by exponentially increasing knowledge, knowledge-intensive work can no longer be handled by one or a few experts. Such work requires many who are knowledgeable across a wide range of areas. Thus, the number of project participants on the technical side has increased dramatically, making coordination much more challenging. The number of other stakeholders has increased just as dramatically. Anticipating that expanded regulatory processes will provide leverage points for broad involvement of nontechnical stakeholders, savvy owners and managers design such involvement into project process. And now that nontechnical stakeholders have greater access to information and knowledge, they are unwilling simply to accept the dictates of experts, or of stakeholders whose interests they perceive to be incompatible with their own. Nontechnical stakeholder interests now include not only the familiar local issue of "not in my backyard," but concerns regarding regional and global impact as well.

Like the comparably transformative impact of the printing press, information technology has set off a thunderous wave of democratization. While the earlier wave produced the Industrial Revolution, modern capitalism, and the movements that have challenged and attempted to modify it, this second wave has unleashed the much more rapid evolution of the postmodern Information Age. One of its effects has been the emergence of a new middle class across 17 developing and three transition nations driven by the most rapid acceleration of consumption ever known.[40] Where this potent new democratizing force will take us remains to be seen. Its connectedness- and interdependence-producing nature appears to have at least the potential to achieve a significantly broader global inclusiveness and, thus, perhaps a more sustainable global system.

In short, a number of global factors are driving the new social complexity, which has in turn created an explosion in the number of participants in built environment projects, their diversity, and the empowerment they have and expect in the process. The global factors are as follows:

- Greater connectedness among people, groups, organizations, and countries across the globe.
- More population, more concentrated.
- More awareness of how human groups and systems affect one another locally, regionally, and globally relative to social and economic inequity, consumption of basic resources, toxification and degradation of the biosphere, and transportation of human and nonhuman disease.
- Technological advances that fundamentally challenge existing social infrastructure and value systems.

- More regulation and expansion of democratic processes associated with regulation.
- More involvement of empowered individuals and groups in democratic processes based on greater access to information, knowledge, networking, and organizing.
- Exponentially increasing knowledge and technological complexity in human activity, requiring more participants, more coordination, and more consensus.
- More cultural and linguistic diversity in many developed countries, including the United States.
- Emergence of new middle classes in China, India, and 18 other developing and transition nations, driving the most rapid consumption boom in history.
- Labor-based business relocation and coming worldwide skilled labor shortfall.
- Greater need for physical security in a global situation destabilized by socioeconomic inequities and cultural conflict.

These are deep and fundamental changes in our social fabric that are not going to go away any time soon. Professionals who want to keep their domains and fields current must embrace these conditions and figure out how to work with them.

One of the two major thrusts of our approach to new business models in this book will be to develop awareness of the social and leadership capacities architects and engineers need to work effectively with this new social complexity and how those capacities can be leveraged to build powerful new value propositions. As a foundation for that awareness, however, professionals must embrace this new social complexity at the very core of their professional mind-set. Such an embrace requires a new worldview—a new set of mental models—capable of describing and illuminating this complexity as well as enabling us to navigate it.

We have begun to take hold of that new worldview in looking at professional firms as living companies rather than solely as man-made constructs to be manipulated for economic ends. Client organizations as well as professional firms are living systems and subsystems nested in complex webs of interdependence. Let us turn now to a mental model for incorporating 21st-century social complexity that we believe can form the basis for a new, more adaptive professional mind-set.

Organization Ecosystems

A year before de Geus published *The Living Company*, James F. Moore published another groundbreaking book, *The Death of Competition*. In it, he introduced the idea of "business ecosystems" based on the metaphor of biological ecosystems with their diverse species and complex interdependent dynamics. A biological ecosystem is a community of organisms, interacting with one another and with their environment. Each organism or species in the ecosystem not only looks after its own survival but also makes unique contributions to the vitality and sustainability of the ecosystem as a whole. Some species extract or use sunlight, water, and minerals to grow. Others consume the first, alive or dead, along with other raw materials, producing wastes of their own. These wastes become food for other organisms in an endless cycle in which everything produced is used by some organism for its metabolism, simultaneously growing themselves and the ecosystem. In Moore's words, a business ecosystem is "an economic community supported by a foundation of interacting organizations and individuals—the organisms of the business world. This economic community produces goods and services of value to customers, who are themselves members of the ecosystem."[41]

Moore's business ecosystem has three domains: the *core business, extended enterprise*, and *business ecosystem*.[42] These domains may be visualized as three nested ellipses, with the core business as the central one, the extended enterprise the next larger one containing it, and the business ecosystem the largest, containing both.

The core business includes first-tier suppliers and distributors as well as any customers whose collaboration is critical to the delivery of its product or service. The extended enterprise typically includes less directly involved customers and customers of customers, second-tier suppliers, suppliers of complementary products and services, standards bodies, and regulatory agencies. The outer ring of the business ecosystem includes investors, owners, trade associations and unions, communities, competitors, and government agencies. Over time, members of an ecosystem co-evolve capabilities and roles and align themselves with the directions set by one or more central companies. In the computer chip industry, for example, a business ecosystem has formed around Intel. In the retail sector, a business ecosystem has formed around Nordstrom. Other organizations that are part of these ecosystems align their strategies and investments around the needs and activities of these central companies.

So important is the ecosystem level in Moore's view that it transforms the nature of competition itself. He convincingly argues that it is not really individual businesses that compete, but business ecosystems. Thus, the primary work of a company must be expanded—beyond building better products or services for sale to customers—to include the building of its ecosystem. That means creating and maintaining those interdependent and mutually beneficial relationships across businesses and organizations involved in the entire process of a product or service being delivered and used. In some cases, these relationships may be with businesses that might have been thought of as competitors in the past. In the largest sense, Moore is looking at what it takes for a whole industry to be healthy, given that it is difficult for an individual business to be healthy in an industry that is ailing or being competed out of existence by an emerging industry.

Moore also sets out an evolutionary framework that can help leaders of a particular business ecosystem develop strategy appropriate to its stage of development. This framework identifies four major stages:

1. The *pioneering stage*, where the ecosystem is formed.
2. The *expansion stage*, as the ecosystem extends to achieve maximum market coverage and critical mass.
3. The *authority stage*, where the ecosystem matures.
4. The *renewal or death stage*, where species must work together to radically improve or reinvent the ecosystem to sustain its ongoing growth.

We will come back to this framework in chapter 18, where we introduce another important new mental model we call "developmental awareness," and in chapter 22, where we explore the advanced leadership capacities involved in providing strategic support for the development of client organizations and their ecosystems.

We make two additions to Moore's powerful model. First, though he includes considerations of the larger environment in his examples and conclusion, this element does not appear in the business ecosystem model itself. His model focuses primarily on business organisms and their relationships and interactions. To reflect the full complexity of business ecosystems within the model, we want to include the larger environmental component that is central to biological ecosystems. In human systems, this larger environment consists of other business ecosystems, social and political systems, and the physical environment—both built and natural. We call this element of our modified model the "Societal and Environmental Ecosystem."

Second, Moore restricts his use of the metaphor to businesses. His model can be used to understand ecosystems that have formed around other types of organizations as well: for example, the San Francisco International Airport's ecosystem or Harvard University's ecosystem. These examples exhibit similar kinds of organizational species, relationships, and dynamics as those that form around businesses. We will use the phrase "organization ecosystem" to refer to this more inclusive concept. Figure 5.1 shows our modified model.

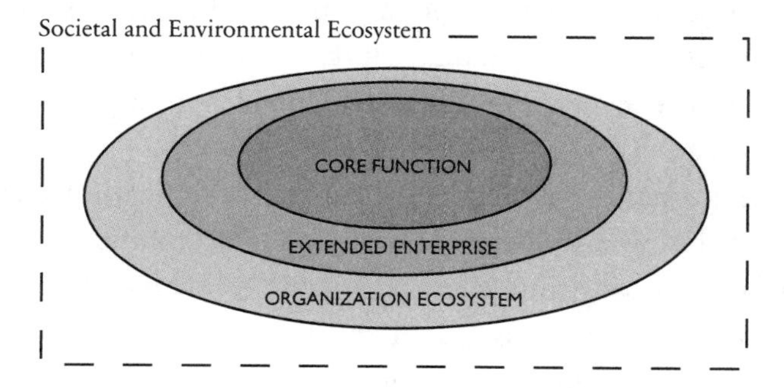

FIGURE 5.1. Organization Ecosystem

This model provides professionals with a framework for understanding the stakeholder complexity embedded in their clients' organization ecosystems. To effectively grasp this complexity, the professional must also understand that to view an organization ecosystem in exclusively economic terms is as reductive and potentially destructive as viewing a business as an exclusively economic entity. The recent economically driven crisis in U.S. corporate and accounting ethics has vividly demonstrated the value of this perspective. Economics certainly play a key role in the exchange of value in organization ecosystems. However, just as biological ecosystems exchange many forms of currency, organization ecosystems involve the creation and exchange of many nonmonetary as well as monetary forms of value. The nonmonetary forms may, in fact, be more significant for the ongoing vitality of these systems. Dee Hock, founder and former CEO of VISA, underscores the significance of the exchange of nonmonetary forms of value:

> The essence of community, its very heart and soul, is the non-monetary exchange of value; things we do and share because we care for others, and for the good of the place. Community is composed of that which we don't attempt to measure, for which we keep no record, and ask no recompense. The non-monetary exchange of value arises from the deep, intuitive, often subconscious understanding that self-interest is inseparably connected with community interest; that individual good is inseparable from the good of the whole; that in some way, often beyond our understanding, all things are at one and the same time, independent, interdependent, and intradependent.[43]

For the human systems we are calling organization ecosystems, then, we need a broader definition of value. We might think of value as anything that contributes to the means by which the system can maintain its vital functions, adapt to changing conditions, and successfully grow or develop through its evolutionary stages. These means can include the following:

- *Resources* that are used as nutrients by the system (energy, materials, time, dollars).
- *Information and knowledge* that support vital functions and projects.
- *Tools and technology* to improve processes and function.
- *Relationships* within the community and with other living systems.
- *Leadership* that focuses attention on and spurs action to deal with opportunities and threats associated with the system's environment and stage of development.
- *Learning* that enables internal improvement, adaptation to major environmental change, emotional and spiritual development, and strengthening of culture and identity.

An architecture or engineering firm has its own organization ecosystem, and from a strategic perspective it is important to understand and build it. However, our interest here is in the value provided by professionals to their clients and to society. Consequently, we want to focus on the client's organization ecosystem and the professional's role in it. In this context, the professional firm's ecosystem is best viewed as a resource-rich part of the client's organization ecosystem—with the firm itself located most likely in the "extended enterprise" ring, but in special circumstances, perhaps even in the "core function."

Though the organization ecosystem model highlights the, at times, overwhelming social challenges that are present in today's built environment projects, it also reveals a basis for envisioning new forms of value that architecture, engineering, and related professions can provide. When clients' needs are viewed in the context of their organization ecosystem, three realms of value creation open up. These three realms are both distinct and overlapping, as shown in figure 5.2.

The first realm is that of value provided directly to the client organization, including not only the familiar technical solutions and knowledge but other kinds of contributions—relationships, tools, funding arrangements, leadership, and learning—that satisfy specific client building, infrastructure, and environmental needs. In making these contributions, professional firms leverage value exchanged within their own organizational ecosystem to enhance value for the client organization.

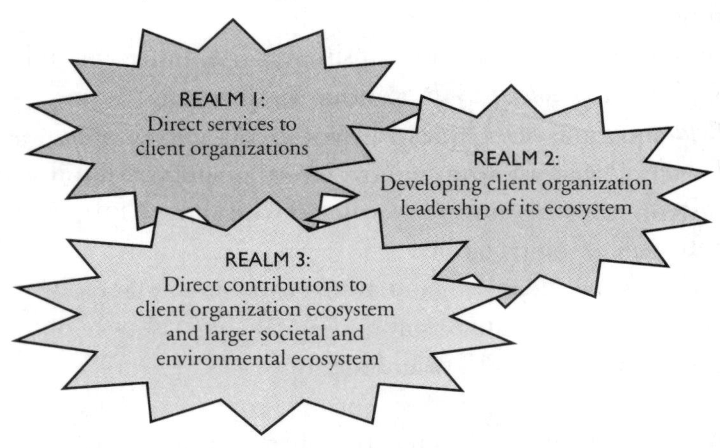

FIGURE 5.2. Three Realms of Value Creation

The second realm of value creation that this model opens up is influencing and guiding clients to respond to or lead changes in their organizational ecosystems. For example, professionals might encourage or assist the client organization to lead a renewal effort in its community in order to revitalize the local economy. Or they might convince a client organization not only to incorporate sustainable building technologies and energy-conserving strategies into their projects, but also to become leaders in their industry by advocating for sustainability-based operations. The value created in this realm is the client's leadership and the beneficial results that flow from it into the organization ecosystem and back to the client organization.

The third realm of value creation that the organization ecosystem model opens up for professionals is making direct contributions to client organization ecosystems, including the larger societal and environmental ecosystem. An example would be providing leadership for a client organization ecosystem problem-solving or renewal effort rather than guiding the client to provide that leadership. Another would be lobbying, advocacy, or public relations on behalf of the client organization's ecosystem. Research and development of a product or process the client organization could incorporate into the products and services it provides to its ecosystem is another. Still another would be speaking engagements at client industry meetings that address emerging industry or related social problems.

Professionals may shape and develop organization ecosystems to make them more sustainable environments not only for themselves and their clients, but for the host of other individuals, groups, and organizational species that live in them. They may raise awareness of problems and opportunities, perturb action, stimulate creative thinking, and share leadership as ecosystems respond to the opportunities and threats emerging in the course of their evolutionary processes. While most such services will be offered in exchange for consulting fees or other types of compensation, professionals may sometimes choose to offer them as unpaid extensions of other services or wholly independent of monetary exchange.

Once professionals are attuned to these three realms, they can more easily see how to work in the first realm in ways that simultaneously contribute to the second or third, working always from a big picture view rather than from a more myopic view. We suspect the greatest leadership power lies in the ability to affect all three realms in the context of the same work—in effect, establishing a modality of providing technical solutions that simultaneously develop the client's leadership and contribute directly to the health of the local and larger ecosystem. However, simply being able to contribute to these realms at different times would substantially enhance the leadership position of these professions.

Viewing a client organization as the center of an organization ecosystem offers the additional benefit of expanding the professional's understanding of the client. Technical analysis of whatever nature now occurs in the context of a web of relationships and current dynamics. These relationships and dynamics introduce many variables, possibilities, and constraints that bear directly on the value of the project outcome. Instead of focusing on the built product alone, the professional can focus on the health of the client's organization ecosystem and how both the process and the outcome of the design or problem-solving process can contribute to that health.

Professionals can thus move away from struggling in tension with clients, who are actually not the primary source of most of the constraints that are in play for projects. Instead, they can partner with clients to address the challenges of thriving in their organization ecosystems, given these systems' current and likely future states. The ecosystem worldview thus provides a powerful basis for healing the client relationship and, as the professional begins to create value for the client's organization ecosystem, for inspiring the client to reauthorize a leadership position for the professional. Imagine a client's reaction to a practitioner who can bring a genuinely more strategic approach to technical work that adds significantly greater value to the ecosystem as a whole, and thus to the organization.

With the help of the organization ecosystem model, the social complexity surrounding built environment projects not only becomes more navigable, but also provides the basis for creating new forms of value that can achieve genuine renewal for the architecture and engineering professions. We will further explore organization ecosystems as a means of evolving the core of professional practice in chapter 8, and they will appear again as powerful guidance for business model innovation in chapter 13.

6 Understanding New Technological Complexity: Value Networks

Even though most managers don't think of themselves as being theory driven, they are in reality voracious consumers of theory. Every time managers make plans or take action, it is based on a mental model in the back of their heads that leads them to believe that the action being taken will lead to the desired result. The problem is that managers are rarely aware of the theories they are using—and they often use the wrong theories for the situation they are in. It is the absence of conscious, trustworthy theories . . . that makes success . . . seem random.

—Clayton Christensen and Michael Raynor[44]

The technological mind-set of engineers and architects has focused in the past primarily on building things, as the comment at a workshop for consulting engineers suggested. This focus expresses the essential technological challenge of the industrial age but not the range of technological challenges that confront society today. Increased technological complexity is one of the new elements of social complexity, and many of the same factors that have led to dramatically increased social complexity have led to dramatically increased technological complexity. These factors, from the perspective of technological challenges relevant to the built environment, include the following:

- More population overall to be supported, and especially a greater need for basic infrastructure in the developing world.
- Larger and denser urban/suburban areas to be served, as reflected by a recent study by the World Bank and World Wildlife Fund-International, which counted 105 so-called megacities.[45]
- Degradation of the biosphere by huge volumes of nonbiodegradable waste based on man-made chemicals, overextraction of toxic metals,[46] overconcentrated organic waste material that cannot be assimilated, and invasive species.
- Depletion of fundamental resources, including clean water, forests, fisheries, fossil fuels, and, in some places, clean air, along with an accelerated loss of biodiversity, which has become the sixth great wave of extinction in the earth's history.
- Climate change affecting energy needs, availability of water for all human activities, including agriculture and industry, and the resources provided by biological ecosystems.
- Greater need to address built environment challenges at a whole-system level—locally, regionally, and globally based on greater understanding of interdependent dynamics.
- More regulation.
- Increased need for physical security of the built environment, including basic infrastructure.
- Water, energy, and transportation infrastructures in developed countries that are inadequate to current loads, aging, and based on old technologies.
- Fundamental changes in the infrastructure of major human activities based on information technology, with more changes coming soon from biotechnology and nanotechnology.
- Greater need to keep up with expanding knowledge and to integrate and coordinate diverse technologies.

In recent decades, architects and engineers have typically taken advantage of new business opportunities provided by regulation, but have ventured to address other challenges only when asked by a client who perceives their importance.

The mind-set of "just building things" has expanded somewhat as natural and socially imposed restraints to growth have caused new construction to diminish as a percentage of overall professional activity. Two pressing sets of needs related to existing facilities have risen in importance: repair and retrofit on the one hand, and redesign and reuse on the other. In *The Restoration Economy*,[47] Storm Cunningham explores the surprising economic scale of this shift, extending its scope to include restoration of natural environments as well as built facilities and infrastructure. Professionals have gradually responded to restoration-related technological challenges, even incorporating new performance criteria of maintainability and adaptability into design for new construction. Despite this activity in their fields, the domains of the built environment professions retain a significant bias toward new construction.

Centering the domains on restoring as well as building, however, would not be sufficient to address the full range of challenges outlined above. Even if we narrow the list to focus just on the developed world, in particular, the United States, we face the needs to

- Provide for population growth in ways that produce livable cities and conserve our natural resources;
- Expand, replace, and update transportation (and other) infrastructure;
- Replace and update our electrical grid;
- Transition to new forms of energy;
- Reduce the impact of toxic compounds and elements on ourselves and the biosphere;
- Reduce the volume of waste not being recycled and reused.

Because these needs are so enmeshed with as yet unresolved social issues, they are fraught with controversy. For this very reason, leadership from built environment professionals who have expanded their capacities to deal with social complexity is especially needed. To more fully grasp the built environment technological issues embedded here, we want to suggest three significant shifts that are driving their expansion: new performance criteria, new scale and complexity of retrofit and replacement, and upstream issues usually considered to be outside the scope of architects and engineers.

New Performance Criteria

Though the built environment has been affected by a slower rate of innovation than many other arenas, that rate has nevertheless been significant. Computer-aided design, Geographic Information Systems (GIS) applications, new products, and new construction processes all require architects and engineers to maintain a brisker pace of technological learning than in the past—just to keep up. Such innovations generally make it easier for designers to meet existing performance criteria. At the same time, however, new types of performance criteria are also emerging, creating a level of complexity beyond acceleration. For example, the increasing integration of information and communication technology into buildings is enabling and requiring types of performance that did not exist even a few years ago. And as we move deeper into the knowledge age, in which value resides more in people than in capital assets, research into physical environment attributes that promote human creativity and performance is driving a new set of performance criteria tied to these aims.

Ecological Sustainability

Given the problems facing society, however, the most important new set of technological performance criteria relates to achieving a healthy relationship with the larger environmental ecosystems that support human life. The way we are currently living on our planet does not appear to be sustainable. The problems cited above will only worsen as our population continues to grow to an estimated 7.9 billion by 2025[48] and 1 billion new consumers accelerate consumption alongside this growth. There is broad agreement among leaders of the sustainability movement that the core problem is one of design. Our current systems, designed and built under the models and methods of the Industrial

Revolution, must be redesigned.

Perhaps the most ambitious as well as inspiring vision of new performance criteria for design is expressed by the architect William McDonough and his chemist partner, Michael Braungart, in *Cradle to Cradle: Remaking the Way We Make Things*. They envision an "eco-effective" rather than "eco-efficient" solution that does not focus on limits but instead mimics nature's abundance, diversity, and redundancy. They summarize their design criteria for eco-effectiveness as the following:

- Buildings that, like trees, produce more energy than they consume and purify their own waste water.
- Factories that produce effluents that are drinking water.
- Products that, when their useful life is over, do not become useless waste but can be tossed onto the ground to decompose and become food for plants and animals and nutrients for soil; or, alternately, that can return to industrial cycles to supply high-quality raw materials for new products.
- Billions, even trillions, of dollars' worth of materials accrued for human and natural purposes each year.
- Transportation that improves the quality of life while delivering goods and services.
- A world of abundance, not one of limits, pollution, and waste.[49]

While discussing the insufficiency and even negative impacts of the more familiar eco-efficient approach, McDonough and Braungart do acknowledge its value as a means of slowing down damage to the biosphere while redesign is proceeding.

The authors of *Natural Capitalism: Creating the Next Industrial Revolution*, Paul Hawken, Amory Lovins, and L. Hunter Lovins, agree that eco-efficiency, which they call *resource productivity*, is not enough. They believe that it must be accompanied by implementation of three other core principles: "biomimicry" to redesign industrial systems as "continuous closed cycles" that enable the "constant reuse of materials"; a "service and flow economy" that shifts perception of value from "acquisition of goods" to "continuous receipt of quality, utility, and performance"; and "investing in natural capital" to restore the biosphere to its full healthy capacity.[50]

The Natural Step, an international organization founded by Swedish oncologist Karl-Henrik Robert, provides a framework many companies are using to guide their sustainability efforts. This framework articulates a scientific consensus on four system conditions required to sustain life on the planet. These conditions include

1. The biosphere is not subject to systematically increasing concentrations of substances extracted from the earth's crust,
2. Or concentrations of substances produced by society,
3. Or degradation by physical means,
4. And human needs are met worldwide.[51]

One company that has used the Natural Step as the basis for its transformation to sustainability is Interface, the largest manufacturer of commercial carpet in the world. Its former CEO, Ray Anderson, recounts the story of his own shift of mind and commitment to this goal in *Mid-Course Correction*.[52] Moving the entire company from selling carpet to selling carpet tiles that it takes back for closed-loop remanufacturing, Anderson exemplifies the transition to a service-and-flow economy. We will return to his story in chapter 13.

If design is the fundamental sustainability challenge, architects and engineers are potentially in a very powerful position to contribute on at least three major fronts. First, the design and performance of buildings and infrastructure govern a large percentage of society's resource consumption and waste production. Second, the construction and maintenance of buildings and infrastructure require an enormous range and quantity of materials. The composition of these materials has a huge effect on the degree of toxicity of our environment. And third, the construction process itself produces the largest single waste stream of any industry. The design of facilities and infrastructure drives waste-producing processes associated with building as well as with demolishing in order to rebuild.

Though engineers and architects have always aspired to ensure the high performance and longevity of the infrastructure and facilities they design, they have been slow to respond to the new performance criteria inherent in the sustainability challenge. One factor appears to be that these new criteria were not part of their original training in the domain—either at the fundamental values level or at the technical level. Another factor, consistent with the short-term focus of the U.S. economy as a whole, has been that most clients and owners over the last 30 years or so have emphasized first cost rather than whole life-cycle costs. Architects and engineers have generally been frustrated in their attempts to persuade clients of a long-term approach that could also meet short-term requirements. Investments in long-term efficiencies have been perceived as simply not affordable with current dollars, or as too expensive relative to their benefits. Given the long-term focus inherent in the sustainability arena, professionals have shied away from attempting to lead their clients, especially given the relationship of diminished trust.

As clients have begun to shift, however, performance criteria for sustainability are finally beginning to enter the field. The primary vehicle has been the LEED (Leadership in Energy and Environmental Design) certification process, which focuses primarily on resource productivity.[53] LEED is a good first step, but if the authors cited above are correct, much more is needed. From a regained leadership position, architects and engineers could jump redesign for sustainability to a whole new level.

The importance of new types of performance criteria will always vary by client and client organization ecosystem. Increasingly, however, built infrastructure and facilities will need to meet not just the familiar functional and aesthetic performance criteria of the past, but also be "smart" with information and communication technology; enhance human well-being, creativity, and productivity in more holistic and demonstrable ways; and contribute to the sustainability of human and all life in the biosphere as a whole. As fundamentally new technologies continue to emerge, new types of performance criteria unimaginable now will likely continue to appear. Professionals need to be prepared to understand, assimilate, and work creatively with them.

New Scale and Complexity of Retrofit and Replacement

Achieving the change-out of Industrial Revolution structures and processes to sustainable ones through all-at-once processes seems unthinkable. The growth of local facility cogeneration and a distributed power model in response to the seriously unreliable U.S. power grid reveals a different process. About eight percent of all electrical capacity in the United States now comes from cogeneration facilities.[54] Mimicking the movement from centralized to distributed computing, this growth may point the way toward a self-organizing approach to large-scale redesign. The degree to which sustainability goals will be achieved through large-scale replacement projects remains to be seen.

Large-scale retrofitting or replacement of water, electrical, or transportation infrastructures in densely populated areas, however, will likely be required. Although these are large-scale structures, they were built incrementally in smaller elements or segments. Even projects that were undertaken at large scale were generally new construction that did not have to take into account existing structures. Today, in many retrofitting or replacement projects, large scale must be achieved while minimizing disruption to existing structures and functions. These are technological challenges of a scale and complexity not quite seen before.

Keeping a hospital running while it is being remodeled is a challenge—as many architects have discovered—but keeping a big city running while its transportation infrastructure is being rerouted under its surface represents a quantum leap in complexity. This is exactly what Boston has attempted in its famous Central Artery, or as it is familiarly known, Big Dig project. Planning for the project began in the 1970s, and its disruptive construction period has spanned 20 years. As might be expected, the learning curve on this level of scale and complexity has been steep. Everyone seems to agree that at least the new tunnels and unique bridge are terrific. However, extreme budget and schedule overruns have produced a great deal of physical disruption and financial pain. Today, bitter political conflict embroils all the engineers involved.

The planning and management challenges of a project of this scale and complexity presented a new technological challenge. At the same time, much of the pain has resulted from underestimating the degree of social complexity that would have to be addressed to be successful. No doubt, more projects like the Big Dig are on the horizon. If architects and engineers want to regain their leadership position in 21st-century society, they must adapt to the new social and technological complexity such projects will present.

Upstream Issues

In large-scale infrastructure retrofitting or replacement projects, a long-term transition process must be designed and planned alongside the design and construction of the physical infrastructure itself. But before any of that, an even more fundamental step must be completed: the envisioning of a viable or optimal future state. For a large-scale project, this envisioning process is itself a huge undertaking, and a hugely important one, for it will drive and guide all planning and design. Whatever the project size, the vision of the client and other members of the client organization ecosystem governs what becomes possible in the planning and design.

At the largest scale, the range of systemic issues to be considered in envisioning an optimal future state is mind-boggling. Population distribution drives where we build, but where and how we build also drives population distribution. The positioning of population relative to resources determines what and how much will have to be transported from elsewhere. The relative positioning of facilities for living, working, shopping, and recreating, as well as for agricultural and industrial production, further determines transportation needs. This positioning also produces environmental impacts that can in turn affect resource availability. While air, water, and food are requirements for human (and all) life, the forms of fuel we use to produce energy are not. Decisions we make about these fuels determine the kind of infrastructure required and impacts on the earth's biosphere and resource base.

No single domain currently combines the issues of land use planning with those of energy sources. Energy source issues have been left to government and the energy companies, while land use planning has been defined as the domain of city/urban, regional, and environmental planners. Planners, however, represent a small segment of the array of professionals addressing the built environment. They have had a significant impact on policy for growth and land use in pockets, but their leverage for whole-system issues on a large scale has been very limited. Instead, land use and the policies governing it have played out in the drama between developers, driven primarily by economic return, and public agencies and advocacy groups. Part of the problem is an enduring division in public sentiment. While some fret about traffic, crowding, and the environmental degradation they view as inherent in urban sprawl, others strongly support the jobs and prosperity they perceive to flow from unplanned growth.

Generally speaking, built environment professionals (other than planners) have left these issues—which we might call "whole-system requirements"—to others, choosing to be the implementers of those others' decisions. Occasionally, professional associations have attempted lobbying efforts on behalf of whole-system issues; for example, the AIA's recent effort to advocate for livable cities and sustainable communities at the state level in California. More often, associations have focused on legislation that directly supports their businesses, promoting the public's identification of architects and engineers with the economic interests driving growth. To the extent that professionals have withdrawn from local and regional whole-system planning, they have lost the point of greatest leverage from which they could fulfill their professional mandate to serve society as a whole beyond monetary value. While architecture and engineering professionals obviously cannot and should not take on these newly complex responsibilities alone, neither should they simply leave them to others.

In the early part of the 20th century, when the issue was only building things, engineers and architects did provide this level of visioning and guiding. As the challenges have become more complex both socially and technologically, these professionals have felt disempowered by too many issues perceived to be outside their discipline, especially in the face of other stakeholders asserting their empowerment. An expansion of the domains to include these upstream issues would focus professional energy and creativity on challenges sorely in need of sound approaches. However, the extent to which each domain is currently identified with its discipline stands in the way of this expansion. Focusing some architects and engineers on becoming planners as well might be helpful. However, the planning discipline is itself not adequate to the complexities, nor is this the best way to address whole-system requirements at a significant scale. Working across disciplines within or across firms and other organizations is a far more effective approach.

The same discipline constraint that shows up at the macro level of whole-system planning surfaces in the micro-level upstream issue that McDonough and Braungart so effectively highlight in *Cradle to Cradle*: materials themselves. The LEED certification limits the professional's responsibility to the selection of materials, which reflects the traditional view of the boundaries of the disciplines. Just as ceding land use planning to others moves architects and engineers out of strategic advising into an implementation role, the selection role excludes professionals from materials development at a time when innovation in this area may be the planet's most pressing need. Again, the point here is not to turn engineers into chemists, but to bring the chemistry of materials into a domain where professionals from many disciplines can work on it together.

The three new forms of technological complexity, like the new social complexity, challenge traditional ways of thinking about professional services for the built environment. This new complexity is a lot to hold in the mind and difficult to organize on a practical level. To help professionals navigate this new level of technological complexity, we want to introduce another powerful mental model, the value network. Complementing the organization ecosystem model, the value network enables us to extend our living systems worldview into the technological realm.

Seeing the Big Picture with Value Networks

Clayton M. Christensen, whose application of the technology S-curve idea we have already referenced, published *The Innovator's Dilemma* in 1997, the same year that de Geus published *The Living Company*. Christensen contends that to really understand the value that is being provided to a client by a product or service, we must understand the larger system-in-use in which that product or service is embedded. For this, he introduces the concept of the "value network." In the case of a product, the value network consists of "the nested physical architecture of a product system [and] . . . a nested network of producers and markets" that is implied by the product architecture. A value network expresses "the context within which a firm identifies and responds to customer's needs, solves problems, procures inputs, reacts to competitors, and strives for profit."[55] Together with the "value criteria" embedded in it, a value network enables us to see how value is created, assessed, and exchanged within the larger system of which it is a part.

Christensen uses the disk drive industry of the mid-1990s to illustrate how specific a value network is to a particular product or service, and to a particular set of customers. The value network for mainframe computers was very different from the one for portable personal computers and from the one for computer-automated design and manufacturing. In the mainframe computer value network, the value criteria for disk drives emphasized capacity, speed, and reliability. In the portable PC value network, the value criteria emphasized lightness, small size, ruggedness, and low power consumption. The value differences between these two networks were so profound that disk drive manufacturers were seldom able to migrate from one to another or compete successfully in both.

Christensen's central point is that companies can be blindsided by a disruptive technology because their view of the value of their own technology is limited to the value network in which it is currently contained. Such a view does not recognize how a product of another value network can evolve to provide a whole new form or level of value to their current customers. In Christensen's example, the product architecture for mainframe computer disk drives places the "disk architecture," with all its elements, in the center of the product architecture. The "disk drive architecture," with all its elements, appears in the next ring out, the "mainframe computer architecture" in the next ring, and the "management information system architecture" in the outermost ring. This outermost ring includes elements as physical as "configuration of remote terminals" and as intangible as "service and repair requirements" and "careers, training, and unique language of EDP staff." This product architecture makes clear that the disk drive has virtually no value without the disk on the one hand, and without the mainframe computer or the management information system in which it is embedded on the other.

This insight parallels the breakthrough leap in thinking that produced Moore's business ecosystem model. In the same way that a business organization does not exist in isolation, neither does a product or service. Its value depends on the larger system of end-use of which it is a part. Christensen's "nested network of producers and markets" closely mirrors Moore's business ecosystem idea. While the organization ecosystem helps us understand the organisms and relationships that form the client's web, the value network specifies the functions and forms of value exchanged by these organisms through their relationships.

Christensen argues that the value network concept applies across all kinds of industries, including services. Because professionals involved in creating or supporting the built environment are linked to the physical products their designs and solutions produce, it is tempting to anchor value-network thinking for their industry in a product architecture—say for a building or a system of stormwater drains and pipes. However, it is the contractors and builders who directly produce the physical products. Strictly speaking, professional services provide just that—services to client organizations and to organizations that directly produce those products. In order to effectively leverage Christensen's value network concept, therefore, we must translate his product examples into a service framework. This requires that we switch our thinking from the physical tangibility of a product architecture to the less tangible notion of a service architecture. Given that the convention is to place the smallest freestanding element of the architecture in the center and build out from there, we could draw a service architecture for interior design as part of new construction as shown in figure 6.1.

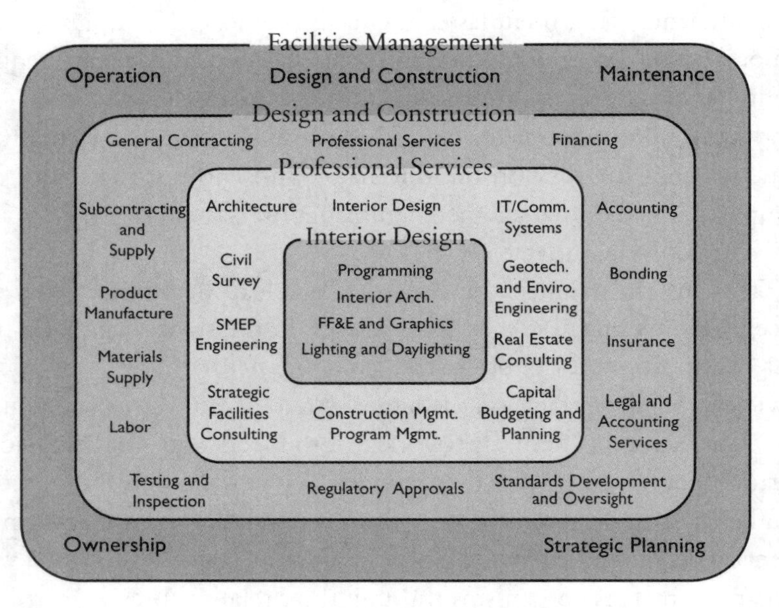

FIGURE 6.1. Interior Design Service Architecture

In the telescoping convention of this way of depicting a service (or product) architecture, interior design is itself one of the elements of the next ring, which shows all the professional services required for interior design to have value for a client in the context of new construction. We could place any of the other professional services in the center of this generic service architecture, but have chosen interior design for this example. As shown in the center, interior design in turn includes interior architecture, programming, FF&E, lighting and daylight, and other functions. Moving to the third ring, professional services appears as one of the larger groups of services and functions required to achieve design and construction as a whole. The outer ring, showing facilities management, is highly schematic, showing only a few of the many functions that make up this level of the system, including design and construction. The interior designer viewing this service architecture is reminded of the common-sense truth so easy to lose sight of: that the value of what he or she provides depends on all of the other functions included in the service architecture.

To add the nested network of producers and markets to the product (or service) architecture to represent the full complexity of the value network concept, Christensen devises another type of diagram, which he titles the value network itself. This diagram presents a set of boxes vertically arranged with the name of the product architecture system level in the top box and the names of the component levels in the boxes below, in descending order. While this diagrammatic form makes space to write in manufacturers and specific value criteria, it allows the representation of only two subcomponents for each level of system, one on each side of the central boxes. Christensen acknowledges that "these depictions are not meant to represent complete structures."[56] In fact, they are less evocative than the product (or service) architecture itself in suggesting the full structure of the value network. Though it requires working with separate lists of providers and value criteria, we prefer the service architecture as a representation of the value network in its full complexity and will use that diagrammatic form in this book.

In translating the generic service architecture for interior design into a value network for a particular market, the interior designer would have to consider the variations in value criteria that characterize a client industry. In retail, for example, the value criteria relevant to a Target store would be very different from those relevant to a Nordstrom. Target would look to support its low-cost, high-convenience merchandising, while Nordstrom would be seeking the appropriate physical context for its higher-end aesthetic values and extraordinary customer service. While being clear about the value criteria differences, the interior designer would also benefit from being aware of the extent to which—in the manner of Christensen's disruptive technologies—low-price versions of designer goods sold at stores like Target have begun to cannibalize the business of Nordstrom and other higher-end stores. This awareness might have significant implications for changing value criteria.

Christensen also explains that firms may structure their range of products or services in a given value network in different ways, more or less integrated or segmented. For example, a design firm that provides planning, program management, architecture, multiple engineering disciplines, and landscape architecture may provide an integrated approach to the entire professional services ring of a value network. For such projects as a new airport or toll road involving international transportation, a large engineering construction firm may take an integrated approach across multiple levels of a facilities management value network in a design-build-finance-operate mode. In a highly segmented approach, many different specialized firms may contribute to a given level in a value network. In a horizontally segmented but vertically integrated approach, a firm can address individual services across levels, such as a consulting engineering firm that provides technical studies, permitting and regulatory support, and community and environmental planning. Decisions about integration vs. segmentation are a critical component of strategy for professional service firms. We will return to this subject in chapter 10 and again in chapter 14.

Christensen's nested network of producers and markets emphasizes only some members of what Moore would identify as the business ecosystem. However, if we play out Christensen's logic fully, the value of different services in a facilities management service architecture does in fact depend upon such entities as regulatory agencies, community groups, professional associations, and trade unions as well as on providers of complementary services and competitors. The business or organization ecosystem is quite literally the counterpart of the value network in this sense, providing a model for understanding the full context in which value is created, assessed, and exchanged.

Christensen bounds his disk-drive value network examples one level up from the discrete computing product, but places no inherent limit on the scope of his outer ring of the architecture. Allowing Moore's concept of the business ecosystem to enrich Christensen's notion of the value network, we can add rings to reflect functions performed at the level of the client organization as a whole and at the level of the organization's ecosystem. And to complete this logic, we can add an outer ring that reflects the larger societal and environmental context we added to Moore's model. The resulting generic value network for any professional service involved in construction is depicted in figure 6.2.

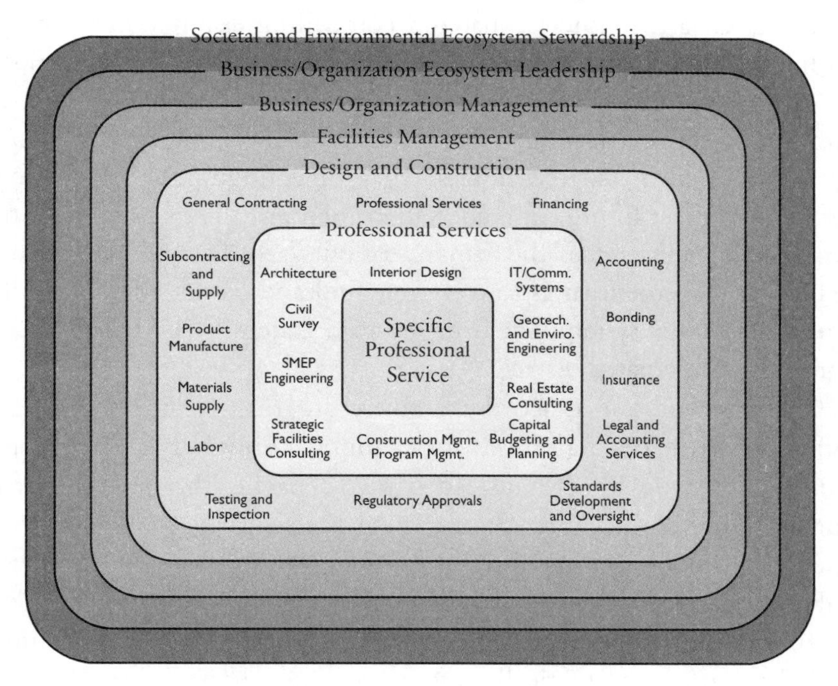

FIGURE 6.2. Generic Value Network for Construction-Related Professional Services

This expanded generic value network provides a conceptual and visual model for understanding the full context in which professionals provide their services. It provides the basis for mapping the value network for a particular type of service for a particular type of client ecosystem along with the accurate identification of value criteria for that service. In so doing, it promotes the pursuit of clarity about how value criteria for a particular service vary between types of client ecosystems. It also helps professionals see better how to provide inner-ring services in ways that not only integrate effectively with other components at the same level, but also support the functions being performed in the outer rings. And, most important for the upstream issues involved in technological complexity, it opens possibilities for providing services in rings beyond the professional's usual level.

One of the firms that participated in our Discovery project, Lucchesi Galati Architects, shared a compelling example involving a utility client. The client was looking for design services to build new facilities. The firm instead suggested that it study the company's utilization of existing facilities. The study concluded that new facilities were not needed and recommended the reconfiguration of existing facilities to meet future demand. Rather than provide its usual suite of technical design services, the firm moved out to the facilities and organization management rings to creatively find and design efficiencies in the management system. The firm was paid gratefully and handsomely for this unanticipated service.

Another of our Discovery architecture firms, Anderson Brule Associates, developed a process for enabling a university and a city to build a common library (instead of two separate libraries), the first such instance in the United States. The processes involved in getting to the design required a melding of two very different—even antithetical—organizational cultures and sets of operational processes. The firm not only provided professional services for design of this facility, but also found a dramatically resource-saving solution in the organization management ring and exercised organization ecosystem leadership to bring these two organizations and their ecosystems together in a completely new working relationship. You will hear more about their story in chapter 12.

Clearly, moving upstream into the outer rings of the expanded value network is one way that built environment professionals can create value and reclaim a powerful leadership position. We will explore the shifts of mind required to begin to leverage this possibility in chapter 8. Leadership can also be reclaimed, however, while staying in an inner ring. This requires solving new problems or solving old problems in new ways, through the technological innovation discussed in chapter 15. Essentially, the expanded value network gives professionals a framework for recognizing ways to work in all three realms of value creation revealed by the organization ecosystem perspective.

To summarize, then, the expanded value network provides a means of understanding and organizing the full range of technological complexity confronting the built environment in the 21st century. The value criteria linked to each service or function in the value network anchor the discipline of attending to new performance criteria, now and in the future. The inner rings provide a means of mapping the increased number of functions and disciplines that must be included and integrated at each level. And the outer rings contain the upstream issues in a way that makes them visible in their full context, particularly when the value network is used in conjunction with the organization ecosystem. Because it contains social as well as technological services, the value network provides a framework for navigating and creating value with respect to technological challenges in their full social complexity.

The new social and technological complexity we have explored in this and the preceding chapter, along with the new models for understanding that complexity, have revealed that the domains of architecture, engineering, and related professions must be updated if they are to support practitioners effectively in the 21st century. The update must include expanded core definitions of the domains as well as similarly expanded sets of values and ethics. We will synthesize ideas for updating the domains in chapter 9. This synthesis will depend not just on understanding the new forms of complexity, but on the new ways of thinking required to address them effectively. The new thinking required to once again exercise societal leadership will draw heavily on the organization ecosystem and value network models.

Before professionals can hope to take this kind of leadership, however, they must first close the gap between their performance and their clients' expectations. Let's turn now to the new ways of thinking we believe are required to close that gap and thereby restore fundamental trust.

7
Shifts of Mind for Catching Up

People only see what they are prepared to see.

—Ralph Waldo Emerson[57]

New ways of thinking rarely come easily to any of us. Our current ways are based in values and beliefs that have long since become assumptions, operating beneath the level of our awareness. To see new things, we must be prepared to see new things, which involves first seeing the constraints of our former way of seeing. As we begin to explore the new ways of thinking required for catching up with clients and stakeholders, we will find that they pose a different kind of challenge from those required for getting out ahead. To get out ahead, professionals must stretch into arenas currently outside of familiar professional competence. To catch up, professionals must instead do familiar things differently. In this territory, shifts may seem deceptively simple or unworkably subtle. And in this territory, changes may feel very personal and potentially identity-altering.

Closing the gap with clients and stakeholders must involve more than simply adopting their mind-set. It must deeply acknowledge them and effectively include their concerns while establishing a strong basis for empowered professional contribution. Table 7.1 identifies the assumptions we believe are creating the most serious gaps. For each, we propose a new way of thinking that we believe can close the gap, restore trust, and provide the basis for doing "good work" once again. In each case, the shift does not entail simply replacing the prior focus, but incorporating its essential elements into a broader orientation.

From...	To...
Task and product focus	Relationship awareness
Autonomy in project work	Whole-project-system awareness
Individual control of preferred solutions	Group process awareness and orientation to emergence
Either-or thinking	A nondualistic, inclusive view
Assumed authority	Self-aware professional contribution

TABLE 7.1. Shifts of Mind for Catching Up

Relationship Awareness

Much attention has been focused on the subject of relationship in professional firm practice over the last 15 years or so, mostly from a marketing and business or strategic planning perspective. Much less attention has been given to the way in which relationship awareness plays into performing the work itself, or into the psyche of the professional performing the work. It is much easier to reposition one's marketing approach or organizational structure than to fundamentally alter the mind-set of a professional whose expertise and sense of accomplishment center on the product or outcome of the work rather than the relationship through which it is accomplished. This change in mind-set is tightly linked to the equally deep shift in role from the expert who is given autonomous authority to do the work to the project participant who must collaborate before and in the process of making his individual contribution.

In traditional organizational development, this awareness accompanies the growth of the individual beyond a technical contribution into a management or leadership role. The individual must now pay attention to both task *and* relationship. The manager's or leader's outcome is not just the work performed at a certain point in time, but the development of individuals or teams capable of performing the work with greater capacity in the future. Because this development depends on the relationship the individuals or teams have with the manager or leader, it must now be considered a primary asset for sustaining present and developing future capacity. Today, the notion of relationship as the means to the end of task is beginning to disappear. From a living system or living company perspective, the human community is viewed as the organization's primary asset, and relationships among individuals are as important as individuals themselves. In that context, task may even be viewed as a means to the end of relationship.

The technical professional's acquisition of the relationship awareness and related competencies that enable him to become an effective consultant parallels the journey from employee to organizational manager or leader. However, translating this mind-set into external relationships with clients and others tends to be challenging for the built environment professional—despite awareness that a good relationship is likely to lead to another project. We doubt that project stakeholders feel that their relationship with professionals is more important than the outcome of the work for which they do indeed seek expert help, but experience tells us that stakeholders put more importance on the quality of relationship than professionals do. We believe that the best mind-set for the professional catching up to stakeholder expectations is to hold task and relationship in equal importance and to understand the degree to which clients see them as inseparable. Client relationships are more than a means to an end. They are fundamental elements of the firm's living system that enable it to thrive, and thus fundamental outcomes of the work itself.

The recently established field of "emotional intelligence" has powerfully reinforced these truths. Research on management success indicates the following:[58]

- Emotional intelligence (EI) is the differentiating factor in success.
- 90 percent of the difference between outstanding and average leaders is linked to EI.
- EI is two times as important as IQ and technical expertise combined.

The seminal researcher behind emotional intelligence, Daniel Goleman, has identified four clusters of emotional competence: self-awareness, self-management, social awareness, and social skills.[59] These clusters reveal strong developmental relationships to one another. Both self-management and social awareness require self-awareness. And social skills require both self-management and social awareness. A manager or leader who does not place significant focus on these areas of competence will almost certainly be limited to low performance.

Goleman's social awareness is the capacity to recognize and work with the feelings of others while recognizing and working with one's own feelings. Our concept of relationship awareness emphatically includes these emotional dimensions, as well as competencies associated with other aspects of relationship. We address these competencies as the social capacities for delivering higher value technical work in chapter 20. Our subject here is the shifts of mind underlying these competencies. From that point of view, the recent research on emotional intelligence drives home the critical importance of shifting to relationship awareness as the essential context for project task, product, or outcome.

Given the overarching importance of relationship awareness, it is important to acknowledge that not all relationships are created equal. In their drive for economic survival, built environment firms have often taken on projects with clients whose values were severely at odds with their own. Every firm learns hard lessons in this arena, ultimately determining the need to part ways with some of these clients. But few firms feel they have the luxury of being selective. With the crop of new business models we envision, more differentiation among firms will be possible. More differentiation will likely provide firms with more opportunity to work with clients who are a good fit.

Whole-Project-System Awareness

Becoming attuned to the integration needs of the project as a whole, regardless of position within that whole, requires two more shifts of mind that may be just as challenging as adding relationship awareness to product focus. First, professionals must fully acknowledge that their role does not correspond to control of the whole project. The degree of this shift increases with project size and complexity, of course. But even on a small project, or where a professional is lucky enough to be in a relatively upstream strategic planner or program manager position, or a prime rather than a sub, the role rarely entails whole-project control. Most professionals will experience some degree of disempowerment in the face of this shift. It cuts at the very heart of their professional motivation—at the ambition and aspiration to make their highest and best contribution. What is required here is the acceptance that individual self-expression will be diffuse at best, diluted in a composite of the self-expressions of many other project participants.

Even in their area of specified scope, professionals do not have complete control. They must integrate with other processes and parts and the people who represent them. Letting go of the project-as-a-whole might be reasonable, if at least the professionals felt some sphere of control where their expertise could fully express itself. These situations are rare in today's complex projects. Undermining the notion of any kind of autonomous authority, this second shift often manifests as the ultimate experience of disempowerment.

When the professional does not feel appropriately valued, his self-esteem suffers. We want to suggest a way to reframe this inevitable experience. The flip side of this sense of diminishment is the experience of being part of something much larger than oneself. As we know in other contexts, it is possible to feel uplifted by this awareness. To feel empowered in this position will require the professional to develop stronger social and leadership capacities (discussed in part IV). Such capacities can enable the professional not only to exercise potent influence within the scope of a limited role, but also to have broader impact. With a sophisticated understanding of the project-as-a-whole and the interrelationships that will drive its successful integration, a professional's influence can ripple throughout the whole-project system as necessary linkages are made. With due humility and respectful role awareness, a role of any size becomes a platform from which to build vital relationships and exercise leadership. This reframing, however, depends on the capacity to gracefully leave one's professional ego at the door in the service of the larger whole.

Group Process Awareness and Orientation to Emergence

For professionals who are deeply attached to wrestling individually with a design or technical challenge, or to seeing their own creative ideas purely expressed in project outcomes, working in a group involves a fundamental shift. First, group process is messy and takes enormous patience. When multiple stakeholders are present, multiple viewpoints are inevitably also present, along with multiple personalities and styles. Even well-led group process, when viewed through the lens of task or topic, is at best nonlinear and at worst completely chaotic. This complexity seems to increase exponentially with the size of the group, to the extent that individuals are empowered to participate.

The professional's natural impatience may result from the doubt that other project participants can contribute at the same level that he can. Or it may be a defensive reaction to the sense of losing control that tends to evolve as group process inevitably challenges attempts to organize it. The complex group dynamics that evolve, however, are not just working on the task. They are working on the group as an entity—the group mind, being, and capacity—and they are working on the relationship of the individuals and group to the task.

When group process goes well, the payoffs are a group of stakeholders who feel good both about each other and about the quality of the outcomes they have achieved together. Furthermore, they feel alignment with and ownership of these outcomes and are ready to steward them through the rest of a long and complex process. Professionals working more autonomously and then trying to sell their solutions to today's stakeholder groups know all too well the experience of having their ideas rejected because those groups have not been sufficiently involved in developing them.

Parallel to awareness of relationship, the professional needs to develop awareness of and respect for groups as organisms in the project's living system. If the project involves stakeholders other than key decision makers in the client organization, project groups will likely be diverse and complex. If the professional can detach from the product and outcomes sufficiently to embrace these more complex social organisms, impatience can evolve into curiosity and respect. An immediate benefit is greater understanding of the larger living system to be served—better input for the professional's own design or problem-solving contributions. Together, members of the client's organization ecosystem know far more about the system that needs to be expressed in the solution than the professional entering the system from the outside. Beyond this immediate benefit, the professional may find something even more exciting: a surprisingly creative collective collaborator.

Our phrase "orientation to emergence" refers to openness to a collective, creative process that is essentially unpredictable. In its technical usage in the new life sciences of complexity, "emergence" is the outcome of a self-organizing process whereby a living system, suspended in, connected to, and interacting with its environment, arrives at a new state substantially different from its prior state. This technical sense draws directly on the more familiar meanings of *emergent*: "rising as if out of a fluid," "arising unexpectedly," "arising as a natural consequence."[60] While we will be making more literal reference to the technical sense later, we refer to it here only in the background of the more generic meanings.

The messy and sometimes stressful group dynamics in play alongside the stew of diverse ideas are part of the mysterious emergent process and the often unpredictable outcomes it produces. Disagreements mark spots where creativity can arise or be exercised. If the professional can embrace a self-organizing group process rather than fight it because it doesn't fit a picture of a more orderly way that things should happen, its richness can provide a new set of possibilities for design and problem solving. This is another way in which the professional—in the context of work that calls for creativity rather than routine approaches—can discover a bigger sandbox.

To leverage this creative power of the group, the professional must first let go of the need for project outcomes to express his own creativity in a pure form. With this must go attachment to prior, preferred ways of thinking about the challenge at hand. This is easier said than done for those who tend to consider favorite, tried-and-true solutions the shortest and safest route to closure. Further, the professional must be willing to live with a lack of certainty that a new solution will emerge either from the group or from himself.

Interestingly, the nature of the process for individual creativity is not very different from that for group creativity. Though the stages of preparation, saturation, incubation, and illumination[61] occur inside the individual mind through a rhythm of accessing its unconscious as well as conscious dimensions, the process is hardly more linear or predictable—or, for that matter, less anxiety producing—than group process. Perhaps the professional's deepest resistance to group process lies in the belief that the primary locus of creative process is the individual, and the desire to engage in individual process for the purpose of his own self-expression. However, genuine openness to and curiosity about the group's capacity give the professional access to even greater creative potential. Individual creativity is not excluded from but flows into the group creativity. And richer, more diverse material flows through more "mind" than any individual can muster.

A Nondualistic, Inclusive View

While professionals have traditionally sought to serve society and take the longer term view, they have typically seen these goals as opposed to the client's organization-centered, short-term concerns. We have all been conditioned to think dualistically: to see different or apparently contradictory options as mutually exclusive rather than as both possible and thus worthy of simultaneous consideration. The ability to register and consider apparently opposing concerns from different stakeholders is a significant skill unto itself, and many professionals are already capable of this kind of pluralistic awareness. The rub comes, however, with the underlying logic we use to organize and work with the pluralistic data and information we take in. We tend to collapse oppositions into dualities requiring simple trade-offs.

One of the best observations of our habits of mind is Robert Keidel's inquiry into organizational structure, *Seeing Organizational Patterns*.[62] Keidel points out that almost all organizational structure reflects a dualistic approach to design. Organizational "architects" play along the axis of control vs. autonomy, or autonomy vs. collaboration, or control vs. collaboration. He suggests that a triadic approach—including all three parameters of control, autonomy, and collaboration—much more adequately reflects the complexity organizations actually deal with.

If we look closely, we can see that thinking about autonomy vs. collaboration, for example, in effect reduces the number of variables we are considering at any one time to one. Autonomy appears in the mind as an absence of collaboration, or collaboration as a lack of autonomy. Thinking in threes produces the interesting effect of exploding out of this polarity trap. When we force ourselves to consider control, autonomy, and collaboration, we recognize that each of these is in effect its own variable, and we jump from considering one to three simultaneously. From this perspective, we now have the capacity to see all three oppositions that are inherent in the triad as consisting of two separate variables that need to be considered in their own right. From here we can say, yes, it is possible to have both autonomy and collaboration in an organization.

Professionals may be familiar with the triadic homily: quality, cost, and speed—you can have two of these, but not all three. Keidel would say that this is still a dualistically based simplification. He asserts that, of course, there are trade-offs in organizational design in the sense that it is very difficult and probably not desirable to put equal emphasis on all three of the core parameters all of the time. He asserts equally strongly, however, the need to consider all three, as well as to consider that the balance between them may shift as the organization adapts to changing conditions. The triadic approach is showing up in emerging approaches to ecological sustainability, such as the so-called triple bottom-line of people-profit-planet or equity-economy-ecology. Each of these sets of needs must be considered in its own right, and the operative word governing their relationship to one another is *and*.

The mind-set required to take the inclusive view, in which all concerns and aspirations are considered in their own right, the short term is viewed in the context of the long term, and the client's needs are addressed in the context of larger societal needs, is the capacity to move beyond dualistic thinking. While this mind-set must be present to some degree to catch up to clients' and stakeholders' expectations that all of their concerns will be included, it is even more important to getting out ahead of these groups in order to lead.

Self-Aware Professional Contribution

With all this focus on the client and other stakeholders, the whole-project system, group process and emergence, and an inclusive view, holding on to a strong sense of professional self is clearly a challenge. Yet this is absolutely critical in order to fully offer and leverage professional expertise and creativity. To the extent that a professional remains unaware that clients and stakeholders expect collaborative expertise rather than autonomous, expert authority, she will tend to view any expression of this expectation as a personal attack on her professional identity. This response results in an unfortunate vicious circle. The professional who feels disrespected will in turn withhold respect and be less able to hear and effectively understand client and stakeholder concerns. A client who feels disrespected and unheard will tend to withdraw from the relationship, which will in turn produce even more defensiveness from the professional.

The way out of this vicious circle involves becoming aware of and shifting several interrelated assumptions. First, the professional must resist the temptation to think dualistically about clients' and stakeholders' response to his way of working. Just because clients and stakeholders reject the autonomous expert mode doesn't mean they reject high-level professional contribution. Second, the professional must understand that the gap between clients' and stakeholders' expectations and her own mind-set does not have its origin at the personal level. Though the professional expresses and embodies the current state of the domain, that state is a function of much larger systemic forces, as we have discussed. At the level of cause, the professional must resist the temptation to frame the gap as a personal issue. And third, the professional must recognize that clients' and stakeholders' experience of the professional mode they don't want is very personal indeed. While not taking questioning or disappointment personally, she must conceive of a new way of being at a very personal level.

As the research on emotional intelligence makes clear, the underpinning for relationship and social awareness is self-awareness. Self-awareness entails being aware both of what is going on inside oneself and of the effects that one's behavior has on others. In order to be effective in the context of more inclusive and collaborative social processes, professionals must pay attention to both of these aspects of self-awareness. They must see this attention not as a threat to their professional identity, but an expansion of it—the very basis for making an even more valuable professional contribution. Expanded and ongoing self-awareness essentially becomes the vehicle for making the passage from disempowerment back to professional self-confidence.

Operating in arenas where there is so much other input—both from other technical professionals and from nontechnical stakeholders—requires the professional to perform one other fundamental shift of mind. At the very core of her professional identity, she must begin to focus as much on what she doesn't know as on what he does. In the context of complex social processes, a professional who can rely on other participants' expertise while maintaining rigorous transparency about the limits of her own is in the best position to make powerful offerings from the core of her professional competence. The professional's focus on what she doesn't know also paves the way for lifelong learning in the broadest relevant contexts. Given the current and likely future pace of expanding knowledge and complexity, such learning provides the only means of maintaining high-value professional contribution.

As we contemplate the profound changes architects and engineers will need to make in order to catch up with client and stakeholder expectations, we must ask how these changes can best be made. Our experience over many years of facilitating change at all levels, from the personal to the organizational, indicates that some people learn by changing their minds first, which then enables them to change their behavior. Others learn by trying new behaviors that produce experience which then changes their minds.

We think that most of you who choose to read this book will be oriented first to ideas, hence we begin with them. Each of the five major shifts of mind discussed in this chapter—as well as the five we discuss in the next chapter—is associated with a set of competencies that translate behaviorally into new value creation. We will take these up in part IV. We hope that those of you who do learn best by first understanding and trying new behaviors will bear with our head-on approach to the underlying assumptions and beliefs. You may want to return to these early chapters to take them in more deeply once you have made your way through part IV.

8 Shifts of Mind for Getting Out Ahead

The significant problems we face cannot be solved

by the same level of thinking that created them.

—Albert Einstein[63]

Things were simpler in the first two-thirds of the 20th century, when engineers and architects exercised societal leadership for the vast building projects that created today's cities and infrastructure. While the world has changed drastically, these professionals have continued to operate from the same fundamental assumptions that applied earlier. As a result, they have fallen further and further behind in the eyes of clients and stakeholders. The awareness of self, relationship, and group, discussed in chapter 7, provides the foundation for more effectively including clients and stakeholders in decision-making processes. These forms of awareness also provide the foundation for exercising the basic forms of leadership that clients are seeking to help address the social and technological complexities of their projects.

Once professionals have achieved the shifts of mind required to close the gap and restore trust with clients, they can then begin to contemplate exercising a more advanced level of leadership. Compelling new leadership will require new thinking that leverages the social and technological complexities of today's built environment projects for new levels of creativity and innovation. To imagine that such new thinking will be easy or comfortable flies in the face of what we know about any serious leadership endeavor. Stepping outside the comfort zone of what is already known will definitely be required.

The shifts of mind described in this chapter are likely to strike professionals as outside of their discipline, as somebody else's job. We believe it is just this kind of expansion that will be necessary for architects and engineers to get out ahead in a meaningful way. We describe these shifts in terms that draw heavily on the new mental models introduced in chapters 5 and 6 as powerful conceptual structures for navigating the new forms of technological and social complexity. Table 8.1 identifies the new ways of thinking along with the corresponding familiar ways of thinking. In each case, the new thinking incorporates rather than replaces the old.

From...	To...
Tactical thinking	Strategic thinking
Focus on parts or entities in isolation	Systems awareness
Organization focus	Organization-ecosystem awareness
Human system focus	Whole-ecosystem awareness
Technical services focus	Whole-value-network awareness

TABLE 8.1. Shifts of Mind for Getting Out Ahead

Strategic Thinking

Implicit in the capacity to provide leadership is the mind-set of strategic thinking. Tactics and strategy both involve thinking about the future, but in fundamentally different ways. Tactical thinking involves asking, given what we have decided to do, how should we do it? or given where we are and what we are already doing, what should we do next? Strategic thinking, in contrast, asks, given where we are, what should be doing and why? or why are we doing what we are already doing and does our thinking still hold? In other words, strategic thinking assumes that we have some power to shape the circumstances in which we find ourselves, and that we can best exercise this power by considering fundamentally different courses of action.

This consideration of fundamental alternatives requires knowing what we want to achieve or where we want to go. It involves being willing to continuously consider what is most important and valuable, and why. Strategy is intrinsically oriented toward a purpose or goal. It also requires our best attempt to understand the relationships and dynamics operating in our environment so that our actions may be based on effective engagement with that environment. We will discuss ongoing disciplines to support these aspects of strategic thinking in chapter 19.

For many professionals who are accustomed to performing technical work, strategic thinking represents a significant shift of mind. Most technical work is governed by prior knowledge and procedures both for understanding a problem and developing a solution. It calls for a professional to operate within these proven processes rather than to ask whether they are the best way to approach a problem. Even when a professional is being innovative on the technical front, the questions being addressed tend not to embrace the most fundamental alternatives. These inevitably entail more holistic consideration of the client system, including its social dimensions. The better a professional is at performing technical work, the more difficulty he may have with the strategic thinking mind-set.

According to some widely used personality style theories, strategic thinking is to some extent a personality trait. In the Myers-Briggs system based on Jungian psychology,[64] for example, the continuum from "Intuition" to "Sensing" describes the difference between gathering information by seeing large patterns and gathering it by observing concrete data. The person with a strong preference for Intuition naturally focuses on the future and the big picture. In contrast, the person with a strong preference for Sensing tends to focus on the present and the tangible details. The Myers-Briggs continuum from "Judging" to "Perceiving" may also be implicated in natural facility with strategic thinking. The Judging personality prefers making a plan, sticking to it, and achieving closure—in short, getting things done. The Perceiving personality, in contrast, prefers to consider alternatives and emerging possibilities—in short, continuing to explore what might be done.

Engineers tend to exhibit a Sensing-Judging preference and architects an Intuition-Perceiving preference, and these differences help explain their challenges in working with one another. However, both know that these differences show up among members of their own disciplines and firms, and that neither of the types can be effective without being balanced by the other. Individuals can show up at any point along each continuum, including a midrange where both preferences operate. Even more important, adult maturation involves developing the capacities of our opposing traits, thus moderating the influence of innate preferences over time.

While we agree that not everyone is cut out for strategic thinking, we believe that its absence is partly learned and that for a majority of professionals in these domains it is learnable—a matter of cultivating the mind-set consciously. Perhaps the most significant challenge is overcoming a felt threat to professional identity and confidence. Because professional confidence tends to be rooted in expert knowledge, strategic thinking may unsettle that confidence in two ways. First, strategic thinking requires stepping outside the professional's familiar area of expertise. Second, the area the professional must step into is intrinsically different. No one can know the answer about the future in the way it is possible to know the answer to a technical problem. Strategic thinking requires a high comfort level with questioning fundamental assumptions, with making decisions in the face of uncertainty, and with real-time learning and ongoing adjustment based on experience and results. Once a professional realizes that certainty is simply not available in the strategic arena, he will likely be able to begin learning to think this way.

Recall the example, discussed in chapter 6, of the firm that helped its client, a local utility, meet future demand by optimizing use of its existing facilities rather than building a new one. The president of this firm has reported that many of its members have had difficulty letting go of the assumption that the solution to the client's problem is a building—the familiar form of their technical work. Over time, however, the firm has increasingly addressed issues at the facilities or organization management level, and it is being recognized and well paid for this more strategic kind of work. Positive experiences such as the one with the utility are gradually building the strategic thinking mind-set in this firm.

Systems Awareness

Ask any built environment professional about the biggest technical problems on a project, and the answer will be about coordination. Construction projects inherently involve many different professionals working on many different elements or systems that must ultimately work together as a coordinated whole. Each professional, however, tends to proceed by focusing inward on the internal integrity of the element or system she is working on. In this case, coordination comes at a later step, when fully developed designs have already made it an expensive challenge.

The tendency to pay primary attention to one's own piece of the puzzle is a natural outgrowth of a mind-set that focuses on the parts or pieces even in conceiving the whole. Systems awareness involves a shift of mind from the parts to the relationships between the parts, which produces a different sense of the whole. The effect of this shift is a bit like the trick drawings that illustrate the notion of a perceptual paradigm. If you look at a famous example with one perceptual pair of glasses, the picture looks like the sculpted light-colored base of a table lamp against a dark background. With another, it suddenly appears to be two silhouetted faces in profile looking at one another across a light-colored open space. In Gestalt psychology, this is called the relationship between *figure*, which is in the foreground of perception, and *ground*, which is in the background of perception. Systems awareness moves the elements we tend to perceive as unconnected into the background; the relationships between the elements now occupy the foreground of our attention. What we now see is totally different.

A professional who has managed the shift of focus from product or task to relationship is already on the way to systems awareness. Along with the human quality of social relationship come the twin notions of connectedness and mutual influence. Because everything is connected, everything ultimately affects everything else. Systems awareness applies relationship awareness in this broader sense to the larger system. But just as the professional cannot throw out product or task while adding focus on relationship, she doesn't have the luxury of throwing out focus on the parts as she adds focus on their relationships. The professional must learn to focus simultaneously on the elements *and* their relationships—to hold a greater degree of complexity than before.

The shift to systems awareness is challenging initially because the relationships between the entities do not appear to our perception as tangibly as do the entities themselves. Introductions to systems thinking tend to begin with a comparison, such as a heap of sand to a cow. Both consist of a whole lot of molecules and atoms, but how are they different? The answer goes something like this: unlike the molecules of sand, the molecules of the cow form a dynamic set of relationships that make an organized whole—a whole that behaves very differently than would a mere aggregate of its parts. But this is an easy case: the cow is obviously a living system because its cohesive identity and boundary are visible. A more relevant analogy for the members or elements of a project system is the barnyard in which the cow lives. The web of interdependencies that make the barnyard a living system does not register in the same visible way in our perceptual processes.

Bringing systems awareness into the design of integrated architectural, engineering, security, and electronic systems for a building produces obvious benefits for construction and facility operation and management. But that is only the first level of systems awareness. Beyond the building as a mechanistic system, systems awareness takes us to the interface between the building and its surrounding natural environment. Trees and other vegetation, soil systems, and the animal life they support are disturbed and displaced by construction. Sun, rain, and wind affect the skin of the building, and dissolved solids from this interaction leach into soil and groundwater, altering their chemistry and ability to support life. Shade cast by the building alters the natural ecosystem in its immediate proximity. The well-being, productivity, and creativity of the individuals who live or work in the building are substantially affected by aspects of the indoor environment: the availability of natural light and visual space, as well as habitable physical space, internal traffic flow, temperature, air quality, acoustics, and aesthetics. These are just the most physically immediate interface exchanges, and do not consider the less local issues of energy use, waste production and management, and transportation, or other kinds of impacts on the surrounding community.

Systems awareness asks us to see connections and dynamic mutual influence not only where they are obvious but also in places where we haven't seen them before and where they may be counterintuitive. This means seeing as many of the parts of the system we are working on as possible. It also means seeing their relationships and understanding that through those relationships, every part and every action affects the whole. In addition, systems awareness asks us to see that we ourselves are part of the system: we affect the system and are affected by it.

As Peter Senge is careful to help us understand in his seminal work, *The Fifth Discipline*,[65] we are never able to grasp the whole complexity of the vast nested systems of which we are a part. Nor are we ever able to anticipate all of the unintended consequences of our actions for those larger systems. However, the discipline of viewing ourselves and our organizations as living systems, which are in turn part of larger living systems, fundamentally alters the consciousness with which we make decisions and take action.

For professionals who have focused on their part of a project in relative isolation, making the shift to systems awareness will indeed require an ongoing conscious effort. The professional must consistently ask, what other subsystems are present in the building or infrastructure system? How do those subsystems affect the one I am working on, and how does the one I am working on affect each of them? What difference do these relationships make to the problem or the design? How does the subsystem I am working on affect the social and environmental system in which it is embedded? How does that larger system affect it? What difference do these relationships make to the problem or the design? What information and knowledge do I need beyond that I already have to address these questions effectively? Where can I get that knowledge and information? With whom do I need to consult and collaborate in order to optimize all these relationships?

Systems awareness is the broad and fundamental shift of mind underlying all the others we will discuss. Organization-ecosystem awareness, whole-ecosystem awareness, and whole-value-network awareness all involve more specific forms of systems awareness. Professionals who make this fundamental shift of mind will be prepared to assist client organizations in developing solutions that respond to and leverage the interdependencies in which they operate. This way of thinking will tend to produce solutions very different from those conceived in relative isolation. Interdependency-based solutions are likely to arouse less opposition, be more efficient and effective in the present, cause fewer problems in the future, and thus provide more value over the long term.

Adopting this fundamental mind-set is perhaps the most powerful single step that a professional can take to lay the foundation for getting out ahead of clients' and stakeholders' current thinking and exercising leadership for society as a whole.

Organization-Ecosystem Awareness

In chapter 5 we introduced the organization-ecosystem model as a means of navigating the new social complexity as well as of defining three realms of value creation that architects and engineers can pursue. The power of this new mental model to move professionals beyond the thinking of clients and stakeholders involves two shifts of mind. The first is the blend of systems awareness and strategic thinking that Moore achieved in formulating his idea of the business ecosystem. The second focuses on the relationship between an organization ecosystem and its larger societal and environmental ecosystem. This second shift is explored in the next section as whole ecosystem awareness.

Organization-ecosystem awareness entails expanding the professional's focus on the client and client organization to include awareness of organizations, groups, and communities that make up the client organization's strategic context. From this viewpoint, the professional sees the members of the organization and their internal relationships in the context of the organization's external relationships and interdependencies. And from this viewpoint, optimally serving the client organization includes promoting the well-being of its ecosystem as well.

The first challenge inherent in making this shift is acknowledging that dealing with the client organization alone can itself be overwhelming. Particularly when more than one major constituency is involved—say, faculty and students as well as administration and facilities managers at a university—a professional can feel that the organization itself lacks the cohesion required for it to be served in a coherent way. To take an even broader view may seem to risk further fracturing and present even greater difficulty in finding a successfully integrative solution.

Just as daunting, such an effort would require a monumental information- and knowledge-gathering effort. Most professionals agree that knowledge of the client organization enhances their ability to win the project and do their best technical work. Practically, however, within the existing business model, professionals are severely restricted in the time they can spend learning about their clients' organizations. Few succeed in expanding their focus beyond the immediate client responsible for overseeing the performance of the technical work. Expanding this effort to the organization ecosystem level may seem nearly impossible.

Making this shift challenges assumptions about what activities qualify as value creating. The willingness to suspend these assumptions depends at least in part on having made the deeper shifts to relationship and group process awareness. A professional who would prefer not to deal with social complexity in the midst of technical service will be unlikely to embrace a view that could increase rather than decrease that complexity. And a professional who would prefer to keep clients and other stakeholders at a distance from the creative process will certainly have difficulty including and encouraging the creativity of the client organization in the larger context of its organization ecosystem's needs and aspirations.

Clients also may initially resist expanding their view beyond organizational boundaries and familiar external focal points. Like most individuals in our society, they have tended to view themselves in relative isolation. Typically, they are willing enough to focus on external groups for whom their need is very tangible, but they may not see the value in other connections. Our relentlessly individualistic American society tends to place the freedoms and rights of the individual above those of the collective. We have difficulty grasping and owning the other side of the coin: the degree to which our actions and expressions affect others and, in turn, how those affected now affect us. Our deep cultural bias is to want to be unobstructed in our freedom to express ourselves. Organizations with this bias, like individuals, tend to pursue their aims in ways that are disconnected from the larger communities in which they operate.

Ironically, today's hyper-competitive context may cause organizations to focus even more doggedly on their own survival without consideration of the larger system on which they depend. While the third realm of value creation—making direct contributions to the client organization's ecosystem—may not require bringing clients along in this consideration, the second realm—enabling a client organization to lead its ecosystem—does.

Expanding clients' awareness of their organization's ecosystem is not primarily an appeal to their altruism. Though Moore titled his book *The Death of Competition*, he admits that he is really talking about the death of a way of viewing competition. His claim, as discussed earlier, is that competition occurs more between organization ecosystems than between organizations—although organizations may compete at a micro-level within an organization ecosystem at various stages of its development. His subtitle, *Leadership and Strategy in the Age of Business Ecosystems*, highlights that the thrust of his book is how to thrive in the world of those dynamics: the pursuit of organizational self-interest in that context. His implication is that many businesses—and by extension, organizations—don't yet see the real basis of their well-being.

If an organization's impacts on its organization ecosystem are healthy, those impacts will promote its own health. If they are not experienced as beneficial by other organizations and groups in their ecosystem, those effects will diminish the support the organization receives from its ecosystem and thus the ecosystem's ability to compete effectively with other ecosystems.

The organization ecosystem model of strategic thinking has emerged in parallel with our understanding of organizations as living systems and businesses as living companies operating in a knowledge-based economy. Given that context, professionals can move ahead of client and stakeholder thinking by using the organization-ecosystem model to think strategically about the client organization's development. The advanced capacities involved in this kind of leadership, including how an understanding of Moore's four stages of ecosystem development can support the effort, are explored in chapter 22. The following example illustrates what organization ecosystem-based strategic thinking can yield.

Imagine a private college in an earthquake-prone area faced with a requirement for seismic retrofit. A structural engineering firm is hired to do the initial analysis and make recommendations. Taking a traditional approach, engineers would perform the technical work without looking into the state of the college, much less its ecosystem. However, organization-ecosystem awareness could reveal that this college relies heavily on the local community for its enrollment and that the towns around it are actually losing population because of several corporate bankruptcies and downsizings. The college will probably need to attract students from other regions in order to maintain its enrollment.

What implications does this situation have for the college's contemplated facilities upgrade? How will the community and alumni feel about a nonresident, transient student population versus the college's ability to thrive? How might these and other impacted constituencies be willing to participate and help? Awareness of these important questions now equips the professional to lead the organization in reconsidering its plans. Would new construction aimed at a new identity be more strategic and cost-effective in the long run? Would it enable both a better technical solution *and* superior leadership of its organization ecosystem? Would a higher powered, nonresidential college ultimately attract new businesses looking for well-qualified graduates, bringing families who would consider sending their children to the college?

This college may or may not be ready to consider a significant shift in its identity. However, by leading this strategic thinking process, the professional would help the college clarify its facilities upgrade project, as well as its broader identity and direction. Through his organization-ecosystem awareness, then, the professional resists his initial desire to jump at self-expression by performing familiar technical work. Through engagement with the clients, he explores their capacity to set aside the need for immediate resolution of the technical problem long enough to consider the college's future in its larger organization-ecosystem context. If the college chooses not to express its creativity through a change in identity, the professional can still perform the anticipated technical work. If the college does choose the more transformative route, the professional may have the opportunity to perform expanded and more interesting technical work as well as offer other forms of value. Organization-ecosystem awareness thus does not tend to suppress the professional's or the client's creativity, but rather to expand the possibilities for both in the context of the larger system's well-being.

While this fictional example involves a monetary exchange, organization-ecosystem awareness might also result in exchanges of nonmonetary value. A recent *Harvard Business Review* article cites a real-life example relevant to built environment professionals:

> In 1996, SC Johnson, a manufacturer of cleaning and home-storage products, launched "Sustainable Racine," a project to make its home city in Wisconsin a better place in which to live and work. In partnership with local organizations, government, and residents, the company created a communitywide coalition focused on enhancing the local economy and the environment. One project, an agreement among four municipalities to coordinate water and sewer treatment, resulted in savings for residents and businesses while reducing pollu-

tion. Another project involved opening the community's first charter school, targeting at-risk students. Other efforts focused on economic revitalization: Commercial vacancy rates in downtown Racine have fallen from 46% to 18% as polluted sites have been reclaimed and jobs have returned for local residents.[66]

This organization-ecosystem-level leadership could have been provided just as well or better by an engineering consulting firm working with a municipality involved in the water/sewer coordination, or by an architecture firm working with a developer of the downtown commercial buildings.

Whole-Ecosystem Awareness

Whole-ecosystem awareness reflects our expansion of Moore's organization-ecosystem model. It focuses on a human system's exchange with the biosphere and attends to how the larger global ecosystem may be affected by local ecosystem dynamics.

Most environmental scientists and engineers would probably say that just managing to comply with regulations at the local level would be a major achievement. Inquiring into the interconnectedness of local and global dynamics would seem less than relevant to these professionals' project challenges. Most other types of engineers and architects would probably say that just managing to balance the dictates of regulations and environmental impact reports with facility and infrastructure needs would be a major achievement. Though the relationships of the human system with the biosphere are addressed in regulatory aspects of their work, the human system remains paramount in their mind-set.

The relatively few architects and engineers who have been early to respond to clients' increasing focus on green building and sustainable development would probably say that whole-ecosystem awareness is inherent in those emerging approaches. We would agree, but more from the perspective of resource efficiency than from the fundamental redesign urged by the seminal thinkers in the field.[67] Even for professionals now practicing with the LEED standards, keeping up with evolving frontiers of redesign-oriented whole-ecosystem awareness may seem daunting, given all the new learning already required of them. Moreover, their clients may not be ready for a more ambitious redesign approach.

Even though some progress has been made in the direction of sustainability, then, the fundamental shift of mind involved in whole-ecosystem awareness has yet to occur in these professions. This shift is essential if exercising meaningful leadership in this arena is a goal. Insights from what the young science of ecology calls the "pattern of ecological succession" will help flesh out the basis for this deep change of mind-set. While these new understandings underlie much of the seminal work in the field, they are not yet widely known. We believe they can deepen our awareness of the relationship of human systems to natural systems.

In her wonderful book, *Biomimicry: Innovation Inspired by Nature,* Janine Benyus presents these new insights in a striking contrast between developing and mature ecosystems. Table 8.2 outlines the key differences between the two stages.[68]

Natural ecosystems exhibit a fundamental pattern of development. They evolve from simple to complex forms. Linear food chains evolve into weblike ones, while low diversity develops into high diversity and specialization based on increasing symbiosis and interdependence. Small bodies with short life spans that drive rapid growth are succeeded by longer-lived, larger bodies that incorporate balancing influences into their growth patterns. As production shifts from quantity to quality, resources and energy are increasingly conserved, and organisms function in the context of more feedback and information. Mature ecosystems are not only more complex but also more stable.

Mature ecosystems remain in this state "until the next big disturbance"—a fire, a storm, a volcanic eruption, a meteor collision, a species reproducing without predators or resource constraints, or a large human development. Such disturbances restart the process: developing organisms (the colonizers) evolve into mature organisms (those who learn to close the loops). With each ecological succession, more creativity and diversity unfold. Organisms that make up developing systems not only provide renewal after a large-scale disturbance but also find little gaps of disturbance in and around mature systems. In this sense, mature systems depend on developing ones for ongoing creativity that supports their stability.

Ecosystem Attributes	**Developing Stages (Type I)**	**Mature Stages (Type III)**
Food chain	Linear	Weblike
Species diversity	Low	High
Body size	Small	Large
Life cycles	Short, simple	Long, complex
Growth strategy	Emphasis on rapid growth	Emphasis on feedback control
Production	Quantity	Quality
Internal symbiosis	Undeveloped	Developed
Nutrient conservation	Poor	Good
Niche specialization	Broad	Narrow
Stability	Poor	Good
Entropy (energy lost)	High	Low
Information (feedback loops)	Low	High

TABLE 8.2. Patterns of Ecological Succession

(Extracts from table: "Ecological Succession" [pp. 252-3] from *BIOMIMICRY* by JANINE M. BENYUS.

The mature ecology of a rain forest is a familiar example of the web of dynamic interdependence. The dense canopy reveals different species to be at least as interdependent as the members of a wolf pack or a bee colony, or even the internal organs of a bee. Though we know that the various species in a rain forest make each other's lives possible, in our own world, we still place our primary focus on individual entities or organisms. Are we, as an ecosystem, in fact less interdependent or less mature? Or are we just as densely organized, just as dependent on resource sharing and exchange, but unable to recognize and work with this interdependence? We suggest that many of the problems of our increasingly complex human systems flow from our inability to work consciously with—and tendency to work in opposition to—our interdependencies.

Moore uses the pattern of ecological succession as a metaphor to describe the stages of business ecosystem development. For example, in discussing the forms of vulnerability that emerge at his third stage, "authority in an established ecosystem," he highlights the role of pioneering species that come in from the outside to challenge existing ones as well as the role of competing ecosystems. More generally, he views leaders who are working with the business ecosystem model as moving their organisms out of the colonizing or pioneering style of the developing ecosystem toward the interdependent style of the mature ecosystem.

In *Surfing the Edge of Chaos,* Richard Pascale, Mark Milleman, and Linda Gioja blend the new science of complexity with biological approaches to self-organizing systems to reveal the relationship between "the laws of nature and the new laws of business." They make the apparently bold but intuitively sensible claim, "'Living systems' isn't a metaphor for how human institutions operate. It's the way it is."[69]

Moore's use of ecological succession to understand how business operates wonderfully captures the living nature of business systems, yet the status of metaphor leaves the living system picture incomplete. To his credit, he reflects on this limitation in his closing chapter, pointing out that the dynamics he describes at the organization-ecosystem level must be understood in the context of relationship to the larger societal and environmental ecosystem. He even hints at the insight that business ecosystems essentially institutionalize mature ecosystem relationships among organizations that maintain the fundamental characteristics of developing or pioneering species with respect to the larger societal and environmental ecosystem. As Benyus observes:

> Back before our world was full, when we still had somewhere else to go, the Type I ["developing"] strategy looked like a good way to stay one step ahead of reality. These days, when we've gone everywhere there is to go, we have to find a different kind of plenty, not by jumping off to another planet but by closing the loops here on this one . . . It won't do to just tweak the current system and hope that we'll

> evolve, just as a common ragweed or fireweed could not be expected to evolve into a redwood. Instead we must replace portions of our Type I ["developing"] economy with portions of a Type III ["mature"] economy until the whole thing mirrors the natural world.[70]

From this perspective, an invaluable template for our collective survival emerges. Given the current and growing scale, scope, and density of our human population, our larger societal ecosystem must be populated by organization ecosystems designed on mature principles—both in their internal relationships and in their relationships with the larger societal and environmental ecosystem. Developing organization ecosystems capable of responding to disturbances and renewing the larger system should play only a supporting role. Benyus's lessons learned from mature ecosystems echo many of the principles cited in *Cradle to Cradle* and *Natural Capitalism*. Taken together, these ten principles provide a powerful, integrative basis for design and problem solving with whole-ecosystem awareness:[71]

1. Use waste as a resource.
2. Diversify and cooperate to fully use the habitat.
3. Gather and use energy efficiently.
4. Optimize rather than maximize.
5. Use materials sparingly.
6. Don't foul their nests.
7. Don't draw down resources.
8. Remain in balance with the biosphere.
9. Run on information.
10. Shop locally.

These principles have not yet entered widely into the thinking of the public. However, the public is becoming increasingly aware that apparently conflicting needs in different parts of the ecosystem should not be viewed in a simply adversarial manner, as noted in our prior discussion of the nondualistic, inclusive view. The professional deeply grounded in whole-ecosystem awareness can lead clients in understanding all parameters for effective decision making in interdependent systems, and in fully appreciating the life support value of decisions that help maintain balance between our human systems and the biosphere.

Given the public's growing concern not only with the ethical responsibility of our institutions but also with their social and environmental responsibility, the goodwill that such efforts provide can be priceless in terms of image, reputation, and investment. When professionals help clients and other stakeholders operate with whole-ecosystem awareness, both local and global benefits can begin to flow. Project opposition may be overcome, project design may avert predictable crises and unnecessary expenses in the life cycle of the built entity, and cumulative and aggregate effects may begin to make a difference for the well-being of the biosphere.

Whole-Value-Network Awareness

In chapter 6, we introduced Clayton Christensen's concept of the value network as a means of navigating the burgeoning levels of technological complexity that fill our 21st-century world. To apply his concept to the world of professional services, we translated his product architecture into a service architecture, which we then used to represent both the technological and social dimensions of the value network. We also expanded the value network's boundaries beyond the facilities management ring, where Christensen would likely have bounded it. To fully mirror the expanded organization-ecosystem model, our value network for built environment services includes rings representing the services and functions involved in organization management, organization-ecosystem leadership, and societal and environmental ecosystem stewardship.

The value-network structure makes clear that just as the value in the innermost ring depends on the value of the next ring out, the value of the inner rings together depends on the services and functions provided in the outer rings. The telescoping structure also reveals the way in which the parts are indivisible from the whole. All the detailed professional services that appear in the professional services ring are implicit in the design and construction ring through the element of professional services, and so on. Since all services present in an inner ring are implicitly present in the next one, ultimately all services, no matter how micro, are implicit in the outermost ring. The impact a problematic soils analysis can have on an organization's bottom line or the geographical layout of its organization ecosystem reveals this intuitive truth about the relationship of the part to the whole. As we asserted in chapter 6, the value network enables a professional to see the entire context of value-creating activities in which the service she is providing is making its contribution.

Like the organization ecosystem, the value network represents a powerful synthesis of systems awareness and strategic thinking. As such, it provides a framework for professionals to get out ahead of current client and stakeholder thinking in order to exercise societal leadership. But the model can only deliver this capacity to the extent that professionals assimilate it as a fundamental model for their thinking and work.

Making the shift to value-network thinking, like organization-ecosystem thinking, requires expanded information- and knowledge-gathering that is undermined by assumptions governing the current business model. Also like organization-ecosystem thinking, it is undermined by the desire to avoid any more social complexity than is absolutely necessary. Like systems awareness and strategic thinking more generally, value-network thinking also requires the professional to overcome tendencies to focus on her own piece or part in isolation and to resist stepping outside arenas of familiar technical expertise—tendencies encouraged by discipline-bounded domains.

Architects and engineers have successfully moved into certain limited arenas outside their domains. Specific types of database management, for example, have seemed a natural extension of the information management capacities they have developed in order to build things in the information age. Specific types of operations and maintenance have similarly built on the people and process management capacities learned in the context of managing projects. Both of these upstream functions reside in the facilities or organization management ring of the value network. Functions outside the domain that would require professionals to step into more unfamiliar territory, however, have been feared, resisted, or at least perceived as out of reach. In many cases, these internal challenges have been reinforced by client thinking, also shaped by current mental models that view the architect's or engineer's role as limited to familiar forms of technical work.

As we suggested in chapter 6, value-network thinking prompts a series of considerations that enhance the professional's ability to make her service truly valuable to the client. First, visualizing the other essential services in the ring where a service is being provided enables the professional to effectively coordinate with those other services. Second, visualizing the contributions that service makes to the next ring out, or understanding how the next ring depends upon that service, enables the professional to ensure that form of value is delivered. Third, grasping contributions further out in the value network enables the professional to shape a single service so that it delivers value in multiple rings rather than just one or two. And fourth, taking this big-picture view enables the professional to identify additional services that could be provided in rings other than the familiar and expected one.

Though most technical services in our generic built environment professional services value network are provided in the inner rings, not all are. Services in other rings may be different kinds of technical service, such as the capacity analysis performed for the utility that turned out not to need a new building, or more social- and leadership-oriented services, such as those provided by the firm that brought together the processes and cultures of a university and a city, making it possible for the two to share a library. Ultimately, whole-value-network awareness enables the professional to perceive unmet needs throughout clients' value networks and to help clients perceive needs beyond their traditional thinking about technical services. This awareness thus provides the basis for innovations that will create higher value for client organizations and their organization ecosystems—both in ways that clients and stakeholders are already expecting and in ways they haven't yet imagined. We will explore the value network as a tool for business innovation in more depth in chapter 14.

For the professional who has overcome the basic obstacles to whole-value-network awareness, one more very challenging barrier may present itself: the lack of formal authority from which to address issues beyond familiar technical work in the value network. If this professional has already made the shift to whole-project-system awareness, she has replaced a sense of disempowerment with recognizing a larger sandbox in which to use influence rather than control. In the value-network context, no individual or group has, or ever could have, authority over the whole system. Though daunting in scope and scale, a client's value network offers abundant opportunities for exercising leadership. In chapter 12, we will introduce a new model for professional work that covers the full range of technological and social challenges present in built environment-related value networks. This model is based on the seminal work of Ronald Heifetz in *Leadership Without Easy Answers.*[72] In addition to providing a basis for understanding and developing the capacity to work in any ring of the value network, Heifetz addresses the core question of how to lead without a position of formal authority.

Building on the five shifts of mind required to catch up with clients' and stakeholders' expectations discussed in chapter 7, the five shifts explored in this chapter provide the basis for getting out ahead of those expectations and once again exercising significant leadership in society. In order to position individual practitioners for this role, the domains of architecture, engineering, and related professions must be fundamentally updated to reflect and address the new social and technological complexity of the world in which they operate. The next chapter summarizes the changes required.

9 Updating the Domains

People who do good work, in our sense of the term, are clearly skilled in one or more professional realms. At the same time . . . they are concerned to act in a responsible fashion with respect toward their personal goals; their family, friends, peers and colleagues; their mission or sense of calling; the institutions with which they are affiliated; and lastly the wider world— people they do not know, those who will come afterwards, and, in the grandest sense, to the planet or to God.

—Howard Gardner, Mihaly Csikszentmihalyi, and William Damon[73]

A professional domain has a sacred quality. To suggest that a domain needs updating is thus a sort of brazen act, the more so if the profession in question has a long and illustrious history of contribution to civilization. Yet, as the authors of *Good Work* make clear, a domain must effectively support its practitioner community. Like that community, it must be a living system that adapts to changes in its environment. Architects, engineers, and related professionals have fallen behind the thinking and expectations of clients and other stakeholders because their domains have not effectively integrated the new social and technological environment that characterizes the 21st century. Given the broad and now longstanding nature of the rift between these professionals and their clients, individual efforts will be insufficient to address the underlying systemic issues. The domains themselves must be updated.

Part of the difficulty is contemplating how the disparate elements of the professions might be brought together to make such a change. On the one hand, firms bemoan the failure of schools to prepare students for the realities of the world in which they must operate. On the other hand, schools, feeling the moral obligation to defend the domains, appear to many to have been digging in their heels. The professional societies are working to adjudicate some sort of middle ground. In the midst of this dynamic, many individual practitioners themselves remain ambivalent.

Another part of the difficulty is that although some elements of these domains have been formalized—for example, in documents produced by professional societies—their essence remains only partially articulated. Core principles are transmitted in indirect ways that deliberately preserve generality. Fuzzy boundaries are good for fundamental beliefs and assumptions. They anchor people in deep alignment while allowing for the diversity of understanding and interpretation required for a domain to be creative. How do we evolve something so vague and entrenched as to be not only second nature but difficult to put in language at all?

Our approach has been to start with the most visible version of the essence of the domains: the dissonance expressed by new practitioners when they find they must work with firms whose values are inconsistent with their training. The values at the heart of the current domain can be seen most clearly as they clash with client, stakeholder, and business concerns. The best hope for overcoming individual ambivalence and institutional conflict, both of which are obstructing evolution of the domains, is a compelling conceptual framework and vision. We hope that our approach has begun to provide a compelling basis for updating the domains. Our suggestions here address only their very core—the aspects that have been most difficult because they are so implicit and entrenched. First, we propose ideas for shaping new overarching definitions. Then, we offer a new set of values and ethics. In both these arenas, the evolution does not throw out the thinking of the past, but includes that thinking in a significant expansion for the future.

New Overarching Definitions

In chapter 6, we noted that the definitions of these domains are rooted in the organizing principle of building things. The boundaries implied by this core thrust are too narrow to contain the new performance criteria related to restoration and ecological sustainability, the new scale and complexity of retrofit and replacement, and the complex scope of upstream issues that now confront built environment projects. In addition, definitions of these domains are tied to a particular discipline. Neither the new forms of technological complexity nor the increasing social complexity now inherent in built environment work can be effectively addressed from within the bounds of a single discipline. To adapt to these new conditions, professional domains must be defined in ways that better reflect the full range of project-related challenges and that embrace a multidisciplinary approach at their very core.

A transformation unfolding at the Berkeley campus of the University of California (UCB)—the foremost public research university in the United States—offers an interesting model for achieving this adaptation. The administration recently challenged faculty "to define the most critical new areas of teaching and inquiry" for the 21st century. Writing in a summer 2003 letter to alumni, Chancellor Berdahl reported:

> Five new initiatives were selected . . . Computational Biology, Nanosciences and Nanoengineering, Regional and Metropolitan Studies, New Media (exploring technology to communicate truth and beauty and how to best incorporate new media into modern life), and The Future of the Planet (studying the Earth's environment, the changes wrought by human intervention, and how we can manage or mitigate those changes). These initiatives are compelling on their own, but what makes them especially exciting is that they are all interdisciplinary, involving a large number of departments and disciplines across campus.

In the wake of the massive Health Sciences Initiative and the new buildings dedicated to housing it, these new initiatives reveal UCB ushering in a new era of academic study based less on individual disciplines than on arenas that require many disciplines. Although these arenas are not domains in the traditional sense, they do seem to be functioning as domains in a new sense: despite their complexity, defining an arena of endeavor that encompasses a core of concerns, values, and ethics as well as methodologies.

Engineers and architects, already overwhelmed with an explosion in the number of technical disciplines required to design and build a large facility, may not want to hear that they need to think even more broadly about working across disciplines. However, recognizing the degree to which discipline-bounded domains have disempowered professionals in the midst of 21st-century complexity opens exciting new possibilities for reclaiming a leadership position in society.

An appropriate expansion of the guiding definitions of these domains will need to include building things within a larger idea. One possibility would be the notion of optimizing the built environment in balance with the natural environment. Another, more in the spirit of McDonough and Braungart, would be creating human infrastructure that integrates and promotes the healthy generative processes of the biosphere. Each profession will need to shape its own language for an expanded overarching definition. These new definitions must encompass the ongoing emergence of new performance criteria as well as all the upstream issues—social and technological, macro and micro. They will need to include the notion of maintaining a sustainable relationship with the earth's living systems. And they will need to acknowledge the essential role of multiple disciplines while preserving the core of a particular discipline and suggesting how that discipline will be engaged in the larger endeavor.

To provide the basis for new practice, expanded overarching definitions must be accompanied by new values and ethics that express the responsibilities and possibilities inherent in the expanded domain. The values and ethics we propose build directly on the shifts of mind we have argued are required to address the new social and technological conditions operating in the field. As with those shifts of mind, one set is required to catch up with clients' and stakeholders' expectations. Another is required to get out ahead to exercise leadership in society.

Values and Ethics for Catching Up

Like the ideas to which they refer, the terms *values* and *ethics* are themselves fuzzy. Dictionary definitions reveal why many people are not clear on the distinction between them. Their meanings do overlap. *Webster's* defines *value* as we use it today, in the plural form *values*, as "something intrinsically valuable or desirable," "worthy," or "important." It defines *ethics* as "a system of moral values" or "the principles of conduct governing an individual or group." One has to look up *moral* to discover that ethics refers specifically to the issue of right vs. wrong behavior. The overlap in meanings occurs because a stated ethical principle can look like a value—something intrinsically valuable or desirable—at the same time that it refers to an ideal of behavioral conduct. *Values* is, then, a broader term that refers to all things held to be intrinsically important, while *ethics* is a narrower term referring to principles of conduct. We will allow these terms to blend here, rather than attempting to make a distinction that would not add significant value.

As with the new overarching definition, the needed evolution in values and ethics should not simply replace the cherished current set. Instead, it should continue to honor current values while adaptively embracing and acknowledging the new conditions operating in the field. The expanded core principles that emerge should not only provide the basis for more effective practice, but also have equal or greater capacity to inspire professional passion and contribution. The clash between the mind-set of young professionals and the conditions they encounter in their firms reveals the dynamics of the evolutionary process. Table 9.1 modifies the headings of the table in chapter 5 to show the values of the current domain alongside the new conditions that need to be embraced in the new values set.

Current values	Current conditions
Autonomy/expert authority	Participatory politics across complex web of stakeholders
Power to shape the built environment	Marginalized role and diminished influence
Access to and ability to preserve integrity of whole project process	Access to small part of project process
Contribution to society/legacy (long-term view)	Current economic and functional realities (short-term view)
Aesthetic or technological ideals	Pragmatism of client decisions
Requirements of design, problem-solving, or analytical process	Tight budgets and schedules
Interesting problems to solve	Relatively predictable needs that are the firm's bread and butter
Opportunity to be creative	Production based on existing procedures and solutions
Large-scale ambition	Small scale of actual work
Desire to keep learning	Requirement to be billable

TABLE 9.1. Basis for a New Set of Values and Ethics

Developing a new set of values begins with substituting a thought process that involves *and* rather than an oppositional *vs.* in all the rows of this table. To achieve optimal results on all fronts, the client's needs and aspirations must be deeply honored *and* the professional's full range of talents and contribution mined. Budget and schedule must be treated as parts of the design or technical challenge, and the perceptions and convictions of stakeholders as essential criteria for excellence of outcomes. At the same time, adequately challenging input from the professional must inform and energize the process.

The shifts of mind we discussed in chapter 7 laid the foundation for the new, more complex values that will enable professionals to catch up with the current conditions prevailing in client and stakeholder systems as well as in the broader culture. Representing a deep assimilation of these shifts, a new list of ethically oriented values might look something like this:

- Deeply respect and regard the client and the relationship with the client as the fundamental basis for the project process.
- Respect other stakeholders and take into consideration their interests, ideas, and needs.
- Value participatory, emergent processes that involve the client and other stakeholders as a primary contributor to design and problem-solving.
- View the professional's highest and most valuable creativity as emergent with and from the self-expression of the client and other stakeholders rather than as shaped by preexisting professional preferences or norms.
- Define excellence in design or technical solutions to include *all* concerns and aspirations of the client—including budget and schedule—as essential design or problem-solving parameters.
- Be aware of and respect integration needs for the whole project, whatever the professional's role, position, or element.
- Fully offer and leverage professional expertise and creativity.
- Serve society in the broadest sense while serving the client and other stakeholders.
- Consider the long-term impacts of decisions for the short-term.
- Provide leadership for the highest and best use of resources—whatever the context.
- Honor the importance and impact of all scales and levels of detail.
- Pursue continuous learning and adaptation to develop one's personal expertise, team or organizational capacity, and profession, including the arenas of technology, materials, methods, and processes.

Some of these values have already appeared as strategies in the more forward-looking firms in the industry, however imperfectly implemented. Sufficiently rapid adaptation, however, will require the migration of all these values to the core of the domain, where they move beyond the discretionary to the fundamental and imperative. Professionals practicing from these principles as their core set of values and ethics would likely inspire their clients' trust and even expand their role in their clients' future projects based on the value they are providing. They would likely begin to reverse the vicious cycle that has displaced them from the leadership they would like to exercise. To fully recapture societal leadership, however, professionals must adopt additional values that drive a new level of leadership.

Values and Ethics for Getting Out Ahead

In order to provide societal leadership, professionals must move beyond just reacting to the changes that have occurred in the built, social, and natural environment. They must fundamentally reframe and redefine the needs and problems they are addressing in ways that offer new insights and power. The professions would be deeply reinvigorated by embracing new social and technological complexity from a whole-systems view consistent with the notion of firms as living companies—the view provided by the organization ecosystem and value network models for strategic thinking based on systems awareness. The shifts of mind discussed in chapter 8 laid the foundation for new values based on this view. A set of ethically oriented values deeply assimilating these shifts might look like this:

- Expand the view of the whole to which you are contributing beyond the project and the client organization to the client's organization ecosystem and to the larger societal and environmental ecosystem.
- Alongside technical capacities and disciplines, develop social and leadership capacities and disciplines that enable you to provide leadership for client organizations and their ecosystems as well as to develop clients' leadership of their ecosystems.
- Exercise leadership for recognizing, articulating, and solving the chronic and emerging problems of the larger societal and environmental ecosystem as part of and beyond remunerated work.

- Pursue innovation both in technology and in social process for addressing those larger societal and environmental ecosystem problems.

These leadership-oriented values are perhaps no more brash or bold than the ones that drove the great wave of professional contribution in the first two-thirds of the 20th century. They are more explicit. And they do perform the necessary integration of an entirely different level of social and technological complexity. Nothing less will be sufficient to enable professionals to once again pursue their deepest ambitions to improve the quality of life for society and to leave a legacy of significant contribution.

To restore sacred trust with clients and to reinstate this kind of leadership position will require professionals to leave behind the model of the autonomous expert generating ideal solutions that are informed more by their own preferences than by the needs, interests, and aspirations of clients and their organization ecosystems. And it will require that they seriously undertake expansion of the social and leadership capacities that support their technical expertise.

Though some consultants to the professional services industry have seen business and practice as separate and important to differing degrees in different firms, we take the view that in today's world, they are inseparable and equally important in all firms. While the primary purpose of this book is to enable professionals to develop new business models, practice is the foundation of professional service businesses. Because a professional's identity resides first in a domain and only secondarily in a particular firm, many of the issues involved in transforming a firm are deeply rooted in the professional mind-set that is grounded in the domain. New levels of creativity and innovation in the business require fundamentally evolving the core of professional practice. With such an evolution now envisioned, let's turn now to the business context.

PART III

Guiding Business Model Innovation

10 A New Ecology of Firms

In fifty years' time knowledge of physics, biology, chemistry, astronomy, and maybe earth science will be immensely different from knowledge today—far more different from today's knowledge than ours is from that of fifty years ago.

—Peter Schwartz[74]

At the turn of the 19th century, steam powered the first industrial revolution, transforming mills, factories, and transportation, reshaping cities and towns, and giving birth to new types of business and commerce. It was during this awakening that engineers and architects made their formal entrance into the world of business and commerce. A hundred years later, the invention of electricity, synthetics, and the internal combustion engine drove the second industrial revolution. Ford invented the assembly line, Dupont pioneered chemical processing, and Rockefeller developed oil and gas production, giving rise to new industrial business models and the modern corporation. The impact on society was equally profound as technology reshaped how and where people lived. This wave of technological innovation was accompanied by a wave of professionalization, as architects and engineers established professional societies and codified ethics, guidelines, and contracts for professional practice.

At the dawn of the 21st century, technology remains a primary driver of social and economic change. We can expect that technological innovation in this century will be at least as transformative as it has been for the past two. On the leading edge of the future, we have the prospect of nanotechnology, offering the promise of assembling materials and goods from the molecular level up. Potential impacts of nanotechnology on our postindustrial society could include new medical diagnostic and treatment tools, super strong materials made from hyper-efficient manufacturing processes, and new electronic circuitry and ever more powerful computers. It may also be used to mitigate many of the negative impacts that society's activities have had on the natural environment: changing the way we extract materials from the earth; how we use energy to heat, beat, and treat materials into the forms we need; and how we deal with the waste products from these processes.

The biotechnology field is already exploding, giving us the power to genetically design food, plants, animals, and even people. The potential benefits of these developments include improvements in our ability to keep people healthy, feed them, and increase their longevity. Biotechnology also offers the potential of environmental gains, by learning and applying lessons from biological systems to produce more and waste less than today's dominant mechanical technologies do. Organizations and inventors increasingly turn to biomimicry—innovation inspired by nature—for inspiration and guidance in how to grow food, make materials, store and communicate information, and even conduct business.[75]

Much more familiar is information technology, which, in the space of a few short years, has linked one billion personal computers worldwide in the astonishing Internet. Progress in information technology follows logarithmic functions: computing power doubles every eighteen months (Moore's Law), the amount of information stored per square inch of disk surface increases by more than 60 percent a year, and the bandwidth of telecommunication systems (wired and wireless) doubles or even triples every twelve months. As Nicholas Negroponte points out in his book, *Being Digital*, we are moving "from atoms to bits," with the digitization of words, sounds, pictures, music, movies, television, drawings and specifications, and countless other forms of data and information.

Rapid technological advances in computers, communications, and software have led to dramatic decreases in the cost of computing, processing, and transmitting data and information. Along with voice recognition software and video streaming, we are moving toward a communications infrastructure that will seamlessly handle not just sound and sight but all five senses. A major element of that infrastructure will be wireless, using airwaves to link people without the resource intensity required by wired connections. Not far behind are forms of artificial intelligence that may rival our own, at least in some ways. We have already seen significant social and economic transformation resulting from the information revolution, and there is much more to come.

Lest we consider the central evolutionary role of technology a late development in human history, it is important to remember that humankind is a tool-making animal. From the very beginning, our development has been driven by a basic reinforcing loop in which the invention of a tool enables people to create a society different from—and more complex than—the one that existed without the tool. The transformed society then prompts the invention of new tools, which in turn shape a more complex, evolved society that prompts the invention of yet newer tools. The invention of the printing press brought about the Enlightenment, which in turn led to the early industrial revolution, and so on. Thus one could say that technology begets more technology, but only by transforming society along the way. This loop moved very slowly in early human history, gradually picking up speed until it became the exponential warp we are experiencing now.

However, in this reinforcing cycle there is often a significant delay between the invention of a new technology and the realization of its full impact on society; in other words, social institutions change more slowly than technology. As futurist Peter Schwartz notes in his book *Inevitable Surprises*, "Historical research has confirmed that it takes at least ten years and sometimes more for a generation of people to come into positions of leadership and organize their companies and societies around . . . new tools."[76] Edison's electric light bulb and electric motor provide classic examples of this lag. Both were invented in the 1880s, but it was another four decades before Samuel Insull brought electric light and power into widespread use in the American home and Henry Ford reorganized the factory to take full advantage of the electric motor.

We are now experiencing serious unintended consequences from this technology-driven reinforcing loop. Past technology choices have accelerated environmental problems, depleted our resources, and widened the gap of health and wealth between those members of the human family who are participating in these processes and those who are not. Social evolution is significantly lagging behind technological innovation.

Major turbulence is the result, with significant parts of our world poised either for collapse or transformation. According to the punctuated equilibrium theory of evolution, chaotic turbulence creates the conditions whereby creative adaptation can cause a non-linear shift, a leap to a higher level of order that contains greater complexity and greater freedom than the prior level. While we can be sure that over the very long term, evolution will proceed toward higher levels of complexity and freedom, there is no guarantee, in the context of a particular historical moment, that this leap will occur, or that temporary devolution won't be the result of that chaos.

Information Technology: Supplying the Conditions for Business Model Innovation

Of the three technological trends outlined above, information technology offers the most potential for reshaping the landscape within which architecture and engineering firms operate. In the long term, firm leaders need to pay close attention to developments in biotechnology and nanotechnology. However, six major trends growing out of the information technology revolution are particularly significant for business model innovation (Figure 10.1). Architects and engineers have been aware of all these trends, but their transformational potential is most apparent when they are considered together, holistically, as a system.

Architecture and engineering firms have lagged significantly behind other businesses in assimilating and adapting to information technologies. Although they are investing in more and costlier computers and communication technologies, many of the ways they use these technologies are still relatively superficial. Firms use CAD technology, but mostly to automate the drawing process previously done by hand. Employees have access to e-mail and the Internet, but the project delivery processes they support remain remarkably similar to the 1950s paper-based processes. Servers and networks connect remote offices, clients, and other industry stakeholders, but little has changed in how the work is organized and managed. Each of the trends we will discuss offers new possibilities for business model innovation, allowing firms not only to catch up but also move into leadership positions within their organization ecosystems and value networks.

Capital-Intensive	⇒	Knowledge-Intensive
Local	⇒	Global
One-of-a-Kind or Mass Production	⇒	Mass Customization
Integration	⇒	Segmentation
Centrally Managed	⇒	Self-Organizing and Collaborative
One Shop	⇒	Network/Alliance

FIGURE 10.1. Transformation Trends Driven by Information Technology

The new possibilities growing out of these trends will not simply *replace* the original state or condition from which they emerged, although taken in isolation, new technologies can and do replace old technologies. When viewed in the context of a living system, the newly emerged possibility usually *transcends and includes* the prior form, to borrow language from Ken Wilber. The human brain is a clear example. It really consists of three brains: the reptilian, source of our instincts and most fundamental emotions; the mammalian, source of our nature as social animals; and the cerebral cortex, our uniquely human source of self-reflective cognition. We are who we are by virtue of including prior forms as well as of our unique evolutionary innovation. We suspect that ecologies of social organizations, such as industries, are more likely to follow this pattern of inclusion rather than simple replacement of prior forms.

Trend 1: From Capital-Intensive to Knowledge-Intensive

From the Middle Ages until midway through the last century, capital was the primary production factor used by individual and organizations to create wealth. Capital was applied to commercial activities to increase both their efficiency and effectiveness. Factories became more capable, railroads more powerful, and ships bigger and faster. Organizations used capital to construct buildings, bridges, ports, and other infrastructure in support of their operations.

As Arie de Geus asserts in *The Living Company*, "sometime over the course of the twentieth century, the Western nations moved out of the age of capital . . . and into the age of knowledge."[77] As the knowledge age matured, capital-based assets became more easily acquired than the more precious human assets. Across our economy, jobs focus, or at least depend on, knowledge and problem-solving rather than just repetitive activities and manual labor. Many organizations operate with an ever-diminishing core of capital assets relative to their human assets—these are the knowledge-based companies.

At the same time, machines have eliminated many forms of blue-collar and white-collar labor. Information technology has pushed this development even further, enabling machines to perform many mental functions as well. Workers are expected to possess and function at higher levels of knowledge than ever before—not only to design, operate, and maintain smart machines, but to handle the functions required to navigate and be productive in the complex world we have created around these machines. In this knowledge-intensive environment, a company establishes its sustainable niche through the quality of the people it is able to attract, retain, and effectively channel into relevant creativity.

Architects and engineers may have difficulty understanding the profundity of this shift. First, they have always thought of themselves as members of knowledge-intensive organizations, dependent primarily on the quality of the people who provide their design and engineering services. Their clients and other industry-related companies with whom they do business, such as construction companies, have been mainly capital-intensive organizations, but they themselves have generally run on little capital. Second, if anything, they are headed in the other direction. Over the last decades, they have had to make substantial capital investments in computer hardware and software. Consequently, capital assets make up a growing share of a firm's total assets.

In reality, the evolution of the architecture and engineering professions has followed a convoluted pattern. The models of practice that emerged in the 19th century were generally knowledge-intensive forms, in some cases, derived from the medieval and Renaissance craft guilds. Quite recently, lagging the rest of our economy significantly, firms moved to a more capital-intensive approach that wrapped itself around the earlier form. Their economic business model, at times reminiscent of early industrial assembly lines, often treats architects and engineers simply as cogs in a production process designed to produce billable hours. Managers have learned to treat many employees, CAD operators and even junior to middle level professionals, more as "replaceable parts" than as essential assets to the organizational system. Now, pressed by new technology-driven knowledge intensity, firms are returning to a greater valuation of their human assets.

Trend 2: From Local to Global

The second trend is the shift from a local, geographically bounded operational world to a global one. Although the trend toward globalization has been going on for centuries, it has recently been transformed by information technology. In the words of Joseph Nye, globalization is now "thicker and quicker,"[78] offering enhanced opportunities for individuals and organizations of all sizes to establish and profit from worldwide networks of interdependence, both socially and commercially.

Previously large corporations worked in many countries around the world and considered themselves global. Communication and travel were slower, more difficult, and relatively expensive, but they were possible for organizations with deep pockets. Today, instantaneous, inexpensive communication and much easier travel, enabled by the information revolution and other technology advances, make a global stance possible for many smaller organizations. This "democratization of technology," according to *New York Times* columnist Thomas Friedman, "is enabling more and more people with more and more home computers, modems, cellular phones, cable systems and Internet connections, to reach farther and further into more and more countries, faster and faster, deeper and deeper, cheaper and cheaper than ever before in history."[79]

Easy to state, but difficult to truly grasp is the magnitude of this change. This degree of connectedness has never before existed, and it is occurring at a time when the human population is skyrocketing beyond six billion. The sheer size and complexity of this global system is unprecedented. In this new world, many architecture and engineering firms may opt to remain local or geographically bounded in their operations, but many others will choose to evolve, to varying degrees, into global operations. Enabled by information and communications technologies, others will respond by shifting operations that can be performed by computers to workforces located in countries where labor costs are lower.

Trend 3: From One-of-a-Kind or Mass Production to Mass Customization

In the past, if we wanted to purchase something we had only two options: an item was either one-of-a-kind or mass-produced. Today, enabled by information technology, we have a third option, mass customization. Mass customization can deliver goods and services that are both low cost and customized for individual preferences.

This seeming oxymoron is actually an excellent example of how the evolutionary process produces greater complexity by creatively resolving paradox. Greater efficiency and lower cost represent one set of needs that have been rapidly evolving in our world. But corporate profits and organizational sustainability are actually only the small picture behind this set of needs. When we grasp the numbers of our global population, it becomes obvious that global humanity, as a whole, needs low-cost goods and services. The development and support of individual or community identity has created another rising set of needs. Our burgeoning population and accelerating globalization pose new challenges for those individuals or communities who are attempting to grow into their unique potential. In the past, these two sets of needs seemed directly opposed, resulting in entirely separate domains of production. In our information technology–enabled world, however, it is possible—at least with many kinds of products and services—to choose a great basic design and then have it customized to fit our individual identity, while still keeping costs down.

Consumers can now opt for a mass-customized car off a Detroit assembly line. Purchasers of Dell computers can have the company build a new PC according to their individual specifications and have it shipped within one day of ordering. Shoppers buying vitamins from GNC can have customized Vitapacs prepared for their individual lifestyles. Mass customization extends beyond manufacturing to many service businesses as well. Progressive Insurance customizes its insurance services using process modules. ARAMARK Healthcare Support Services provides customized, integrated, nonclinical support services to hospitals, with clients designing specific process modules in areas such as food service, distribution, linen, and maintenance services.[80] All of these are done without a significant cost premium over traditional mass-produced goods and services.

Of course, many organizations will continue to produce one-of-a-kind designs, including, prominently, much of the design and construction industry. But as clients shift operations toward mass-customized solutions, demand will increase for designers and engineers who can help them deliver goods and services that cater to particular customer desires and are not subject to the costs and time requirements of one-of-a-kind processes. Other firms are beginning to offer clients their own versions of mass-customized design and engineering services.

Trend 4: From Integration to Segmentation

This trend, moving from integrated to segmented operations, is the result of the maturing of production technologies and the commoditization of automated functions that can be repeated with increasing efficiency. The gap between the birth of information technology and promised productivity and efficiency gains is rapidly closing, accelerated by differences in labor costs across the globe. Many global corporations are relocating their operations and jobs to other countries. As Friedman notes, "India is rapidly becoming the back office for the world."[81]

Architecture and engineering firms are moving in both directions on this trend. Some firms are beginning to separate production processes (particularly CAD and GIS) from design and problem solving, construction administration, and client service activities, moving toward increasingly segmented service offerings. This segmentation has been further reinforced by increased liability concerns. As firms have shed responsibilities, other entities, such as program and construction managers, have taken over aspects of the project delivery process formerly handled by architects and engineers.

At the same time, other architecture and engineering firms are attempting to move in the opposite direction, claiming segments of the process that traditionally have been outside their domains. Front-end services include strategic facilities planning, feasibility analysis, real estate services, and project financing. On the back end, they include construction, commissioning, other post-occupancy services, and ongoing maintenance and operations. Some firms are trying to position themselves as integrators, reclaiming control of the design and construction process as a whole.

Trend 5: From Centrally Managed to Self-Organizing and Collaborative

The fifth trend being driven by information technology is the shift from achieving management through a centralized process to achieving it through self-organization and collaborative. This trend is closely related to the shift from capital-based to knowledge-based organizations. In the industrial age, workers did routine tasks at the direction of managers who were deemed to know everything about what needed to happen and how to solve any problems that might arise. However, as workers have transitioned to increasingly knowledge-intensive activities, and as the complexities of the technology-driven world have increased, the knowledge required to run organizations no longer resides principally with their leaders and managers. Information technology facilitates the generation of more and more information and knowledge, spread across greater numbers of employees. Front-line workers in many organizations now possess much of the knowledge that represents the essence of a company's value proposition; top management possesses only a portion of this knowledge base. And much of the information that once served managers has become either inaccessible or obsolete. As a consequence, knowledge-based organizations have become much flatter, with fewer levels of hierarchy.

A corollary to this lesson, that it is impossible for any one individual to be on top of increasing complexity, is that most levels of work require expertise from a variety of disciplines, functions, and perspectives. Knowledge-based organizations are typically more team based, using collaboration as the basic mode of getting work done. As this trend has evolved, collaboration has, to some extent, also become the basic mode of providing leadership. Self-managing teams chart their own courses, and leadership teams, rather than individual leaders, make important strategic decisions. Information and communication technologies are increasingly deployed to facilitate these forms of collaboration and self-organization.

Although some firms will choose to work with traditional hierarchical structures that allow central control of projects, others will choose to develop the skills needed to take advantage of this shift, empowering self-managed project teams and collaborating with similar teams from clients and other stakeholder groups.

Trend 6: From One Shop to Network/Alliance

The sixth trend that is transforming the way architects and engineers work is the shift from providing products and services as a single organization to providing them as part of a network or alliance. Before information technology really began to drive this emerging condition, a longer history of organizational restructuring created conditions that fostered its development.

For much of the industrial age, companies preferred to perform all the necessary functions for their operations themselves, through horizontal and, often, vertical integration, doing whatever was required their own way whether it was the best way or not. In the late 1970s, as the total quality movement gathered steam in the face of Japanese competition, the idea of forming almost exclusive relationships with an organization's supply chain extended the idea of controlling all the relevant functions for companies that had not chosen vertical integration. As global competition began to heat up in the 80s, companies began to outsource functions they had formerly held internally, paring back to their core competencies. In the 90s, outsourcing of staff and support functions began to extend to significant portions of entire business lines. Together with downsizing and enabling information technology, these organizational shifts also produced the phenomenon of free agents: individuals who make their living by contracting their services to organizations on a temporary basis.

The deregulation of the telecommunications industry, the rise of distributed computing, and the explosion of the Internet made this organizational refocusing even more complex. Now companies were experiencing, at the whole-organization level, what was happening to them internally: their work was becoming too complex to be taken on alone. Companies had to join with other organizations that brought complementary competencies to the table. In many cases, this need continues to lead to mergers and acquisitions. However, these are now more strategically focused than those in the past. Other companies, focused with laser-like precision on their particular area of expertise, have formed alliances with others to produce an integrated product or service.

Architecture and engineering firms are facing similar pressures from this trend. Will they maintain a one-stop stance that attempts to perform all of the required services for a client or will they establish networks and alliances that allow them to focus only on what they do best?

Mapping the New Ecology of Firms

Taken together, these six trends form a multidimensional ecology within which architecture and engineering firms can invent and build new business models—but only if these trends are understood, embraced, and adapted to. Firms that ignore these trends and retain their old economic business model will, at best, find themselves relegated to smaller and less attractive markets within this new ecology, or at worst, be forced out of business as their positions become unsustainable.

This new ecology offers a wide range of habitats (niches) for firms, each shaped by different responses to the six trends. As a guide to this terrain, we offer a map drawn from the Applied Strategic Planning (ASP) process devised by University Associates (now Pfeiffer & Jones) several decades ago.[82] The ASP process emphasizes three critical dimensions: WHO the customer or client is, WHAT the business is providing to the customer or client, and HOW, in general, the business is providing its product, service, or other type of outcome. Specifying these three dimensions provides an excellent foundation for elaborating any business model: to whom, what, and how firms will be delivering value.

The six trends driven by information-technology can be paired in a helpful pattern across the three ASP dimensions. The first two fall in the WHO, the second pair in the WHAT, and the final two trends in the HOW. The pairs produce three four-box models that, when taken together, provide an analytical framework that firm leaders can use to think more strategically and comprehensively about how their current and future business models are responding to emerging trends in the marketplace. Although presented as discrete choices, each of the six trends represents a continuous spectrum of choices.

Who do you work for?

The shifts from capital-intensive to knowledge-intensive and from local to global both fall into the domain of WHO does the business work for (see Figure 10.2).

A capital-intensive client might be a utility, an oil company, an automobile manufacturer, the U.S. armed forces, a city, a hospital, a bank, or a large retailer. A knowledge-intensive client might be a cellular phone company, a software company, a semiconductor manufacturer, a biotech firm, an Internet company, a law firm, a nonprofit environmental organization, or a government think tank. Some firms are developing their ability to bring capital to the table to serve the needs of capital-intensive clients by expanding beyond traditional design services into design-build-finance-operate capabilities. This will be particularly critical for many government agencies struggling to finance needed improvements in capital-intensive infrastructure. Opportunites can be both large and small, ranging from a massive new toll road project outside of Denver financed, built, and operated by the Washington Group to a performance management contract for a small electric generator executed by Wendel Energy Services for a hospital in western New York. Others serving capital-intensive clients, particularly cash-strapped municipalities and utilities, will focus on technological innovation and value engineering to optimize the limited capital available to such clients. For example, David Evans and Associates, an engineering firm based in Portland, Oregon, is using advanced materials analysis and design expertise to help the Oregon Department of Transportation target high-priority bridge repair and replacement projects.

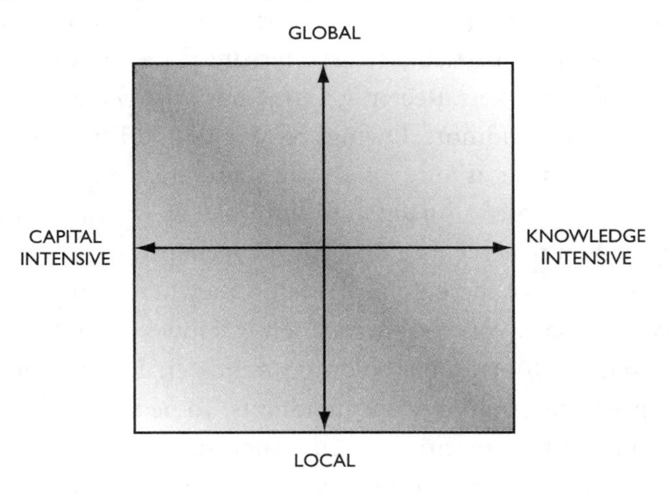

FIGURE 10.2. Who Do You Work For?

On the other hand, firms working with knowledge-intensive clients will focus their efforts on how built environment solutions contribute to the productivity and creativity of a client's knowledge workers. Anderson Brule Architects, in San Jose, uses a process-mapping technique to help libraries, health-care clinics, and other knowledge-based clients redesign work processes, build teams, and align their organizational vision and culture. RMW Architecture + Design, in San Francisco, emphasizes the design of high-productivity workplaces for high-tech clients in Silicon Valley.

A firm might decide to work only locally, serving its immediate community, city, or county. Or, the firm may choose to operate regionally, nationally, only in certain countries, or wherever opportunities arise around the world. The majority of architecture and engineering firms are still positioned primarily to do work locally. Many architecture and engineering firms expanded to regional and national status during the last half-century, either through a single office or multioffice structures. The number of global firms is still limited, dominated by large engineering/construction firms like Bechtel or CH2M HILL. Some midsize engineering firms have made the leap to a global practice. For example, Stanley Consultants, an 800+ person firm headquartered in Muscatine, Iowa, has an active global practice with offices around the world. Other firms, particularly architecture firms, have followed specific hot global markets. Recently, China has emerged as a major market for firms like Skidmore, Owings & Merrill and Kohn Pederson Fox Associates.[83] Gensler is one of the few architecture firms to break through to a true global practice, with offices in London, Amsterdam, and the Far East to serve its global business clients.

Clearly, many of these examples fall somewhere along one of these axes, not necessarily on the extreme ends. However, charting a firm's position relative to each of these two trends enables leaders to place their firm in one or more of the quadrants. To be truly effective, architecture and engineering firms have to understand and, to some degree, model themselves after their clients. Each of these quadrants will have a somewhat different kind of value network, requiring a significantly different kind of firm and providing opportunities for different business models and pricing schemes. If you find your firm working in more than two quadrants, you may be insufficiently focused to be strategically viable long-term. You may also be contributing to that lack of differentiation between firms that has in part been responsible for the undertow of commoditization.

What kind of work do you do?

The shifts from one-of-a-kind/mass production to mass customization and from integration to segmentation fall into the domain of WHAT kind of work we do (Figure 10.3).

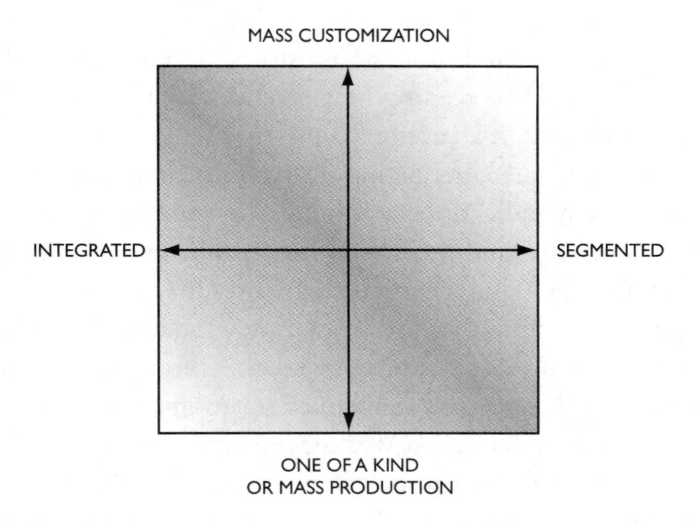

FIGURE 10.3. What Kind of Work Do You Do?

While there is a long history of architecture firms that focused on the design of prototypes for structures such as gas stations and convenience stores, and more recently, cellular phone installations and big box retailers, these firms have often been seen as the second-class citizens of the industry. Engineers have been more amenable to reusing their solutions, but even here, the romance of the unique, tough problem still beckons as the great ambition.

One-of-a-kind solutions are expensive. And good design is good design, regardless of how many times that design is used. Uniqueness may, in many cases, stand in the way of more important forms of value. The design and production of mass-customized solutions—designs that are both more accessible and less expensive—can yield significant value creation opportunities. Kaufman and Meeks, a Houston-based architecture firm, has sold rights to patented subdivision layouts and apartment designs. These cloned solutions for low-income housing developments provide a level of amenity and sustainable design that are often only available to higher-end projects. Large-scale rollouts of standardized store designs, now being managed by Gensler for retail clients, reduce design and construction costs, enhance the retailer's brand by presenting a consistent image across stores, and improve sales and operations by quickly spreading best practices for layout, furnishings, and display throughout an entire fleet of stores.

Opportunities for participating in mass-customization enterprises are spreading as manufacturers expand beyond simply supplying components for one-of-a-kind projects to the design, delivery, and installation of complete building or infrastructure solutions. For example, the office furniture manufacturer Steelcase is now offering buyers a kit-of-parts for offices, call centers, and even schools through a joint venture formed with a real estate management company and a real estate investment trust. The product, called Workstage, combines cost and time savings generated by factory-produced modules with the convenience, predictability, and one-stop shopping that the company offers through its mass-customizable furniture systems.

Similar propositions are also showing up in higher-end markets once thought to be immune from industrialization. A Swedish company is now delivering mass-customized luxury condominiums for erection on top of existing structures in central London. Produced in Sweden, the modules are finished to the specifications of individual owners and shipped across the channel to London, where they are lifted up and assembled on rooftops. Just as some design firms choose to work with manufacturers to create new furniture and products, some architecture and engineering firms may conclude that their future lies in supporting these mass-customized solutions for buildings and infrastructure, rather than continuing to pursue one-of-a-kind projects. By reducing the focus on unique solutions, firms may be able to liberate creative energy for larger, more complex problems, such as making communities more livable, solving pressing environmental problems, or creating new models for sustainable communities that could be customized for use around the world.

Currently, the landscape of architecture and engineering firms spans the spectrum from integration to segmentation. Firms have generally followed an evolutionary pattern that begins with a single-discipline practice and matures into a multidiscipline, integrated organization. Horizontal integration, across disciplines, and/or vertical integration, by extending services at either end of the project delivery process (for example, adding strategic facilities planning to the front end or construction management to the back end), provided a means of growing revenue by capturing a higher percentage of the professional fees available on any given project. This strategy also held the promise of increased client satisfaction and profits through the performance improvements and service innovations that are possible when a single organization controls all of the parts and pieces of a process or technology.

A new wave of integration is moving forward as architecture and engineering firms and contractors move toward design-build (D-B) or even design-build-finance-operate (D-B-F-O) project delivery processes. Organizations like the Beck Group, a Dallas-based contractor and real estate consulting organization, merged with an architecture firm to create an integrated service capability that could offer clients turnkey projects, from site acquisition through programming, design, and construction. The company has made a considerable effort to implement proprietary project delivery processes, databases, and communications software to make its service attractive to potential clients.

Unfortunately, as more firms move along this same developmental track, pursuing roughly similar integration strategies, they all begin to look alike, even though they may have started out very differently. This merging of practice types has ultimately reinforced client perceptions that many design and consulting services are simply commodities to be purchased on the basis of price.

Further movement toward increasing integration may be less attractive than segmented service strategies. As first steps, some firms are outsourcing CAD, GIS, and other production services to low-cost providers, often based overseas. Others are limiting their work to specific pieces of the existing project delivery process; for example, only doing programming or design work. Still others are focusing on new specialty services, such as strategic facilities planning, sustainability consulting, or facilitation services for complex projects. Rather than growth through integration, these firms pursue growth strategies centered on extending these specialties to new client types across wider geographies.

How do you do the work?

Finally, the shift from a centrally managed to a self-organizing and collaborative approach and the shift from one organization to a network or alliance of organizations (or individual agents) fall into the domain of HOW, in general, you do the work (Figure 10.4).

Project management has been a core business discipline of architecture and engineering firms. As generally practiced, it has been a variation on the central management theme. But as communication technologies, including e-mail and project Web sites, have become more common, the self-organizing capacities of project teams are increasing, catalyzing collaboration among all members at all times rather than flowing only through the project manager. Firms need to assess where they fall on this continuum from a project perspective. And where do they fall at the practice level? How self-organizing and collaborative are the processes for managing and leading the firm?

Many architecture and engineering firms have formed joint ventures or partnerships with other firms who are at the same time contracting directly with the same or similar clients. This kind of complexity in an environment of segmented services has been the norm for sometime. Most firms prefer to be the prime, however, and consider the subcontractor position undesirable. At the same time, they dread situations with multiple prime contractors because they tend to lack the leadership and collaboration that is necessary for effective self-organization to occur. Networking and allying are a given in the industry, and yet they remain quite unresolved for most firms. The design-build process has so far brought only new versions of these issues.

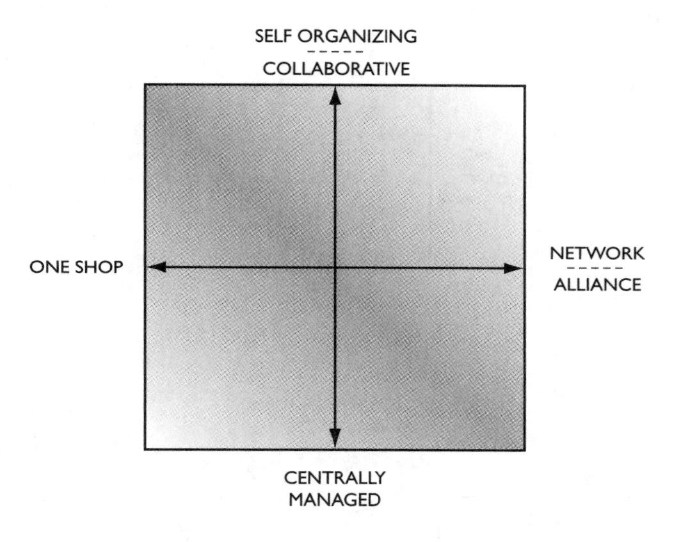

FIGURE 10.4. How Do You Do the Work?

Moving to the far extremes of both these spectrums, combining self-organizing project approaches with a rich network of collaborators, potentially takes project delivery into new territory, one that looks much more like the self-organizing effort that has created the LINUX operating system in the computer industry. In situations where issues are too complex for either a single firm or a centrally managed process (global sustainability issues, for example), expanding the depth and breadth of a firm's networks and alliances and working in a self-organizing fashion may be the most appropriate strategy as well as the only viable strategy. The network or alliance can share the considerable risks that come with these efforts and self-organizing can allow for the emergence of solutions that couldn't be imagined by any one party.

The New Ecology of Firms

Putting these three models together yields a rich, multidimensional landscape within which to understand potential vulnerabilities of existing business models (Figure 10.5). Plotting a firm's position (shading one or more quadrants in the WHO, WHAT, and HOW models) gives a visual picture of how it is responding to the six emerging information-technology-driven trends.

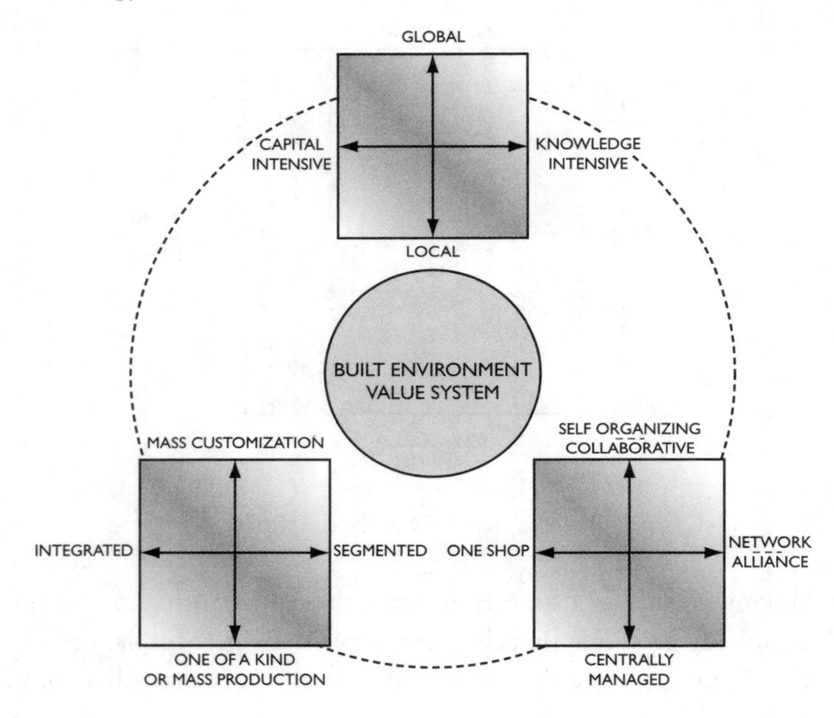

FIGURE 10.5. The New Ecology of Firms

If it has done relatively little adaptation, a firm might ask, what are the reasons? Does the firm's existing market remain resilient and robust? Do leaders believe that it will continue to do so? If the firm has adapted across one or more dimensions, how consciously was this done? What are the implications of what the firm has done relative to the other trends? What quadrants make the most sense to combine?

Architecture and engineering firms can also use the new ecology diagram for business model innovation, to establish the underlying fundamentals of to whom, what, and how they will be delivering value with new business models. They can be more intentional and strategic when selecting specific niches. Some firms will undoubtedly choose to remain in-place. But around them a new diverse ecology of firms will form, operating with different business models and offering different value propositions to their clients and communities. This diversity will begin to diminish the head-on competition that has played a major role in commoditizing services in the eyes of clients. Opportunities for more interdependence and collaboration among firms within this ecology will also emerge. With more interdependence and collaboration, architects, engineers, and other built environment professionals will have a much better chance of reasserting their leadership positions in society, improving the value they offer society and receiving fair compensation for the value they create.

II Experience

Designing new experiences is usually about figuring out a way to connect with people.

—Tom Kelley[84]

"The ultimate candy store for design-technology-creativity buffs." That's how Tom Peters describes the offices of the industrial design firm IDEO. Widely regarded as the leading industrial design firm in the world, IDEO consistently wins design awards and accolades for its work. Driving this success is IDEO's commitment to "creativity and innovation," which is embodied in the way it works, the people who comprise its staff, the way those people interact with clients, and the physical environment they work in.

Tour IDEO's work space in Palo Alto, California, and you'll experience the tumultuous energy of its staff working on a wide range of new product designs, from new computers and cellular telephones to the latest in ski goggles and snowboards and next year's consumer electronics. Along quiet, tree-lined High Street, IDEO created its own urban campus by converting a series of warehouses. These disconnected buildings are interspersed with sandwich shops, cappuccino stands, and other neighborhood stores. IDEO staffers can be seen working at the street-side tables and benches or strolling on the avenue trolling for inspiration.

Inside the buildings is a chaotic mixture of workstations arranged into small neighborhoods, large team work spaces, and support spaces ranging from meeting rooms to machine shops. Each studio space is filled with an eclectic blend of furnishings. In one, the wing of a DC-3 airplane is suspended from the ceiling. In another, plastic beaded curtains serve as doors, and Christmas tree lights decorate the space all year round. Umbrellas sprout as shade devices, bicycles hang suspended from ceilings by elaborate pulley devices, and carts are parked overflowing with "cool junk" to inspire a stalled brainstormer. Everywhere there are caches of toys, gadgets, and prototypes from past projects mixed in with murals and displays illustrating the firm's current work.

For IDEO's community, it is an environment dedicated to fostering innovation and building community. To corporate America, it may look like another planet, the antithesis of the buttoned-down, cube-filled environments many have created for themselves. But many leading companies now turn to IDEO not only for innovative product designs but also for the experience of working with them. In particular, they value the experience of people, process, and place that synergistically support highly creative team efforts at IDEO. Besides a great product design, these companies hope to bring some of IDEO's creative DNA back to their own organization.[85]

To satisfy their clients' aspirations, IDEO developed and implemented an explicit set of experiences that bring clients "behind the stage curtain," allowing them to observe, participate in, and learn from the design activity surrounding projects. These experiences range from guided tours of the firm's offices, attendance at Monday morning staff meetings where IDEO staff show and tell their latest designs, and active participation in a "deep dive" (IDEO's unique brainstorming process). The firm also invites client representatives to move into IDEO's offices and actively participate in the design of their own products over an extended period of time.[86]

One European vehicle manufacturer has sent teams to work with IDEO on its new products for weeks at a time in one of IDEO's famed "war rooms." Matsushita, the Japanese electronics company, sends staff to Palo Alto to become immersed in the IDEO culture. A key executive from Steelcase, the office furniture manufacturer, spends the night cutting foam-core boards for a concept model on the floor of an IDEO studio during a "blitz" project. Samsung, the Korean electronics giant, opened a joint IDEO-Samsung office in Palo Alto so that its designers could live and work with IDEO people. They designed more than 20 products in a three-year period.

In addition to recognizing the benefits they received directly from IDEO's designs, testimonials from clients speak of the significant value of their IDEO learning experience. Matsushita allocates portions of the costs of its on-site projects at IDEO to training budgets in recognition of the value of this learning. Samsung executives believe their three-year venture significantly enhanced the company's overall skill at product development. Validation of this investment came when one of Samsung's internal designs, a banana-yellow television set, appeared on the cover of *BusinessWeek* promoting a story about "The World's Best Products."

In addition to monetary rewards, IDEO also gains significant value from the close encounters between its staff and clients. Clients contribute new knowledge and ideas, continually refreshing IDEO's knowledge about design solutions used in different industries around the world. Collaborative experiences also provide safe settings where engineers and technical team members can refine their client relationship skills, including breaking down stereotypes that can exist between clients and designers. And these experiences help foster deep personal relationships between IDEO staff and client representatives, offering the potential for a fertile downstream flow of future assignments.

The Experience Economy

In *The Experience Economy*, authors Joseph Pine and James Gilmore argue that our economy is shifting toward a new class of economic offerings—memorable experiences. Experiences stand atop a hierarchy of economic value that moves upward—from commodities to goods, from goods to services, and from services to experiences.[87] The authors argue that the evolution of this hierarchy represents a fundamental shift in our economy.

Commodities were the fundamental economic output of the agrarian economy, which peaked in the United States in the 18th century.

Animal, mineral, and vegetable commodities were extracted from the natural world, stored and transported in bulk, and ultimately consumed. Because they couldn't be differentiated, the invisible hand of the market, supply and demand, determined the price or value of any given commodity. A new threshold of value emerged as the economy shifted from agrarian to industrial: commodities gave way to goods (products and merchandise). Prices of goods were set on the basis of the costs of production and differentiation. Useful goods commanded significantly higher prices than the costs of the commodities consumed to produce them. Eventually, as the supply of goods grew and competition among sellers increased, the forces of commoditization set in, and the value of many goods, in the eyes of buyers, declined.

By the 1950s, services began to overtake goods at the top of the value hierarchy. In 1985, with more than 80 percent of all jobs accounted for in the service sector, magazines and business pundits declared the reign of the service economy. As Pine and Gilmore note, "Clients generally value the benefits of services more highly than the goods required to provide them. Services accomplish specific tasks they want done but don't want to do themselves; goods merely supply the means."[88]

Despite their growing dominance, services have inevitably fallen prey to the same value-eroding dynamics of commoditization that befell their predecessors. Service offerings have proliferated in every sector of the economy, forcing service companies into price wars. Whole sectors of the service economy, from telecommunications to air travel to fast-food restaurants, have had their value propositions reshaped by fierce cost-based competition. Gilmore and Pine assert that providing services is no longer enough; companies need to find ways to create new value for their customers and escape the forces of commoditization by staging memorable experiences.

C.K. Prahalad and Venkat Ramaswamy spotlight experience in their recent book, *The Future of Competition: Co-Creating Unique Value with Customers.* They observe that "the future of competition is being shaped by changes in the meaning of value, the roles of the consumer and the company, and the nature of their interactions—changes that are profoundly altering the value creation process . . . the most fundamental change is in the nature of value. Rather than being embedded in the products and services that the firm offers, value is now centered in the experiences of consumers."[89]

Memorable experiences have moved above services on the value hierarchy. As services have been commoditized, their value to customers has diminished. Customers have begun to choose particular products or services on the basis of the type of experience they provide. They are also willing to pay a premium price for that experience.

To illustrate this point, Pine and Gilmore use the example of coffee beans. Sold as a *commodity*, coffee beans earn pennies on the pound. In the supermarket, when merchandised as a *good*, ground coffee sells for a dime a cup. At the diner, when offered as a *service*, a cup of coffee costs 10 times that amount. Finally, when coffee comes as part of an *experience* at a trendy espresso bar, with jazz on the sound system, comfy seating, and the *New York Times* on a brass newspaper rack, the price rises to more than 30 times the cost of the beans alone.

The emergence of the experience economy has profound implications for architects and engineers as they consider new business models for their practices. Professional services, like other forms of service, have been commoditized. One way out of this trap is to become expert at using experiences to create value.

While it is true that some architecture and engineering firms periodically stage value-creating experiences (for example, participatory design charrettes or squatters sessions), these events are generally small add-ons to larger professional processes. Seldom have these experiences been visible enough, or celebrated in ways that command premium fees from clients. Pine and Gilmore's portrait of an emerging experience economy offers firms the opportunity to fundamentally embed experience into business models and value propositions—to create new types of value and to earn attractive returns for doing so.

IDEO has already done it. The experience of working and learning with IDEO provides clients value far in excess of the services that another industrial designer might provide. The memorable experiences IDEO stages help clients acquire and internalize the knowledge and skills needed to continually improve and adapt to changing conditions in their markets and organization ecosystems.

Architects and engineers can leverage experiences to create value in at least four different ways:

- *Experiential learning:* offering learning experiences in conjunction with project delivery processes and client relationship activities.
- *Collaborative creativity and new digital modeling tools:* engaging clients and other project stakeholders in collaborative design and problem-solving processes, enabled by new modeling, simulating, and prototyping technologies.
- *Individual co-creation experiences:* building points of interaction into mass-customization processes where individual clients co-create their design solution, monitor the production of their chosen model, and participate with a larger community of clients who are also solving their problems in the same way.
- *Staging experiences for the client's customers:* designing and staging experiences that clients will subsequently offer to their customers directly and as part of other product and service offerings.

Experiential Learning

Like IDEO, architecture and engineering firms can build learning experiences into their project delivery processes and into the ways they work with their clients. Learning can focus on specific technical topics of interest to the client (sustainable design, new materials and methods, etc.), or it can support the growth of new social and leadership capacities (creativity, team skills, leadership, etc.). Clients can be invited to attend classes, seminars, and roundtable discussions and to observe professionals at work (attendance at IDEO's Monday morning staff meetings). Learning can also occur by directly involving the client in the project's technical work, which traditionally has been viewed as the exclusive domain of the professional (participation in IDEO's brainstorming sessions), or by the rub-off effects that come with co-location of staff members (IDEO's residential program for clients). Enabled by new technologies, firms can also construct dramatic new types of learning experiences.

Collaborative Creativity and New Digital Modeling Tools

Huddled in a darkened space, explorers strap on special headgear and prepare to make a descent into an underground environment beneath the North Sea, off the coast of Norway. For the better part of a day, they tunnel their way through this geologic environment looking for the best way to tap the natural resources to be found there. Led by a skilled guide, they descend miles below the seafloor, stopping periodically to chronicle and analyze the significance of their discoveries. The following day, the explorers complete their journey, having developed a shared understanding of the territory's rock formations, faults, surfaces, and reservoirs; and they have made a series of decisions that will ultimately save their organization millions of dollars. The expedition has also helped them become a team in the truest sense of the term. They learned to appreciate the knowledge and insights of each member of their group, to communicate and collaborate in new ways, and to take advantage of the creative synergy that teamwork makes possible. Moreover, they did it all without leaving the office park in suburban Houston where they started.

This expedition took place in a "Decisionarium," a unique immersive environment designed by Landmark Graphics Corporation, a Houston-based developer of integrated oil and gas exploration and production software. Inside a Decisionarium, teams of geophysicists, geologists, engineers, drillers, and managers from oil and gas companies gather to explore the subsurface conditions of a particular site, evaluate prospective reservoirs, and decide on drilling strategies aided by Landmark Graphics 3-D modeling and analysis software. But a Decisionarium is not just another office or conference room; its setting provides a unique experience, designed to change the way oil and gas companies use their employees' expertise, insights, and creativity.[90]

Founded in 1982, Landmark Graphics was established to develop and sell 3-D modeling software that would enable oil and gas companies to analyze exploration and drilling decisions. By the mid-1990s, the market for this type of software application was becoming increasingly commoditized. To escape this trap, Landmark Graphics decided to move beyond selling software to providing a solution to a much more fundamental problem that their customers' faced: how to make better decisions, faster.

Traditionally, the people involved in making drilling decisions in oil and gas companies have worked in relative isolation. Individual geologists, geophysicists, and engineers worked in their own offices, handing off their analysis and recommendations to the next person down, or up, the organizational hierarchy, until finally a decision was made. When the process was successful, the profits were plentiful, but the cost of a poor decision—drilling a dry hole or tapping a reservoir in inefficient ways—could amount to tens of millions of dollars. Landmark Graphics' Decisionarium offered a new way to make these decisions that could significantly improve results.

In a Decisionarium, all of the members of a drilling team gather, put on 3-D goggles, and enter a large room dominated by a concave, 9.5-foot tall by 20-foot wide computer screen. There they join a group of Landmark Graphics' staff, including geoscientists, a computer operator, and a facilitator. The geoscientist brings to the table knowledge of the digital model that Landmark Graphics created of the particular site under consideration. The computer operator manipulates the model at the team's request, simulating possible actions and enabling the team to visualize site data. And the facilitator guides the team through its decision-making process. The process is designed to leverage the synergy that is possible when groups of diverse people work together in a collaborative decision-making process. Throughout the session, the process stops to capture models, simulations, and algorithms that were particularly helpful in gaining insights about the decision being explored. That knowledge is then made available to future teams exploring the same field or facing similar situations.

The Decisionarium has won over the oil and gas industry. It facilitates better decisions by allowing all of the stakeholders to work together, sharing the same model in the same place. The ability to visualize the immense volume of data describing a particular site provides a common language that can be used by all team members. The decision-making process, which used to occur in a series of meetings or phone conversations, often strung out for weeks at a time, is accomplished in less than a day. The experience builds trust and grows teams. It not only helps the participating team learn but also captures that knowledge and makes it available for future decision makers.

Landmark Graphics projects that customers will double their success rate in drilling decisions and cut development costs by 25 percent by using a Decisionarium. Customers agree. They are not only lined up to spend time in Decisionariums but are also purchasing their own facilities from Landmark Graphics, at costs ranging from $500,000 for a basic package to as much as $2 million for a theater-size environment with a 40-foot screen.

Landmark Graphics changed the rules of value creation for its business, moving away from selling a product (3-D modeling software) toward providing memorable experiences for its customers, experiences that transformed the way those clients worked, learned, and made decisions. The software, and the programming that went into creating it, was only a small component of a much richer value proposition. A Decisionarium creates value by reducing drilling and exploration costs and increasing operating efficiencies for fields under development. Collaboration and decision-making skills honed inside Decisionariums prepare team members to be more effective back in the workplace. Value also accrues in the form of learning and knowledge, stored both in the minds of participants and in digital forms, ready to pay future dividends for customers.

Just as Landmark Graphics creates value for oil and gas companies by inviting exploration teams to experience a computer-aided decision-making process inside a Decisionarium, architects and engineers can offer their clients the experience of a digitally enabled collaborative design and problem-solving process. Traditionally, design professionals have worked in relative isolation from their clients; designs and solutions were developed independently by the professionals and unveiled for client input, review, and approval at designated milestones. New computer modeling, simulating, and prototyping tools offer the potential of changing this dynamic.

In the 1980s, the digital spreadsheet revolutionized the world of finance, triggering a tidal wave of financial innovation that reshaped the global economy. Spreadsheets, produced first on VisiCalc, then on Lotus 1-2-3 or Microsoft Excel, became the dominant medium of communication and collaboration in the business world. This disruptive technology opened up opportunities for people throughout an organization to collaborate on the development of financial strategy. The what-if analyses that spreadsheets made possible enabled organizations to fine-tune financial strategies, reduce risk, and increase value. These financial wind tunnels, where organizations could test financial strategies risk free, enabled the boom in securities that has fueled the global economy over the last two decades. The early adopters of digital spreadsheet technology—investment bankers, venture capitalists, and mutual fund managers—ascended to preeminent positions in the economy.[91]

Innovation and creativity thrive on the interplay between people and the expression of their ideas. Change the medium, and you change the way people think. That is the tale of the spreadsheet: a new medium for modeling and simulating financial ideas sparked a wave of financial innovation and changed the way the financial world worked. Architects and engineers can harness new digital media (computer models, visualizations, prototypes, and simulations) to transform the way they work and create significant new value for their clients and other project stakeholders.

Digital models can be used to re-imagine and gain new insights about design problems by test flying potential solutions in new digital wind tunnels. In much the same way that what-if analyses grew up around spreadsheets, 3-D modeling software can be leveraged to explore numerous alternatives and iterations. Firms can add rapid prototyping technologies to further enhance the experience. Rapid prototyping has become a mainstay of both the software and industrial design communities. Rather than follow a linear product development path, these organizations use rapid prototyping technology to develop many solutions early. They don't expect the client to pick one of these solutions, but rather use the models and prototypes to elicit customer needs and expectations that often go either unexpressed or are only realized when it's too late to address them. Fortunately, digital abundance (faster, better, cheaper computer processors, storage media, and communication technologies) means that teams now have the ability to try out many variations of any design or solution at a reasonable cost.

The value of digital modeling tools can be further enhanced by deploying them as part of new collaborative design and problem-solving processes. For the design of their new California Adventure theme park, designers at Walt Disney created their own version of a Decisionarium. They built a 4-D virtual park using Silicon Graphics computers and three large projection screens assembled in a U shape to create a viewing center. Time was added to a 3-D model by assigning different colors to each phase of the construction work. This virtual theme park operated as a "giant flight simulator" for the Disney team and other project stakeholders. Designers were able to climb aboard a virtual roller coaster to see what riders would see as they rounded each curve. The model allowed the project team to explore potential designs as well as examine alternative construction sequences and methods, showing what would be built at each stage of the construction process. The simulator was also deployed to explore critical code compliance issues with city officials and obtain necessary buy-in and approvals for the project. For example, one simulation included a virtual fire truck that was "driven" into the virtual park. Then, its virtual ladder was extended to the top of the California Screamin' rollercoaster to ensure that it would reach the top.[92]

Firms can also use digital modeling to help clients discover their program requirements as part of the design process, rather than rely on traditional front-end program documentation efforts. Traditional up-front programming often fails to uncover all of the client's expectations and priorities. However, modeling and simulation processes can be used to reveal these critical requirements.

Cambridge Technology Partners (CTP), an e-business software development company, built its software design approach around this idea. CTP invites its customers to participate in a rapid prototyping process it calls the "Rapid Solutions Process." The process features an intense collaboration between CTP software designers and customer users, including participation by senior management from the customer organization. Rather than starting work by developing a program of requirements for the software, the CTP team begins by building software prototypes almost immediately. Key design and implementation issues surface as customer representatives begin trying out and reacting to these prototypes. By involving customers directly in the design and prototyping process, the CTP software designers also have the opportunity to see how their customers behave around the software, prioritize requirements, and make choices between options. According to CEO Jim Sims, customers rarely end up building the applications they initially said they wanted. "What usually happens is that the team discovers what they really want as they build their prototypes . . . clients end up surprising themselves."[93]

CTP's rapid prototyping process compresses the traditional front-end software design process from six months to three weeks, a tenth of the normal development time. More important, CTP's software applications accomplish what customers really need them to do. Instead of satisfying a potentially off-target formal statement of requirements, the applications deliver the business results companies want.[94]

Digital modeling tools offer the opportunity to democratize the design process. Appropriately deployed, these tools can shift design away from being the exclusive property of a favored few professional firms. Simulation and modeling processes can place design in the hands of a wide range of project stakeholders who serve as co-creators of the solution. Unique relationships and teams coalesce around these digitally assisted design experiences as more people become involved, learn to work together in this new way, and communicate using the graphic language of these models and simulations. Client and stakeholder expectations can be managed as explorations of design "what-ifs" are used to educate and inform participants about the impact of alternative design choices. Clients are much more reluctant to throw out or criticize work that embodies their own direct input.

Overall, new digital modeling and simulation tools enable new experiences that can transform the social process of design and problem solving, setting off a virtuous cycle in which innovative models stimulate innovative conversations, which in turn point toward more innovative solutions and new value creation opportunities.

Individual Co-Creation Experiences

Sumerset, the world's largest builder of houseboats, leverages the power of experience to create significant forms of value for its customers.[95] The company offers houseboat buyers a co-creation experience that combines the benefits of mass production and custom design. This includes opportunities for continual dialogue (in person, on the telephone, and over the Internet) with Sumerset staff during each stage of the design and production process. The company emphasizes its belief in the power of "informed choice" for its buyers and has structured its co-creation experience to take advantage of that power. "To foster this dialogue, the company not only provides customers with access to the firm's employees and knowledge, but also clearly explains the risks and trade-offs of every customer choice throughout the process."[96]

New boat buyers are invited to collaborate with a team of Sumerset professionals in the co-design of their boat, a process that can range from making slight modifications on standard designs to developing highly customized designs tailored to fit individual needs and preferences. During production, the company posts photographs on its Web site daily, allowing buyers to see their boat as it is being built. When the boat is complete, Sumerset's delivery team transports and assembles it and also provides the necessary instruction so that buyers feel comfortable with the operation and care of their new boat. From start to finish, buyers are invited to participate in Sumerset's community of houseboaters, a virtual forum where past and present customers share ideas, information, and helpful advice.

In *The Future of Competition*, Prahalad and Ramaswamy argue that this type of individualized "co-creation experience" is one of the most significant opportunities for value creation in the emerging experience economy, particularly for companies involved in the design and production of mass-customized products and services. They note, "the quality of the experience involved in co-creation differs markedly from that of a traditional product purchase. The basis of value for the customer shifts from a physical product (with or without ancillary services) to the total *co-creation experience*, which includes co-designing as well as all the other interactions among the consumer, the company, and the larger community."[97] Architects and engineers can learn to create value through experience by learning to orchestrate and support similar types of co-creation experiences for mass-customization clients.

Staging Experiences for the Client's Customers

As Pine and Gilmore point out, many service industries—including retail, restaurant, resort, and hospitality markets—are actively wrapping experiences around their service offerings. This shift opens up the fourth way that architects and engineers can create value—by designing and staging experiences for clients to subsequently offer to their customers.

For these markets, professionals learn to act like set designers, creating new experiential settings for their client's operations. Walt Disney pioneered this approach. Disneyland's guests were treated to "a living, immersive cartoon world" experience. Today, the Walt Disney Company has extended this experience-based strategy beyond the amusement parks, to Broadway theaters, retail stores, cruise ships, and even its own office facilities. Restaurants such as the Rainforest Café offer diners an experience of dense vegetation, cascading waterfalls, and tropical birds and fish while they enjoy their meal. Outdoor equipment retailer REI presents shoppers with climbing walls, indoor biking trails, and simulated rapids that can be experienced while shopping for the latest in outdoor sports gear and clothing.

Beyond set design, designers can help clients create more robust experiences combining the physical with the virtual. A consortium of 20 firms led by IDEO and the Office for Metropolitan Architecture (OMA) helped retailer Prada create its first "epicenter" store in 2001. The New York City store offers shoppers "an ongoing experiment to enhance the shopping experience through interactive technology." This experiential environment includes a variety of cutting-edge in-store technologies: dressing room walls can switch from transparent to translucent, providing a shopper privacy or allowing her to show a companion how she looks; controls allow shoppers to view their selections under a variety of lighting conditions; touch-screen displays show alternative colors, fabrics, styles, and sizes; and a video-based Magic Mirror allows a shopper to see herself from all angles. According to co-CEO Miucca Prada, the store delivers a new retail experience that "melts into the architecture of the store itself."[98]

12 Technical, Collaborative, and Transformative Work

Companies today face adaptive challenges. Changes in societies, markets, customers, competition, and technology around the globe are forcing organizations to clarify their values, develop new strategies, and learn new ways of operating.

—Ronald Heifetz[99]

"It's so rewarding to watch the expressions of people as they enter and explore the building," said San Jose Public Library Director Jane Light at the August 2003 opening of the new Martin Luther King Jr. Library in San Jose. "There's a sense of awe and excitement that lets us know we're on the right track. And in the long-term, we are confident that the combined strengths of city and university library staff will deliver what the building promises."[100] The library promises to become a center for lifelong learning and an information hub for San Jose residents and for students and faculty at San Jose State University (SJSU). The newly opened facility combines the collections of a major university with the resources of the city's main library and its 17 branches.

The idea of combining resources to create the nation's first co-managed academic and university library in the nation was conceived in 1997 by San Jose Mayor Susan Hammer and SJSU President Robert

Caret. Six years later, in August 2003, the $175 million facility opened, to the acclaim of its diverse community. The design and construction of the 8-story, 470,000-square-foot structure, the largest library ever built west of the Mississippi, was a significant accomplishment.[101] Equally impressive was the simultaneous transformation of two distinct library organizations, one public and one academic, into a single, unified organization, motivated by a powerful, shared vision.

A joint-use library seemed like an obvious solution to a critical problem that faced both the city and the university. Both communities needed to replace existing library buildings that had run out of space and had no potential for further expansion. Because SFSU occupies a prominent location bordering downtown San Jose's redevelopment area, any library built by the university would inevitably entail significant input from, and involvement with, the city and surrounding community stakeholders. By sharing funding and avoiding the duplication inherent in constructing two buildings, the resulting facility could be bigger and better equipped than either party could afford separately.

Hammer and Caret's idea went beyond simply pooling funding and co-locating the two library programs. They saw an opportunity for a new type of library, staffed by a single organization, taking advantage of the creative synergy that could flow from combining the best of the public and academic library worlds. Public Library Director Light further articulated this vision as the creation of a "seamless-service library" where patrons would be served by city and university employees. These leaders also recognized that the joint-use facility could provide a new urban commons, anchoring San Jose's downtown redevelopment area, where citizens and academics, college students and children could come together, interact, and learn from each other.

The vision was both attractive and easy to grasp at the conceptual level, but the reality of merging these two very different types of organizations into a single operating unit was daunting. The formation of this new organization had to happen concurrently with the design and construction of the building. To help city and university leaders in this transformation process, the San Jose Redevelopment Agency (the authority formally responsible for managing the design and construction of the new facility) engaged the San Jose architecture firm of Anderson Brule Architects (ABA). ABA had successfully tackled several similar efforts for other clients in the San Francisco Bay region.

As a first assignment, the Redevelopment Agency asked ABA to facilitate a process for completing a feasibility study that would provide the means for deciding whether to move forward with the project and serve as the basis for a "memo of understanding" between the two parties about how a new library would be jointly owned and operated. Working with Light and key university library staff, Pamela Anderson Brule and Sam McBane, principal and senior strategist, respectively, for ABA, designed a process that assumed a positive outcome but still allowed the committee responsible for the feasibility study to search for possible deal breakers. Members of the committee were challenged not to create barriers and find reasons "why it [a joint-use library] isn't possible," but rather to determine "how it can be made to work." ABA helped the committee clarify the overarching vision for the new library, examine critical details and issues that would be covered in the memo of understanding, record and document meeting results and key decisions, and move toward completion of a published document. In the end, their optimism paid off and the project was given a green light.

ABA's next assignment raised the bar for its facilitation and collaboration skills. While the Redevelopment Agency began the process of selecting an architectural design team for the project, ABA was asked to work with city and university leaders to design and facilitate a process for developing a strategic operating plan that would describe how Light's seamless-service library would actually function—to provide the details needed to craft specific operating agreements covering all aspects of this new public-public partnership, from library services to administration to maintenance.

The strategic operating plan would also serve as an important predecessor to the traditional architectural program that would be done for the facility. There would be little point in trying to develop a list of spaces, adjacencies, and equipment for the facility until the two organizations had comes to terms with how they were going to operate as one.

Completing the strategic operating plan would be no easy feat. Leaders anticipated enormous resistance internally and externally. Combining the public and academic library organizations was in many ways like mixing oil and water, merging distinct cultures, values, and operating practices. Members of both existing organizations would be challenged to accept the decision to create a joint library as a given, even though they might have personal misgivings, and to work together to figure out how this new institution could function.

Realizing the magnitude of the change confronting the two library organizations, Anderson Brule and McBane proposed a process that included dozens of meetings involving hundreds of staff members from both organizations over a twelve-week period. ABA staff would facilitate the meetings, manage agendas and attendance, document decisions and understandings, and generally keep the process moving so that participants wouldn't get bogged down in storming, or arguing over details. A key attribute of ABA's process design, according to Light, was that it "allowed a large number of library staff from both sides to participate intensely, but not feel that it would go on forever."[102] According to Anderson Brule, a key goal for ABA was to make it easy for the library staff and other stakeholders to put together their pieces of the operations plan, "to provide a strong backbone and process . . . to keep them on track and moving forward."

At the end of the three-month effort, results far exceeded expectations. In addition to creating an operations plan, ABA's facilitation helped members of both organizations become a real team. This team, according to Light, went on to create and implement many of the innovative programs housed in the competed facility, and that teamwork continues to pay dividends in the operation of the new library. Participants learned about each other, compared and contrasted the different ways they worked, and identified mental models and assumptions that drove behavior and expectations. In the process, the people in each section of the two libraries, from stacks to circulation to reference to administration, came together as one organization. According to Anderson Brule, "they did it all, we just made it easy for them to do it."

Commenting on the central importance of these facilitation efforts, Light observed, "ABA's process made the potential of a combined facility believable for staff members from both sides." The collaborative mind-set that ABA facilitators modeled from the beginning not only helped library staff work and make decisions together, but also helped them build collaboration skills that were used even when ABA wasn't present. The process improved communication and "created stronger relationships that have only begun to serve us well."[103] Barbara Leonard, a key staff member for the university, also noted that ABA's facilitation "allowed staff to think outside the box, enabled participants to reflect more on real possibilities, and provided tools to help understand staff concerns on each side."[104] Overall, the process generated buy-in at all levels from both library staffs and supported rapid evolution of the new organization.

For the San Jose library project, ABA stepped out of the traditional role of the architect. The firm was asked to help the client design a new type of library *organization*, not just the building to house it. The new organizational design needed to address the merger of the two operations, resolve the attendant technical details required for unified operation, and also—critically—acknowledge and deal with the social complexity of the situation and potentially wrenching personal changes for staff on both sides. People needed to change their ways of working and job descriptions, as well as their basic values and beliefs. As professionals, Anderson Brule and McBane didn't attempt to offer technical solutions as expert organizational consultants. Rather, they offered facilitative leadership to support city and university leaders as they engaged members of their two communities in a collaborative effort to create a new library.

ABA's facilitative leadership is representative of a powerful trend sweeping the design and construction industry. More and more projects require collaboration and facilitative leadership, not just technical solutions, to reach a successful outcome.

This trend was underscored at a panel discussion we facilitated for senior representatives from client organizations. We asked the city manager of a midsize San Francisco Bay Area city and the director of facilities for a large public university, "What is the most important thing you need from the architectural and engineering firms that you work with?" Both of them answered, with not a moment's pause, leadership. The city manager cited a seemingly straightforward project that involved laying additional pipe for the municipal water system. The engineering and construction work was easy; dealing with the aroused concerns of the surrounding community was not. Community members were currently obstructing this important project, exhibiting classic NIMBY (not in my back yard) syndrome and voicing other concerns, many of which were only loosely connected with the water project. The city manager wanted the engineering firm to step up and help the city deal with these concerns, win community acceptance and support, and then get the project done. The university facilities director told similar stories of projects whose success hinged more on navigating the arcane political and social dynamics of the campus than in wrestling with budget and time constraints or with typical architectural design issues of form and function.

Satisfying these types of client demands requires the mastery of the social and leadership capacities discussed in part II. In turn, mastery of these leadership skills prepares firms to deliver new forms of value to their clients, their organization ecosystems, and society as a whole. It also sets the stage for the evolution of an important set of business models that architecture and engineering firms can choose from.

Adaptive Work and Professional Practice

The value a firm creates is directly related to the type of work that it undertakes; different types of work offer different opportunities for value creation. During the first three-quarters of the last century, architecture and engineering firms defined their work on the basis of their professional disciplines. Firms did architecture, or civil engineering, or structural engineering, or some blend represented by the disciplinary backgrounds of the firm's principals.

In the early 1980s David Maister recommended that professional firms move beyond this definition. He equated "type of work" with a professional service firm's choice of project type. Maister's project types ranged along a spectrum from "brains" to "gray hair" to "procedure."[105] He argued that a firm's mix of project types shapes both the economic and organizational structures of the firm. Over the next 20 years, led by management consultant Weld Coxe, many architecture and engineering firms adopted this perspective as a fundamental element of their business model and part of their strategic planning.

In the mid-1990s Ronald Heifetz, in *Leadership Without Easy Answers*, proposed a new definition that links the type of work professionals do to the leadership challenges that clients or communities face.[106] His definition significantly expands the possibilities for value creation inherent in the work of professionals. It also provides a framework that can be leveraged to build new business models around these possibilities.

Heifetz makes a fundamental distinction between technical work, on the one hand, and adaptive work, on the other. He defines technical work as work that a group, organization, or community already knows how to do. Not everyone knows how to do it, but someone, usually someone who has been given the authority to perform the work, knows how to do it. For Heifetz, technical work exhibits two characteristics. First, the problem to be solved has, or can be given, a clear definition. And second, the solution and its implementation are also clear—previously known and tried in other cases.

When professionals are involved with clients doing technical work, the weight of both problem-definition (diagnosis) and problem solving (treatment) rests with them. Professionals exercise the authority that comes with their role to prescribe a technical solution to the problem. Minimal leadership is required from the professional. Heifetz draws on his personal experience as a physician to illustrate technical work, which could include treatment of an infection or a simple outpatient procedure.

In contrast, the need for adaptive work arises when the complexity of a situation precludes both clear definition of the problem and straightforward solutions. These kinds of situations arise naturally as individuals, groups, organizations, and societies evolve. New complexities are produced by new social phenomena, such as increasing population, increasing diversity in the population, or new laws to deal with increasing economic or environmental stress. They are also produced by new technological developments, such as instantaneous electronic communications, or economic shifts, moving from boom to bust. Natural disasters or other large-scale disruptions can also give rise to situations requiring adaptive work. We are constantly evolving at all levels of our human systems, and new problems constantly emerge at the edge of that evolution. The faster the evolutionary change, the more situations we face that require adaptive rather than technical work.

Adaptive work involves more than developing a new kind of technical solution. Simply imposing a known technical fix on this more complex type of problem won't result in a real or lasting solution. Adaptive challenges invariably involve growth in the social system's capacity to handle increased complexity. People's values and beliefs tend to be based on realities experienced in the past, and on past ways of making sense of those realities. Consequently, emerging realities tend to challenge existing values and beliefs. As Heifetz further explains: "Adaptive work consists of the learning required to address conflicts in the values people hold, or to diminish the gap between the values people stand for and the reality they face. Adaptive work requires a change in values, beliefs, or behavior. The exposure and orchestration of conflict—internal contradictions—within individuals and constituencies provide the leverage for mobilizing people to learn new ways."[107]

If leaders are orchestrating adaptive work, they are no longer "influencing the community to follow the leader's vision . . . [but] . . . influencing the community to face its problems."[108] As Heifetz also points out, "The dependency on authority appropriate to technical situations becomes inappropriate in adaptive ones."[109] In adaptive work, then, the definition of the problem as well as its solution and implementation requires deep learning—double-loop learning rather than single-loop learning. This learning is something only clients and other stakeholders themselves can do.

When confronting situations involving adaptive work, the professional's role changes significantly; he helps the client do the adaptive work. He can't do it himself on behalf of the client, as he can often do when he is engaged with technical work. The professional helps the client learn more about the situation, understand the nature of the adaptive challenge involved, and accept the need for change. Then, the professional facilitates the client's efforts in adapting to the situation, in the best way possible. Heifetz illustrates adaptive work through the story of a young carpenter diagnosed with terminal cancer. The physician had no easy answer for this patient. The patient needed someone (the doctor) to help him and his family confront the harsh reality of this situation, adapt as best they could, and live as high a quality life as possible during his remaining days. Technical work still needed to be done, in the form of treatments or prescriptions designed to prolong life and make the patient as comfortable as possible, but the highest value provided by the physician in this situation was in the nontechnical leadership he offered the patient.

ABA's work for the San Jose joint-use library falls squarely in the realm of adaptive work. The situation the client faced was not simply technical in nature. The client needed a professional who could make the difficult process of becoming a single organization as easy as possible. A typical architectural programming and design process, approached as technical work, would have papered over the deeper social and technological issues that needed to addressed and resolved. The process facilitation effort placed ABA in a role of providing adaptive leadership, collaborating with university and city leaders to help members of both communities establish a common vision and set of values and begin the learning and change process that was necessary to successfully merge into one organization. With ABA's guidance, the new library staff also built a foundation of effective communication and collaboration skills that not only served them well as they moved through the design and construction process, but continues even today as they work to fulfill the mission of this groundbreaking institution.

Three Types of Work

Heifetz is not describing two discrete types of work but rather a continuum that runs from technical to adaptive work, with leadership difficulty ascending from the former to the latter. Heifetz emphasizes this leadership dimension by calling out a third type of situation that falls between technical work, which he labels "Type I," and adaptive work, which he labels "Type III." "Type II" work, which he gives no other special name, involves a combination of technical and adaptive work—the problem can be clearly defined, but the solution requires learning and, crucially, collaboration between the professional and the client (and potentially other stakeholders). We have named Type II work "collaborative" to reflect the nature of the social and leadership capacities the professional needs to perform it. We have also renamed Type III work, labeling it "transformative" rather than "adaptive."

By labeling Type III work transformative, we embrace more than the very particular social dimension of leadership in Heifetz's adaptive work. From a whole-system viewpoint, leading into new territory often involves adapting to both social and technological innovations. Humans are generally much better at adapting to technological changes than dealing with social innovations; many of our most serious problems stem from the failure of our social capacity to keep pace with our technological capacity. Heifetz's emphasis on social innovation (changing values, beliefs, and competencies), embodied in his definition of adaptive work, recognizes both the significance and nature of this gap. However, it is also important to recognize that technological innovation can emerge in response to adaptive challenges. Technological innovation can be as adaptive as social innovation. Consequently, we include both in our definition of transformative (Type III) work. In should be noted that innovations that emerge purely in the realm of technology, in isolation from social issues, fall within the category of technical (Type I) work.

Traditionally, architecture and engineering professionals have been content to focus primarily on Type I work. Whether the professional was doing plans and specifications for a new school building, or providing recommendations for remediating a contaminated site, the work was easy to define and the solutions could be developed with little active participation by clients. Type I work can span all three kinds of projects defined in David Maister's widely used model.[110] It could include a cutting-edge water treatment process that relies on the "brains" of technically innovative staff, a complex expansion of an existing sewage treatment plant that needs the "gray hair" of experienced seniors, or a simple sewer pipeline project that leverages a firm's facility with "procedure."

In the last 20 years or so, a significant portion of the work that used to be considered Type I has migrated to Type II, driven by the growing complexity and interdependence of client organization ecosystems. In fact, we have witnessed explosive growth in Type II work. An airport expansion provides a good illustration. Though the nature of the work is generally well understood, the work is complex and involves a multitude of interdependent stakeholders. Beyond contributing technical design solutions for new terminal buildings or runways, architects and engineers can more fully meet client expectations by helping the client prepare a plan for involving and coordinating all the stakeholders, as well as facilitating their collaboration and decision making.

As turbulence and change in our society increase, the need for Type III work by clients and communities expands. The circumstances surrounding the closure of a military base provide a good example of Type III work. Loss of jobs and downstream economic impacts come with reductions in payroll and spending. Community self-esteem teeters as a major aspect of civic identity vanishes. Often, no alternative use can be readily identified. Major environmental problems may need to be addressed before any new use of the site can occur. Stakeholders in and around the community become active and voice conflicting demands.

In this situation, the community doesn't need a technical solution. In fact, none exists. The situation demands adaptive leadership from someone who is willing and able to guide community members through a difficult change process—accepting the loss, understanding the nature of the adaptive challenge they face, creating a new vision for the community, and conceiving an acceptable transition process. The community needs a leader who can mediate the competing interests surrounding the closure. Though technical solutions may be offered to support desired changes, in a Type III situation such as this, a professional's primary focus must be on providing leadership for this social dimension of adaptive work, and perhaps adding technological innovation to the mix.

New Business Models: Technical, Collaborative, and Transformative

The economic business model of architecture and engineering firms has been built, for the most part, around doing Type I work, selling technical solutions to client problems with little or no need for leadership. Although there is still a demand for this type of work, a significant portion of Type I work has now migrated to Type II. Clients still need technical solutions, but in many cases they also want leadership help in getting those technical solutions implemented.

Much of the work remaining in the Type I category has fallen victim to the corrosive forces of continued commoditization. The value proposition of that economic business model is in jeopardy; i.e., most of the project type niches are filled with established firms competing for the work using mature project delivery processes. Many firms are equally capable of providing the required technical services and offer clients the same value proposition ("we sell hours") for performing them.

To be successful in the long run with a business model focused on Type I work, firms need to develop innovative new technical solutions that differentiate them from others doing similar work. They need to choose organization ecosystems that value the technical capabilities they offer, and to understand how value is created and assessed within those ecosystems and their corresponding value networks. Within those value networks, firms need to clarify which levels of the client system offer the most attractive positions from which to create value.

Firms that focus on Type I work must be able to tackle the increasingly complex technological challenges that surround client problems, not merely repeat solutions that may have worked in the past. They need to improve their abilities in developing and delivering technical innovations, whether in form or function, material or method, outcome or process. As we discuss in chapter 14, firms can use value networks to identify and develop new and improved technical offerings of value to clients. In chapter 15, we explore how firms can transform Type I work with a deep commitment to technical innovation and creativity and create significant new forms of value for clients and project stakeholders. We also believe that higher value technical work will only be possible if firms are willing to develop social capacities that allow them to work effectively with those clients and stakeholders. We will examine this territory specifically in chapter 20.

Despite the recent explosive growth of Type II projects in the economy, architects and engineers have lagged behind in developing the leadership, facilitation, and collaboration skills needed to do this work. From a learning perspective, the tyranny of utilization has eaten up critical time that firms should have devoted to developing these competencies. Clients have turned to other professionals (e.g., management consultants, construction managers) for this assistance. As a result, architecture and engineering firms often find themselves restricted to the delivery of technical solutions. To catch up and move ahead of their clients and other industry competitors by providing Type II services, architects and engineers must adopt the shifts of mind outlined in part II. They also must have a staff skilled at building teams, helping clients and project stakeholders develop shared visions, and facilitating group learning and decision-making processes.

The need for transformative (Type III) work is also growing rapidly. Companies, institutions, and communities face more situations involving their physical environment for which there are no easy answers, and they need leadership to help them adapt. Again, if firms want to take on the leadership challenges that come with Type III work, they will need to learn to operate with the mind-sets outlined in part II. They will need to prepare leaders who can help clients adapt to these difficult conditions, change values and mental models, and learn new ways of being.

Both collaborative and transformative work bring opportunities for architecture and engineering firms to create exceptional value for clients and their organization ecosystems. Firms that take on these leadership challenges will also be taking a significant step toward reestablishing their leadership role in society, providing a basis upon which professionals can update their domains and help society adapt to the growing social and technological complexity. However, preparing a firm to deliver on the value proposition offered with Type II and Type III work will require significant learning and development within the firm. Part IV describes steps that firms will need to take in order to develop the social and leadership capacities required for each of these types of work.

New business models focused on these three types of work, whether technical, collaborative, or transformative, also need to be accompanied by new pricing strategies. Continuing to sell hours is neither fair nor viable. It isn't fair because existing rate schedules and multipliers are not high enough to return a fair share of the value created with these new business models. Clients would keep a disproportionately high share of that value. It isn't viable because revenues generated by selling hours would be insufficient to compensate the firm for its investment and the risks it takes in developing these capacities.

Chapter 23 surveys alternative pricing strategies. When appropriately matched to a value creation strategy, these pricing mechanisms can repay firms for their investments in building new capacities, compensate for risks that they take, and provide adequate retained earnings to support the ongoing growth and development of their living firms.

13 Organization Ecosystems

"Doing it nature's way" has the potential to change the way we grow food, make materials, harness energy, heal ourselves, store information, and conduct business.

—Janine M. Benyus[111]

Chapter 5 introduced two fundamental mind-sets—organization ecosystems and value networks—for thinking about value. Organization ecosystems and value networks are really two sides of the same coin, twin representations of the same reality. The former describes the nature, behaviors, and relationships of the interacting organizations, people, and physical environments that produce goods and services of value to customers or communities. The latter describes how that that value is created, assessed, and exchanged. Parallel to the hierarchy of relationships expressed in an organization ecosystem is a nested set of technologies, goods, and services that describe its counterpart value network.

Taken together, these two mind-sets provide a powerful, integrated framework from which the leaders of architecture and engineering firms can imagine, design, implement, and guide new business models. They provide a new language for talking about value and new value propositions embedded within those business models. In this and the following chapter, we explore organization ecosystems and value networks more fully, with a particular eye toward how firms can use them in their own transformative efforts.

In August 1994 Ray C. Anderson embarked on a journey that would transform him as well as the company he had founded 20 years earlier. That company, Interface Inc., is the world's largest producer of contract commercial carpet. Anderson's wake-up call happened as he prepared to deliver a keynote address to an Interface task force charged with responding to customers who were asking what Interface was doing for the environment. At first, he approached the question from a conventional manufacturer's viewpoint that focused on compliance, what Interface was doing to ensure that its operations conformed to environmental regulations. However, in the midst of his preparation, he happened upon Paul Hawken's book, *The Ecology of Commerce*. Reading that book "changed my life," he notes. He conceived a new mission for both himself and his company, "to do something to correct the mistakes of the first industrial revolution."

His speech served as a call to action for the organization. It set forth both a new vision, to become "the first name in industrial ecology, worldwide, through substance, not words," and a new mission, "to convert Interface into a restorative enterprise, first to reach sustainability, then to become restorative—putting back more than we ourselves take and doing good to Earth, not just no harm—by helping or influencing others to reach toward sustainability." The achievement of these goals would require a major transformation in the company's way of working, doing business, and relating to the world.

During the following years, Interface began to implement a strategy designed to accomplish this transformation by 2020. Internal actions aimed to reduce, reclaim, recycle, and redesign its products and the way that it manufactured and distributed them, advancing and sharing best practices with other manufacturers. Externally, the company challenged its suppliers to do the same and worked to influence its customers and other stakeholders in the design and construction industry to follow its example.

The company conceived and began implementation of a new business model, which Anderson broadly described as, "We will do well . . .

very well . . . by doing good."[112] The model was accompanied by new value propositions, including its innovative "Evergreen Lease." For this new value proposition, Interface committed not only to make carpet with state-of-the-art recycled content, but also to take responsibility for installing, maintaining, and disposing (recycling) of it at the end of its useful life. Customers no longer buy the carpet, but instead sign up for lease payments that compensate Interface for the benefits and services of their carpet.

Over the next decade, Anderson and Interface continued to pursue this mission, establishing a new web of relationships linking thought-leaders in the sustainability movement with industry participants (customers, builders, regulators). The company's promotion of learning about sustainability issues and sharing ideas and best practices spread its influence significantly beyond the boundaries of its core business.[113]

Anderson's story provides a sterling example of what happens when the leader of an organization looks beyond the immediate concerns of the core business and acts on unmet needs and emerging conditions that may be present in client and societal ecosystems. Dealing with these issues can inspire new business models, create new forms of value, and position the company to receive increased monetary and nonmonetary rewards for its efforts. As Anderson observed, "At the very least we will give our people and our company a higher cause and long range reason for being."[114]

By 1994 Interface was highly successful at delivering value to its customers through its core business. The company was providing hundreds of million dollars worth of carpet, carpet tile, and related products to customers. It had assembled its own organization ecosystem, growing relationships with suppliers, builders, regulators, and other stakeholders working in Interface's extended enterprise to enhance the value of these products.

The transformation that Anderson envisioned involved dramatic adaptation of the way Interface worked, shifting its focus from its own organization ecosystem to active leadership and care for client and societal ecosystems. It provided new forms of value directly to its customers' core businesses, satisfying their emerging demands for green products. The company reengineered products and production, distribution, and installation processes, applying its growing understanding of eco-efficient design and manufacturing. It also offered both leadership and learning to customers to raise their awareness of sustainability and transformation issues, sharing best practices that could be applied to core functions. For example, in 1994, Interface started an annual summer event in which 100 of its customers came together in Texas for three days of outdoor adventure and classroom learning that mirrored organizational development processes being applied at Interface. According to Vicki Devuono, director of creative development at Interface Americas, "We had benefited so much from this kind of experiential learning that we thought our customers would benefit, too. Honestly, we were a little scared about what they would think. But we went ahead anyway. Now the *Why?* conference is in its fifth year, strengthening our ties with customers in ways that completely transcend the typical business relationship."[115]

But Interface did not stop at this first realm of value creation. Anderson also aspired for the company to influence customers and members of their customers' extended enterprises (designers, builders, and even competitors) by helping them create a "third industrial revolution," this one driven by environmental sustainability. The Evergreen Lease was originally devised as a solution for the Southern California Gas Company's landmark Energy Resource Center. The utility wanted the design of the Energy Resource Center to demonstrate the merits and opportunities of green construction to the construction industry and its larger base of commercial and residential customers. Interface staff worked with industry consultant John Picard to "devise something he and the Gas Company really wanted in this building, the first-ever in the history of the world . . . perpetual lease for carpet." The utility was not just interested in this concept as a technical solution for floor coverings at the center. It wanted to change the attitudes, behaviors, and even purchasing patterns of the communities it serves by demonstrating the feasibility of this and many other innovations included in the project.

Interface's efforts helped the company exercise that type of leadership for the good of both its organization ecosystem and broader society. This was a solid example of the second realm of value creation, helping clients respond to, or lead, changes in their organization ecosystems.

Interface's transformation also led it to exercise leadership in the third realm of value creation. Doing good, for Interface, meant extending its leadership beyond the boundaries of its core business and direct customers. The company has hosted conferences and symposia that have shaped the industry's and society's understanding of the adaptive challenge posed by sustainability issues. The concept of doing good has also led to extensive collaborations with industry thought-leaders like Paul Hawken, Amory Lovins, and William McDonough. Their influence led to the creation of the company's PLETSUS Web site for gathering and sharing "Practices Leading Toward SUStainability" (sharing Interface's ideas freely) and to the reconceptualization of its annual report to serve not only as a vehicle for communicating financial results to shareholders, but also as a means of educating readers about the issues surrounding Interface's new mission.

More broadly, the company set out to act as a role model for other businesses and organizations. Anderson notes that, "The target group to influence, *other businesses,* are also our customers or potential customers. If we do well enough through creating goodwill and becoming resource-efficient, to the point that we are kicking tail in the marketplace, then that is the example other companies will see and want to emulate."[116] Evidence of that spin-off effect is provided by the experience of Nobuyasu Ishibashi, president of Daiwa House Industry Co., a large construction company headquartered in Japan. After learning of Interface and hearing Anderson speak, "Mr. Ishibashi admired Mr. Anderson's commitment and, as a result, felt compelled to begin moving his own company toward sustainability."[117]

What is the return on investment for these value-creating actions in the higher realms of a company's organization ecosystem? Anderson identified three major ways the company does well by doing good: First, the company earns the goodwill of its customers and builds its brand image in the marketplace, earning their trust and predisposition to increase their trade with Interface in the future. In 1998, four years after setting out on its transformational path, Interface was named one of the most design-driven companies in America by *I.D.*, the international design magazine, alongside the likes of Apple Computer and the Walt Disney Company. The magazine cited one of Interface's greatest achievements as "reducing material consumption, carbon monoxide emissions, energy consumption and non-recycled content in raw material by substantial indices over the last four years."[118] That same year, *Fortune* magazine ranked it as one of the "Most Admired Companies in America," and *Fast Company* magazine profiled Interface as one of four companies creating the "Future of Business."

Second, the company is rewarded by the savings that come through resource efficiency, fueled by the knowledge gained from and relationships built with those in its larger ecosystems. By 2002, the savings realized by designing waste out of its processes was more than $200 million. Savings dropped right to the company's bottom line, accounting for almost 25 percent of pretax profits in some years. Between 2000 and 2002, Interface reduced the amount of carbon dioxide emissions from company sources by more than 25 percent and the amount of nonrenewable process energy used to make carpet companywide by more than 20 percent. At the same time it achieved a 10 percent reduction in nonrenewable yarn consumed per square yard of carpet produced.[119]

Third, Anderson notes, value flows out of the positive feedback loop that occurs as other organizations and investors join Interface in its quest. As more and more do well by doing good, Interface benefits both directly, through increased business, and indirectly, as its organization ecosystem and society as a whole are healthier, wealthier, and more resilient. The company has done well, and Anderson and Interface have made major contributions toward the creation of a more sustainable future. By 2000, sales had risen to over $1.2 billion from roughly $600 million in revenue in 1994 when the transformation began, falling back during the construction recession of 2000–2002 to just under $1 billion. Meanwhile, the company's push toward its goal, "to be sustainable by 2020, accounting for no net reduction in the earth's resources in the conduct of our business and doing no harm to the biosphere," is on schedule.[120]

Transforming Your Business Model Through Organization-Ecosystem Awareness

Just as Ray Anderson did with Interface, architecture and engineering firms can take advantage of the opportunities that come for shaping new business models through their awareness and understanding of organization ecosystems. Firms can adopt a more proactive stance of providing leadership and creating value directly for a client's core function, its extended enterprise, and its larger organization ecosystem. The following are examples of value-creating strategies:

- Bringing capital to the table, as part of new performance management or design-build-finance-operate project delivery processes.
- Identifying and negotiating new water supplies for water districts in Southern California.
- Helping industrial companies dispose of underperforming real estate assets that are weighing down their stock prices.
- Developing software, materials databases, and training programs that teach clients green design principles and help them develop and implement new internal standards and practices.
- Hosting client roundtables, conferences, and training programs to share best practices that will help organizations become more successful, improving their physical environments and management and business practices.
- Testifying at congressional hearings to promote increased funding for basic research about earthquakes.
- Bringing together inventors, entrepreneurs, and financiers to spark new enterprises.

Anderson's vision of a transformed Interface emerged during a time of personal reflection. For the leaders of most firms, a more deliberate strategic inquiry can identify similar opportunities for new business models and threats to existing practices. The following questions will help guide this strategic inquiry:

- Which organization ecosystems does our firm currently inhabit (or want to inhabit)?
- What do we know about those ecosystems? (This is a critical question, because in many firms this knowledge is dispersed and consequently relatively unleveraged.)
- Which of those ecosystems are the most attractive for an evolved business model?
- Which organization(s) leads the ecosystem? Where are they leading it? What is their vision of the future? What are their goals?
- Who populates those ecosystems and what are their underlying dynamics? Which of those organizations or individuals do we have relationships with?
- What relationships need to be developed within the organization ecosystem for our firm to be successful? What existing relationships, either inside or outside this ecosystem, would be valuable to others?
- What is the stage of development of that ecosystem? (Refer to Moore's stages of development, outlined in chapter 5.)
- What are the characteristics of the ecosystem along such dimensions as the following?

Organized	vs.	Disorganized
Authority-driven	vs.	Dispersed
Open	vs.	Closed
Turbulent	vs.	Slow changing
Politicized	vs.	Nonpolitical
Affluent (resource rich)	vs.	Poor (resource starved)
Capital-based	vs.	Knowledge-based

- What is the health of the ecosystem? Is it growing or shrinking?
- What major trends are (or will be) affecting it?
- What are the ecosystem's most pressing needs with respect to the

physical environment? Which of those needs are either unmet or underserved?

- How can our firm contribute to the ongoing success and development of the ecosystem?

The experience of SEA Consultants, a civil engineering firm based in Cambridge, Massachusetts, with a major university client confirms the benefits of this type of strategic inquiry. The firm has moved from a subconsultant role to a prime consultant role on a variety of university projects, and members of its staff have attained trusted-adviser status with this client, providing counsel to university staff on critical infrastructure problems.

The original spark for this transformation was a firm reorganization that replaced SEA's discipline-based structure with client-focused groups, including one for higher education. Art Spruch, the newly appointed client manager for the university, realized that if the firm wanted to become truly client focused and win more work from this client, SEA would need to do things differently. It would need to change the way the firm typically approached projects and clearly differentiate itself from other professional firms serving this client.

Spruch's first step was to gather everyone in the firm who had worked with the university and facilitate a brain dump of everything they knew about the client. Different perspectives and knowledge emerged as civil, environmental, geotechnical, structural, and other discipline leaders talked about their experiences with university projects. They quickly realized how much they didn't know about the client and the client's universe. Their knowledge was, for the most part, confined to particular disciplines and limited to technical aspects of particular projects.

To move toward a deeper knowledge of this client, the SEA team began a strategic analysis of the university's organization ecosystem. They documented the university's broader goals and strategies and mapped both its core functions and extended enterprise. They identified the firm's existing relationships in that ecosystem as well as the ones that needed to be established. They identified major forces and trends that were reshaping the university's ecosystem and significant problems that it was facing, beyond the bounds of infrastructure and facilities. Spruch, as client manager, began attending project meetings, not to assert his authority but simply to listen and learn more about the client. As time passed, the SEA team cataloged a new set of opportunities for serving the client and repositioning the firm, codified in a document they called their "business development knowledge plan."

Spruch also began to guide the shift from simply doing technical (Type I) work to approaching projects as collaborative (Type II) undertakings. Key to this new mind-set was coaching project managers and engineers to listen and learn from the client and to allow university staff to participate in the engineering design process. Instead of just showing up and presenting a recommended technical solution to the client, they learned to describe the problem, offer alternatives, and involve client representatives in collaborative deliberation and problem solving. If three alternatives were presented, the likely solution would be a fourth of fifth one that had been developed jointly. University representatives were delighted to be engaged as valued partners in this way.

During these interactions, the SEA team began to take advantage of their expanding knowledge of the university's world. They were able to identify approaches and alternatives that otherwise would have remained obscured and apprise the client of threats and opportunities that needed attention beyond the bounds of current projects. At the same time, Spruch and other SEA staff members worked to establish new connections and relationships that exceeded the bounds of current assignments at the university and within its ecosystem.

The trust earned through these changes paid off quickly. The university asked SEA to move into a prime role on a major project on which it had previously been a minor player. The firm was awarded several major new infrastructure projects as a prime and, for those projects, was given roles and responsibilities that previously had been held either by the university itself or contracted to outside program managers or construction managers. In addition, SEA was asked to do work in areas that previously had been restricted to other entrenched competitors working for the university. Work for this client is likely to be SEA's largest revenue source next year, up from only a fraction in previous years.

More important, SEA has applied what it has learned—about the firm, its individual staff members, and its teams—to build new mindsets and capacities that can be exported across the firm to other client-focused groups. The strategy of being client focused is now closely linked to an expanding awareness of organization ecosystems and a growing ability to do collaborative work.

Becoming a Keystone Species

Architecture and engineering firms may choose to contribute to their client's organization ecosystems by creating value incrementally, targeting value-creating strategies to specific needs or opportunities. However, by taking a bolder approach, firms can assume a role in these ecosystems and society that they long ago forfeited, operating as what biological scientists call a keystone species.

In the world of biological ecosystems, keystone species are the major contributors to the health and vitality of their particular ecosystems. Without their presence, the very existence of their ecosystem would be threatened.[121] For example, as elephants move through the African savanna, trampling the grass and foraging on bushes and small trees along their path, they create habitat for hundreds of species of birds and small mammals. Or, along the Pacific Coast, starfish feast on mussels scoured from the rocky seashore. Freed from colonization by these aggressive mollusks, the rocks provide shelter and mooring for a teeming community of limpets, barnacles, and other marine life.

Interface's new mission challenged it to move toward keystone status, to make a difference on a global scale. Commenting on the magnitude of this challenge, Anderson noted: "In a global economy of $40 trillion ($40,000 billion) it's very presumptuous, isn't it, for a little company headquartered in Atlanta, that started from scratch in 1973 and took 24 years to reach one billion in sales, to think it can intervene in so large and complex a system as industrialism itself? What makes us think we can make a difference on a global scale and on an issue of such overwhelming magnitude? I don't know that we can, but . . . others have convinced me that we should try. Unless somebody leads, nobody will."[122]

Interface's innovative new business model transformed its position from that of a supporting species (carpet supplier) to a new status as a keystone species (leader of profound change) in its customers' organization ecosystems. Its actions reshaped the environmental attitudes of many of its customers, and society more broadly, and helped those organizations and their ecosystems move toward a healthier future.

The early industrial age is filled with examples of architecture and engineering professionals acting as keystone species, shaping their clients' organization ecosystems and the larger societal ecosystem. They advocated for, facilitated, and inspired critical activities, including building canals, railroads, and bridges, creating new building types, and conceptualizing and implementing public health solutions. In the 1950s, President Eisenhower's plan for the construction of an interstate highway system was shaped by a kitchen cabinet that included consulting engineers. But in the last half century, as we have discussed, design professionals have ceded much of this responsibility to others. Their economic business model did not support this role; they couldn't justify it as a short-term strategy for selling more hours. Instead, they chose to act like a supporting species that makes only minor contributions.

In transforming their business models, architecture and engineering firms face the same challenge that Anderson articulated for Interface. Will leaders of firms be content to focus their value proposition on their core business or their clients' core functions as a supporting cast member in a larger drama being staged and played out by other members of the design and construction industry? Or will they aspire to become keystone species, leading actors who shape the way the drama plays out?

14 Value Networks and Technology S-Curves

Flight became possible only after people came to understand the relevant natural laws and principles that defined how the world worked: the law of gravity, Bernoulli's principle, and the concepts of lift, drag, and resistance. When people then designed flying systems that recognized or harnessed the power of these laws and principles, rather than fighting them, they were finally able to fly to heights and distances that were previously unimaginable.

—Clayton M. Christensen[123]

Multnomah County, the city of Portland, and the Oregon Department of Transportation (ODOT) had a problem. The Broadway Bridge, a historic structure connecting parts of Portland's downtown across the Willamette River, was in need of major repairs. Built in 1912, the bridge was designed with a movable double-leaf center span that could be lifted when tall ships needed to pass along the river. That design was made possible by the use of an open-grid steel deck for the roadway to minimize the weight of the structure. Unfortunately, when it rains, as it often does in Portland, the poor skid resistance of the open-grid surface, combined with the geometry of the span, spawned numerous car accidents and fatalities.

Although a planned rehabilitation could bring the bridge up to current structural standards, there was no obvious solution for the safety issue. Also, the repair work was constrained by the bridge's designation as a historic structure. Replacing the open-grid surface with a solid deck was not considered feasible. The existing structure could not accommodate the additional loads that were required to accommodate the weight of available solid decking technologies.

Engineers from Portland-based David Evans and Associates (DEA) stepped up to the plate with an alternative approach to the decking problem. It hinged on the use of new advanced composite materials for the solid deck to meet weight restrictions, maintain the existing visual profile of the bridge span, and offer improved skid resistance on the surface.

The design was the brainchild of Dr. Frieder Seible, a recent addition to DEA's staff. Seible is also dean of the Jacobs School of Engineering and professor of structural engineering at the University of California at San Diego (UCSD) and head of the its Structural Research Laboratories. He knew how to apply cutting-edge materials research being done at structural materials testing laboratories to create breakthrough designs, such as the one DEA proposed for the Broadway Bridge, and could marshal the technical resources necessary to prove out the concept.

Seible also brought instant credibility to the team. Through his academic research work, Seible had developed an international reputation in the structural design of bridges and transportation structures. When he joined DEA, Seible opened a door into the world of advanced materials research and structural design for transportation structures for the firm and its clients.

Multnomah County and ODOT were convinced. As a first step, they engaged DEA to complete a feasibility study to validate the design, which used advanced composite decking materials with a special high-abrasion polymer topping. As the design progressed through a series of approvals by funding authorities, DEA staff worked with the Federal Highway Administration to flag the project for special "Innovative Product Funding." Those funds were awarded to supplement the budget for rehabilitating the Broadway Bridge, offsetting a significant portion of the added first costs. More important, the FHA funds mitigated a major risk factor for Multnomah County and ODOT by covering additional maintenance costs that might be incurred as a result of the implementation of this innovative solution. Following approvals, the design for repair and renovation of the bridge incorporating the new decking materials proceeded, with construction scheduled to begin in 2003.

DEA's innovative technical solution for the Broadway Bridge created significant value for the firm's clients and the community served by the bridge. It reduced the potential for future injuries and loss of life caused by accidents on the bridge, abating a set of major liability issues for the county. In addition, the Innovative Product Funding minimized operational risks related to future repair costs.

The Broadway Bridge project is one of a series of cutting-edge projects undertaken by the bridge design group at DEA during the last three years. Those efforts have established the firm as a leader in high-end bridge engineering and design and a trusted adviser in the boardrooms of large transportation clients. That accomplishment is particularly significant given that the firm had no capability for doing bridge engineering and design as recently as eight years earlier.

The story begins in 1995, with the arrival of David Moyano, a senior bridge engineer and manager, at DEA. He had been recruited away from a much larger engineering firm. He was attracted by DEA's vision, values, and culture, and an offer that provided a degree of freedom for building a bridge design practice that didn't exist in his previous firm.

Moyano's first step was to build a critical mass of bridge design staff capable of doing standard, technical work. The firm focused on transportation agencies in the states of Oregon and Washington, trying to win work in what was a highly commoditized market, where price was the driving force of competition.

To differentiate DEA's bridge design services, Moyano offered clients a collaborative design process that emphasized the importance of relationship building and team-based decision making with client staff and other project stakeholders such as permit agencies. He visualized this strategy as a leap from a price-dominated technical engineering production curve to a new collaborative process production curve, where clients would select the firm based on performance, not price. The strategy worked, and over the next four years the firm built a highly profitable bridge engineering group of about 30 professionals. However, by the end of that period competing bridge design firms had caught on to DEA's collaborative approach and adopted it themselves, shifting the basis of competition back to price.

It was time, in Moyano's terms, "to jump to the next curve." This time the leap would be much more dramatic. Moyano wanted to catch the curve of advanced technological innovation for bridge design, positioning the firm to move into the signature bridge design market. He wanted to be invited into the boardrooms of transportation clients when critical up-front decisions on plans and projects are made. His new strategy required new types of people with recognized technical and creative skills and proven track records as trusted advisers. As Moyano noted, "You have to have 'A' level players to attract 'A' level players."

To accomplish this, Moyano tapped into what he calls his "nodal network" of personal relationships in the bridge design community. He recruited Steve Thoman, a senior bridge designer in California whose presence would immediately move DEA to the next level of capability. Moyano captured the entrepreneurial spirit of this engineer with the vision of building the type of bridge design group that DEA had in mind. Thoman brought to the firm his own nodal network, which included UCSD's Seible, a person they jointly identified as operating at the highest level of specialty expertise in the structural design of transportation structures. They knew that if they could recruit Seible, he would bring yet another nodal network of relationships and open up even wider possibilities for filling out the bridge design group with additional A-level players.

Working with Thoman, Moyano was able to entice Seible with a vision of assembling a team of respected colleagues capable of doing truly innovative engineering work. DEA also offered a custom-designed employment arrangement, allowing Seible to maintain his position with UCSD and the Structural Research Laboratory. Seible signed on. In turn, he brought another set of relationships, including engineers with high-end structural analysis and modeling capabilities. It had taken a little over two years, but Moyano had assembled his team, and DEA had successfully made its jump.

Competition on this new production curve is based on technical skills and creative abilities, not price. DEA's bridge group is set to ride this curve with limited competition; few competitors can offer a similar depth of expertise. As a result, DEA has shifted the marketing stance for its bridge design group. Rather than remaining in a reactive stance, waiting for transportation clients to establish their capital spending plans and send out requests for proposals on engineering work, Moyano's group now proactively identifies problems that it is uniquely suited to help clients solve. Sitting with clients in their boardrooms and helping them make critical decisions have also allowed the firm to shift away from selling hours to being paid for the technical creativity and innovation it brings to the table.

Value Networks and Technology S-Curves

Technological innovation has always been a source of potential competitive advantage and a means of escaping the difficult straits of price competition. The potential for evolving new business models through technological innovation is as real for architecture and engineering firms as it is for companies that make products.

Moyano's decisions to jump from one production curve to the next demonstrate an intuitive understanding of this potential. This strategic territory and the underlying competitive dynamics of technical innovation have been more explicitly illuminated by Clayton Christensen in his work on value networks and technology S-curves. Whole-value-network awareness helps leaders make sense of the growing technological and social complexity confronting their firms. These concepts can also be used to better understand the nature of value creation and exchange within an organization ecosystem and to design business models that are suited to specific competitive environments.

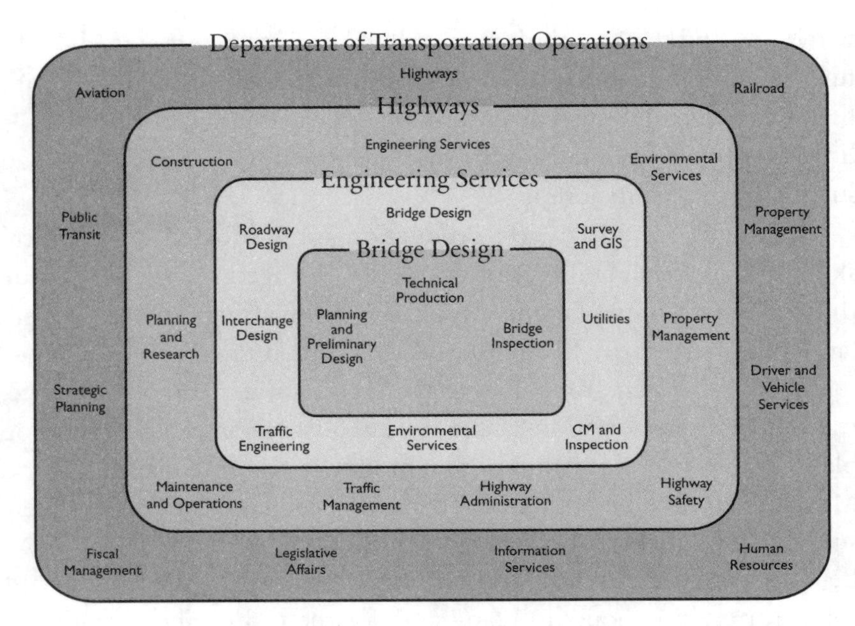

FIGURE 14.1A. Department of Transportation Value Network

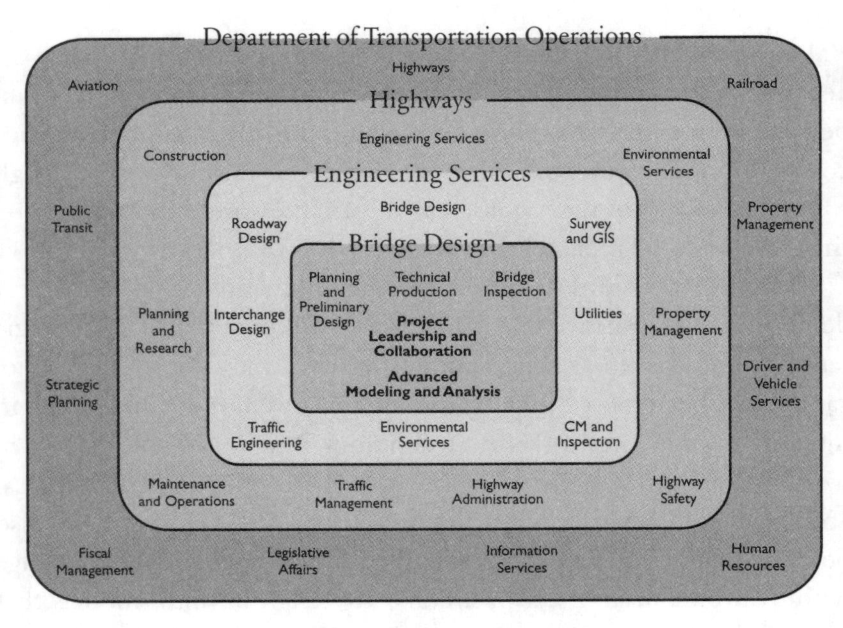

FIGURE 14.1B. Department of Transportation Value Network: DEA Innovations

DEA's decision to start up a bridge design capability represented an extension of the firm's posture within the DOT value networks of Oregon and Washington (Figure 14.1A). DEA was already an active member of the organization ecosystems of these DOTs, providing transportation-related engineering services, including roadway and interchange design as well as traffic engineering, construction inspection, and survey services. Bridge design represented a niche in the value system that, although not unoccupied, appeared to offer significant opportunities for providing engineering services to these DOTs. Moyano correctly read the situation with respect to the value criteria used within the existing DOT value network: cost and convenience, or competition between engineering firms centered on price. We can better understand the strategic situation DEA faced by looking at the technology S-curve for the DOT value network. Technology S-curves show the change in performance over time with respect to particular value criteria used within a value network as a whole or of specific component technologies within a value network. "In the early stages of a technology's development progress in performance will be relatively slow. As the technology becomes better understood, controlled, and diffused, the rate of technological improvement will accelerate."[124] As technologies mature, the rate of progress slows, eventually reaching a natural limit where little further improvement is possible (Figure 14.2).

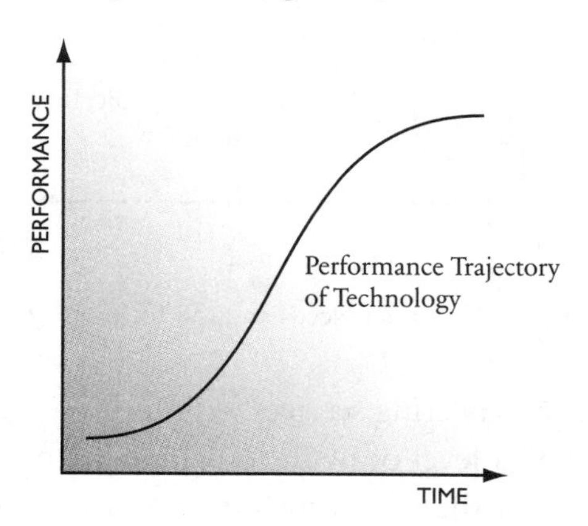

FIGURE 14.2. Technology S-Curve

The technology S-curve for engineering services within the DOT value network has reached the mature stage. To understand the dominance of price as the basis for competition in this type of situation requires the addition of another key insight from Christensen. Technology S-curves do not stand in isolation; they exist in relationship to the performance needs of potential customers for the technology. Customer needs can be charted along with the technology S-curve (Figure 14.3). Christensen argues that as technological performance overshoots customer needs, the nature of competition shifts. When the technology is still not good enough (below the line), technology suppliers compete on the basis of performance. When performance passes the point where customer needs are basically met (above the line), the basis of competition shifts to cost, convenience, and customization—the technology begins to be treated as a commodity.[125]

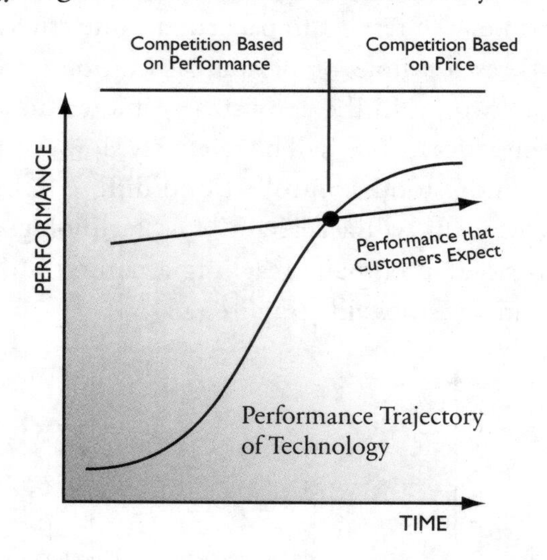

FIGURE 14.3. Customer Needs and the Technology S-Curve

Many of the engineering services within the current DOT value network have reached a level of maturity where performance, in general, exceeds the needs of the client; most engineering firms working with the DOT offer services that are good enough. Consequently, competition within the engineering services portion of that value network is dominated by price and customization (tailoring services to meet the needs of specific clients). To escape these competitive dynamics, David Evans and Associates offered clients two additions to the existing value network (Figure 14.1B). The firm's strategic shift began with Moyano's reshaping of the traditional technical design and documentation process around new project leadership, relationship, and collaboration skills (Type II work). Then, he recruited several key staff that brought with them advanced materials research and analysis and high-end bridge design expertise. Both steps helped the firm move to a position where cost pressure was not as fierce and new value propositions were attractive to clients.

Christensen provides specific advice to firm leaders about how they should deal with different phases of a market's maturity. When performance is still not good enough, Christensen observes that companies that adopt an integration strategy—pulling all the elements of the product or technology's architecture (the way the components interact within the design of the technology or value network) under one roof—will dominate. An integrated operation provides these companies a twofold opportunity for performance gains by improving specific component technologies, and by innovating the way components are assembled and work together. Christensen uses the early days of the computer industry to illustrate this dynamic. When mainframe computers were not yet powerful or fast enough to satisfy customer needs, IBM, the most integrated computer company, dominated the marketplace with a 70 percent market share and 95 percent of industry profits. Christensen notes that similar dynamics have characterized the markets for many products. "Ford and General Motors, as the most integrated automakers, dominated their industry during the era when cars were not good enough. RCA, Xerox, AT&T, Alcoa, Standard Oil, and U.S. Steel dominated their industries at similar stages. Their products were based on the sorts of proprietary, interdependent value chains that are necessary when pushing the frontier of what is possible."[126]

In the design and construction industry, we can identify the 25-year period following World War II as a time when architecture and engineering services were in many ways not yet good enough. Many of the project delivery practices that are still the heart of professional practice today were invented and codified during those years. The architecture and engineering services market witnessed the growth and flourishing of practices integrated vertically along the project delivery process (from site investigation and programming through construction administration) and horizontally across disciplines with the growth of multidisciplinary practices. Firms competed on the basis of performance, expressed in a variety of qualifications-based selection processes. Nonintegrated companies, i.e., single discipline or single-service firms, had difficulty competing with larger, integrated firms because they couldn't offer the same level of performance across the whole set of needs that clients expressed.

However, the pace of technological progress eventually extends beyond the needs of customers, and companies are forced to change the basis on which they compete and how they do it. Christensen notes that as a product or technology matures, companies begin to adopt modular architectures that allow them to make significant improvements in specific subsystems, to achieve cost savings or increase the fit with the requirements of specific customers, without redesigning everything. Once that modular architecture is in place, the advantages of being an integrator fade. In some cases, an integrated posture, with its attendant overhead and structural rigidity, actually becomes a liability.

Christensen identifies two generic strategies for succeeding when a technology has become mature (above the line). First, a company can become what he calls an aggregator. Aggregators pull together component technologies, subsystems, or modules provided by suppliers and assemble them into a single package that offers customers a lower price or tighter fit to their specific needs. The success of Dell Computer in the 1990s sprang from this type of strategy. Dell's PCs were not necessarily better that those sold by Compaq or IBM, but by outsourcing all of the critical subassemblies and software, Dell was able to offer customers cost-competitive computers assembled to a particular specification and delivered at lightning speed. Compaq and IBM, weighed down by their proprietary subsystems, couldn't keep up.

Aggregators, such as Dell, achieve success through growth, leveraging small margins from large volumes. They concentrate on the design of the overall system, customer relationship management, shopping the best suppliers, and management of the assembly process.

Second, companies can target the component technologies or subsystems that are not yet good enough in the eyes of the aggregators. They can design and deliver technologies to be assembled by the aggregators that provide significant cost and customization gains. Aggregators are willing to pay a premium for the performance gains offered by these specialists, even if the ultimate customers still purchase the overall product or technology on the basis of price. In the PC market, Microsoft (operating systems) and Intel (microprocessors) have played that role. Christensen's analysis suggests that the bulk of industry profits for mature products will flow to the subsystem providers. And, in fact, Microsoft and Intel have racked up a disproportionate share of PC industry profits in the last 15 years. The high-end bridge design and advanced materials research and analysis capability of David Evans and Associates represent this type of response. Although the firm can certainly leverage that expertise as part of larger prime engineering contracts with DOT clients, a more attractive alternative may be to market them to a wide range of industry partners that may have need of this expertise.

Maintaining a posture as an integrator after a product or technology has matured is precarious. Outsourcing on a piecemeal basis to compete with aggregators is seldom sufficient to reestablish a real competitive advantage. Instead, those outsourcing arrangements often sap critical financial resources as the price of acquiring their subsystems. Alternatively, firms can adopt a disintegration strategy, searching their current operations for superior subsystems with upside performance improvement potential that could be sold to former competitors or new aggregators that have entered the market. This requires a willingness to give up the prime position with a customer, but offers the potential of more attractive returns in the future.

As design firms examine the status of their particular organization ecosystems and their corresponding value networks, it is critical to assess the maturity of the value network and its component technologies and respond appropriately. Firms in many sectors of the design and construction industry face situations where the dominant project delivery processes, including both design-bid-build and design-build, are mature and subject to the price pressure Christensen describes.

Many integrated architecture and engineering firms have done little to respond to the maturing market condition that they find themselves serving. Although this can be a legitimate response, firms choosing that option will need to make peace with their services being treated as commodities and should prepare for the shakeout that will inevitably occur when profits from this market are no longer sufficient to support the large number of competitors serving those markets. Even if they find a way to live with commodity status and survive the shakeout, they must also be ready for even more dramatic change when new disruptive technologies lure these clients away.

Some formerly integrated firms are escaping mature markets by following a disintegration strategy, and are dismantling their organizations to focus on specific competencies. For example, a number of architecture practices have given up the construction documents and construction administration phases of a project in order to concentrate on design.

An alternative strategy for firms is to become an aggregator, pulling together the contributions of a number of technology suppliers under a single prime contract. Traditionally, many architecture firms have served as a type of aggregator, combining integrated architectural services (inner layer) with subconsulting engineering services to control the next layer up. Firms pursuing this strategy will either outsource more of their existing operations, or move up a level in the value network to offer, through subconsulting and alliance relationships, other functions in the project delivery process traditionally performed by other types of organizations. Designer-led design-build and performance management contracting are examples of this strategy.

Other firms are attempting to reposition to an emerging market niche, such as the hazardous waste market in the 1980s, the telecommunications market of the 1990s, or the homeland security services today. Unfortunately, finding an unoccupied niche is hard to do. And, as the experience of the hazardous waste and telecommunications markets showed, these niches can rapidly mature or become crowded. In either case, price sensitivity can escalate quickly as oversupply chases demand.

Disruptive Technologies

Rather than learning to live within the constraints imposed by a mature technology, organizations can use technical innovation to develop and deploy what Christensen calls disruptive technologies. Disruptive technologies are innovations that introduce a value proposition that is very different from the one offered by an incumbent technology serving a particular market.

Although a disruptive technology initially underperforms established technology used in a mainstream market, it has features that are attractive to a few fringe (often new) customers. Nourished within these fringe niches, the disruptive technology climbs its own developmental curve. As its performance improves it can reach a point where the technology also becomes competitive within the mainstream markets. At that point, the disruptive technology rapidly overthrows the established order and reshapes the value network (Figure 14.4).

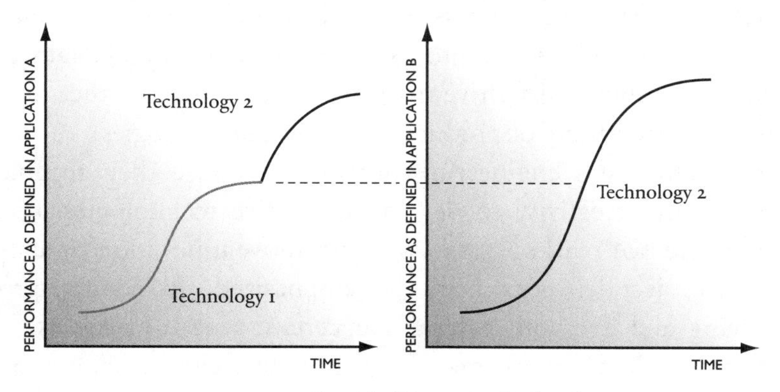

FIGURE 14.4. Generic Disruptive Technology

(Reprinted by permission of Harvard Business School Press. From *The Innovator's Dilemma* by Clayton M. Christensen. Boston, MA, 1997, p 41.)

Christensen illustrates the concept of disruptive technologies through the evolution of computer disk drive technologies.[127] In the first half of the 1980s, the original desktop personal computer value network was built around 5.25-inch disk drives. The capacity, speed, and size of this technology fit the performance needs of desktop machines, and the 5.25-inch disk drives dominated this market. As the decade progressed, continual improvements in capacity and speed more than satisfied the needs of mainstream desktop PC customers.

Then, in a relatively short period in the late 1980s, the 5.25-inch disk drive technology was almost completely displaced by a new disruptive technology, the 3.5-inch disk drive. Desktop PC makers had rejected the 3.5-inch disk drive technology when it was first introduced in the mid-1980s. The 3.5-inch disk drives significantly underperformed the 5.25-inch technology in the critical dimension of capacity (20 megabytes versus 40–60 megabytes). Instead, the smaller disk drive found a home in the emerging portable/laptop/notebook market, which valued the ruggedness, reduced size and weight, and low power consumption of 3.5-inch disk drives, attributes that were largely irrelevant to desktop PC customers.

Nourished by the portable computer market, the 3.5-inch disk drive technology moved up its performance S-curve until it was able to deliver capacity that was good enough for the desktop PC makers at an attractive price. By the late 1980s, the 3.5-inch disk drive had almost completely displaced the established 5.25-inch technology. Christensen goes on to describe how this process is repeated again and again, not only in the computer disk drive market but also in many other industries with a wide variety of product and service technologies.

Architecture and engineering firms can leverage their technical innovation and creativity to develop disruptive technologies. Faced with a mature DOT market, DEA made a disruptive innovation by adopting a new project delivery process that emphasized collaborative problem solving and decision making. For certain types of projects, this process redefined the performance needs of DOT clients (Figure 14.5). Instead of competing on the usual performance criteria of experience, technical quality, conformance to standards, and capacity, DEA emphasized new attributes of relationship, collaboration, and project leadership. Those attributes were attractive enough that the new process quickly replaced the old way of working and gave DEA an opportunity to win substantial work in this established DOT market.

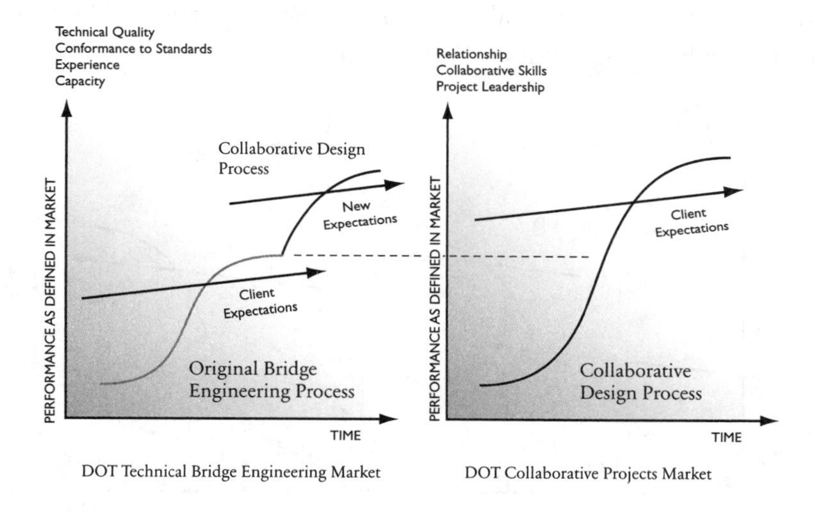

FIGURE 14.5. Collaborative Project Leadership as a Disruptive Technology

A potentially disruptive technology has emerged in the bridge design market with the development of Michael Baker Engineering's BRADD (Bridge Automated Design and Drafting) software for use by DOTs. The software automates the design and drafting of common bridge types. The traditional technical engineering process for doing conventional bridge design has matured and is subject to fierce price competition. Baker's BRADD software offers performance attributes that are attractive to DOT clients in the focused niche of simple span bridges and culverts. The software offers ease of procurement and use (an ability to do the engineering without having to procure outside engineering services), mass customization abilities (tailoring basic prototype designs to specific conditions), and speed. The adoption of the software unseats engineering consultants that traditionally provided design services for such projects. The state of Pennsylvania was the first to purchase the software, and its use by other state DOTs is spreading. More significantly, as the performance capabilities of the software increase, it has the potential of capturing ever more complex bridge design projects. Over time, BRADD software, or similar competitive software that may enter the market, could take over a significant portion of the mainstream bridge design market (Figure 14.6).

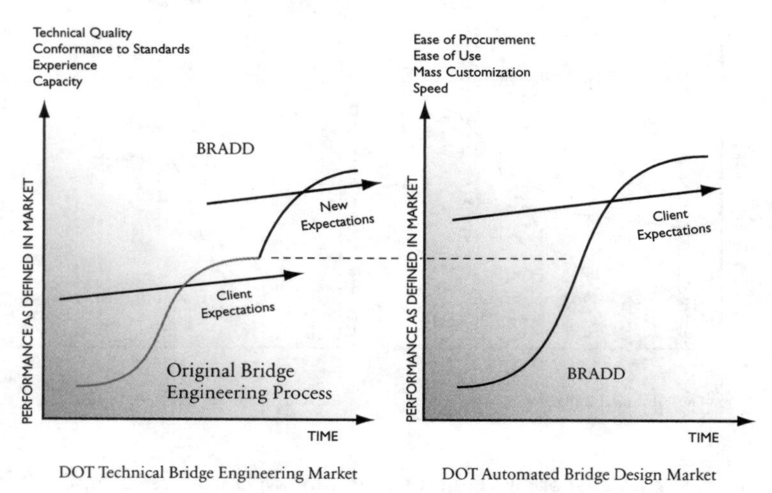

FIGURE 14.6. Automated Bridge Design as a Disruptive Technology

Architect Frank Gehry's new project delivery process, developed to support his work on high-end museums and cultural facilities (a very small, specialized niche of architectural clients), presents another potentially significant disruptive technology. As we will discuss in chapter 15, Gehry's staff needed to reinvent the traditional architectural design, construction document, and construction process in order to bring his distinctive architectural vision into reality. This new process, leveraging computer modeling software and custom-built parametric analysis capabilities, allows Gehry and his team to virtually build his expressionistic designs in the computer before actual construction begins. The result is a major realignment of the entire process—potentially eliminating the need for construction documents at some point in the future—and a significant reduction of risk for building subcontractors and fabricators.

Gehry speculates that the new process will eventually spread outside this small niche to mainstream architecture practice:

> We're on the verge of revolutionizing the way architecture is practiced. Jim [Gehry's key lieutenant responsible for this initiative, Jim Glymph] is starting to write software that other people will be able to use. I told them to go ahead and do it. They're meeting with the lawyers and the accountants. I've provided them with all the legal stuff. And I don't even understand it. You know it's not for me to do that. They're doing it. If they make a lot of money, that's fine. I may become the Bill Gates of architecture![128]

Transforming Your Business Models Using Value Networks and Technology S-Curves

Value networks provide a second powerful mind-set from which to conceive and implement new business models for their firms. Along with organization ecosystems, they provide fundamental tools and language for understanding how firms can create new forms of value for their clients as well as establish new strategies for earning a fair return for their efforts.

Parallel to the process described in the last chapter for examining organization ecosystems, firm leaders can engage in a strategic inquiry using value networks and technology S-curve analysis. Key questions to examine include the following:

- How is value created and assessed within the firm's value network(s)?
- What role does the firm currently play in those value networks? What technology (services, products, project delivery technologies, etc.) does the firm contribute? How does the firm's technology link to other's technologies?

    Construct a map of the value network (service architecture) that corresponds to a client's organization ecosystem. Place the firm's technology in the middle, and work outward through the technologies in each tier of the value network.

- What value criteria are most important within those value networks?
- What is the level of maturity of both the overall value network and the firm's technology?

  Compare technology S-curves with customer performance expectations.
- Is the performance of the value network (or technology) below or above client needs? Is the firm's value proposition correctly aligned to this competitive dynamic? Is the firm adequately responding to this status?
- If the technology is below client needs, is the firm selling value, not hours?
- Are there opportunities to fill in gaps in the service architecture of the value network or to develop innovations within the inner rings that the firm currently serves?
- Are there opportunities for expanding the breadth of technology offered by the firm? Can it expand within its current tier(s) or move upstream into higher levels? Should the firm retrench from technologies that are commoditized and don't offer attractive returns?
- Within a given value network, where are the greatest opportunities for earning attractive profits? What can the firm do to take advantage of these opportunities?
- Are there opportunities to fundamentally shift the firm's business model, adopting integrated, segmented, or aggregator strategies?
- Are their opportunities to create disruptive technologies? Which client niches would find the performance attributes of these innovative technologies attractive? Which value networks could the firm invade with an existing disruptive technology that it has developed?
- Can you identify value networks that the firm currently doesn't serve that would appreciate the performance attributes of your technologies? What would you need to do to enter those organization ecosystems?

15 Technical Innovation and Creativity

We see a world of abundance. Not limits.
In the midst of a great deal of talk about reducing the human ecological footprint, we offer a different vision. What if humans designed products and systems that celebrate an abundance of human creativity, culture, and productivity. That are so intelligent and safe, our species leaves an ecological footprint to delight in, not lament?

—William McDonough and Michael Braungart[129]

When the new Guggenheim Museum Bilbao opened to the public on October 18, 1997, it unexpectedly ushered in a dramatic renaissance for what had been a backwater industrial city on Spain's northern coast. This titanium clad "miracle" was the brainchild of architect Frank Gehry. His bold artistic vision fulfilled the aspirations set by the Guggenheim Foundation for a museum that would not only showcase its modern art collection but also be breathtaking on its own.

The museum's opening signaled the emergence of the Guggenheim Foundation as a global superpower in the art world. And it vaulted this languishing port city onto the world's cultural map, making it a must see on a new art-world Grand Tour.

The creativity of Gehry's expressionistic design draws accolades from visitors and critics alike. Visitors are showing up in record numbers, urged on by volumes of favorable reviews from art critics and travel guides. In its first year, the Guggenheim Bilbao attracted over 1.4 million visitors, triple the number expected and more than the attendance at both of the New York City Guggenheim museums combined. From its soaring atrium to the expansive exhibition spaces and smaller conventional galleries, the museum provides a range of physical settings that allow visitors to appreciate works ranging from Richard Serra's immense steel sculptures to those by modern masters such as Andy Warhol, Mark Rothko, and Robert Rauschenberg.

But the spotlight at the Guggenheim Bilbao is not only on the art. The building itself is a star attraction. Critics note that the structure asserts a presence as strong as the Eiffel Tower in Paris or Frank Lloyd Wright's original Guggenheim Museum in New York City. Gehry's gleaming ensemble has been hailed by architect Philip Johnson as no less than "the greatest building of our time." Spanish novelist Manuel Vazquez Montalban describes it as a "meteorite" of light and titanium, come to earth in the ruins of Bilbao's failed shipyards and steel mills. *New York Times* critic Herbert Muschamp simply calls it a "miracle."

The Guggenheim Bilbao Museum stands as both a cultural landmark and a new architectural benchmark for innovative design. But the true miracle of Gehry's conception is revealed in what has now been called variously the "Bilbao effect" or the "Guggenheim effect." Located in the heart of Bilbao, the museum pumps people, money, energy, and spirit into the city and the surrounding region. It has become the principal transforming agent for the economic and social renewal of the entire Basque region. Rarely a tourist destination before the opening of the museum, Bilbao has become so popular that the director general of the museum reports with glee that officials of the French resort city of Biarritz have speculated that they should change their tourism slogan to, "We're only 1-1/2 hours from Bilbao."

The economic impact of this flood of tourists has been gigantic. Within months of its opening, the government of the region recouped its entire investment. The success of the museum also spurred a major reconstruction of Bilbao's civic infrastructure, including a new airport, subway system, and other improvements. Each year the airport welcomes more than one million visitors who would otherwise never have come to Basque country, filling local restaurants, hotels, and shops. The management-consulting firm KPMG estimates that in 2001, the museum generated almost $150 million in gross domestic product for Spain's Basque region (0.5 percent of GDP), contributed $24 million to the Basque treasuries, and created more than 4,400 jobs.[130]

As significant as these economic impacts have been, the Bilbao effect has also endowed Spain's Basque region with a brilliant new persona and given a massive boost to Basque self-confidence and self-image. The museum has reenergized a local populace weighed down by past recessions and the struggles of the Basque separatist movement, and has bestowed a new sense of pride and identity on the community.

Gehry's design for the Guggenheim Bilbao creates significant economic value directly in the form of ticket sales at the museum, and indirectly in spin-off economic gains in the surrounding community. The design also creates significant nonmonetary value. Patrons appreciate the emotional and spiritual experience of the art as well as the structure. The Guggenheim Foundation moved a giant step forward toward its goal of becoming a global powerhouse in the art world.

Gehry achieved these results through the innovation and creativity he brought to the technical work of architecture itself. His technical creativity and innovation found expression in three different aspects of his work at Bilbao: the artistry of the form he designed, his mastery of new materials, and his transformation of the process used to design and build the project.

He gave free rein to his personal creativity as a designer and form-giver at Bilbao. The museum represents a high-water mark in his experimentation with architectural form, incorporating shapes and spaces of unprecedented kind and configuration. According to critic Michael Sorkin, "Bilbao surely marks the mature phase of Gehry's cubist sensibility, in which he returns cubist two-dimensional depictions of three-dimensional space (his sketches) to the actual realm of volume. It is a masterpiece."[131]

His expressionistic style, with its exploration of increasingly daring sculptural shapes, has been enabled by his innovations in two other areas of his practice. Throughout his career, Gehry has emphasized the innovative use of materials. Early on, he established his reputation as a leading designer with his use of chain-link fencing and galvanized metal as positive design elements in both his Santa Monica Place parking garage façade and in the remodeling of his own Santa Monica bungalow.

A series of projects leading up to Bilbao provided opportunities for Gehry to explore the use of metal skins to clad the complex geometric shapes that were emerging in his sketches and models. One degree of freedom was achieved in the Frederick R. Weisman Museum in Minneapolis, where he used stainless steel panels to create shimmering elevations. By the time he reached Bilbao, aided by a shift in the commodities market, Gehry was able to choose a pure element, titanium, giving the museum a unique appearance that fits magically into its post-industrial environment. According to key Gehry staff member Randy Jefferson, "The titanium surface on Bilbao is about 0.38 millimeters. It's not a whole lot thicker than several sheets of paper . . . what is beautiful about it—on a very windy day you actually see the titanium sheets flutter. The pressure across the surface creates alternating positive and negative pressures. The wind pressure pushes and pulls on the surface. You can see a pillowing that is a product of the thinness of the metal and the system of connecting it to the wall construction."[132]

Beyond innovation in the materials of construction, Gehry and his staff have been equally creative in transforming the process for translating his concepts into information that other members of the design and construction team can use to engineer, fabricate, and build the project. The engine of this transformation has been CATIA (Computer Aided Three-Dimensional Interactive Application) software, designed by Dassault Systemes, the French aerospace company, to represent complex, three-dimensional objects. CATIA was the same software that Boeing, the aerospace giant, used to reinvent its design and manufacturing process for the production of the 777 airliner.

In the late 1980s Gehry and his technical expert, Jim Glymph, saw the potential of this software and set about tuning and tweaking it to the needs of their architecture practice and adapting their architectural process to take advantage of the power of the software. They recruited an aerospace engineer from Boeing who was familiar with the software and knew how to use it.

Gehry's staff folded CATIA into an iterative design process that includes handmade models, laser measuring and digitizing devices, and computer-generated rapid prototype models. By the time of the Bilbao commission, their efforts had yielded a tool and process that made it possible for Gehry to create the sculptural shapes that emerged from his sketches. Commenting on Gehry's process, Michael Sorkin notes: "Thanks to the computer we can, within the limits of materials and gravity, now build any shape. But the computer also provides another liberation. Secure in the knowledge that anything can be produced, drawing—sketching—is itself emboldened, offering a license that gives the sketch validity not simply as a source but as the final technical authority. The computer enables the representation and manipulation of that which cannot otherwise be drawn."[133]

CATIA enabled Gehry's artistically derived sketches and manually constructed models to be quantified, digitized, and mapped as continuous surfaces, providing the information the engineers needed to design the structure and the fabricators to build the structure and skin. It also provided real-time feedback during design on costs and constructability. Gehry notes: "With our new equipment, shapes can be transferred to the computer in fifteen minutes, and now we know how much it's going to cost per square foot to build those shapes, because we've had the necessary experience. Now we can budget jobs in the earliest design phases. Also, we know that if we use flat materials it's relatively cheap; when we use single curved materials it's a little more expensive and it's most expensive when we warp materials. So we can rationalize all these shapes in the computer and make a judgment about the quantity of each shape to be used."[134]

Gehry points out that the software has also provided a new way to communicate with the other members of the building team. "The new computer and management system allows us to unite all the players—the contractor, the engineer, the architects—with one modeling system. It's the master builder principle."[135] The software also allows them to create a virtual building, ironing out potential problems and issues that otherwise might not have appeared until the building was under construction. As a consequence, at Bilbao, "Gehry's technique at construction was able to almost match the amazingly bold artistic concept."[136]

The wellspring of the Bilbao effect lies in Frank Gehry's technical innovation and creativity. The synergy of his bold architectural designs, experiments with materials and methods of construction, and reinvention of the traditional architectural design process establish an important benchmark for all architecture and engineering firms. They can adopt a similar stance with respect to innovative design and problem solving, using technical innovation and creativity as a principal means for creating value for clients and inventing distinctive new models of practice for themselves. Architects and engineers can make technical innovations in the following three areas: innovative form and function, new materials and new uses for old materials, and new approaches to project delivery. We will examine them next, followed by a discussion of strategies for technical innovation that firms can use to create value within them.

Innovative Form and Function

The ability to develop innovative forms and functional solutions that satisfy client needs and solve problems is a primary value that architecture and engineering firms offer. For the Guggenheim Bilbao, Gehry pushed the boundaries of museum design and architectural form. In similar fashion, other firms can move beyond the simple accommodation of programmatic requirements and deliver innovative designs and functional solutions.

Green/sustainable design offers fertile ground for this type of innovation. Architect William McDonough is on the leading edge of this movement, pushing the boundaries of sustainable design, creating projects that move beyond resource efficiency toward eco-effectiveness. For McDonough, eco-effective design means not just reducing the consumption of natural resources but also creating projects that actually return resources to the ecosystem. First, for a new facility for the furniture manufacturer Herman Miller, and then for the restoration of Ford Motor Company's massive River Rouge manufacturing plant, McDonough's team pioneered the development of whole-system storm-water management systems that combine green roofs (roofs covered with a thin layer of topsoil and plants) with porous parking lots, and constructed marshlands and swales to absorb, store, and naturally treat rainwater. The system works so well that when storm water reaches the nearby river it is clean, clear, and moving gently enough to avoid washouts and erosions. This technical solution saves money by reducing the need for storm-water treatment infrastructure (concrete pipes and treatment plants), while at the same time establishing new habitat and enhancing the beauty of the surrounding site.

Creative consulting engineers focus on the development of new technologies that combine scientific advances with practical applications. The importance of this mission was spelled out by Craig Goehring, CEO of Brown and Caldwell, an environmental and civil engineering firm based in Walnut Creek, California, in a message to his staff: "Because environmental technology is a critical part of our business of delivering client success, we work at it—searching for breakthroughs and innovations, developing, pushing, and applying technology to achieve environmental results in ways that are faster, smaller, cheaper, and better."[137] An example of the payoff that comes with that commitment was Brown and Caldwell's innovative approach to secondary clarification technology for wastewater treatment. The firm used its knowledge of particle dynamics and hydrodynamics to show clients how they could completely eliminate the need for this costly treatment process, saving upward of $1 million in clarification construction costs per treatment facility affected.[138]

New Materials and New Uses for Old Materials

Firms also create value through their willingness to approach the selection and use of materials in innovative ways. Gehry pushed the boundaries of metal skins on buildings by his selection of titanium to enclose the Guggenheim Bilbao. William McDonough and Michael Braungart's MBDC group tackles the need for eco-intelligent building products by collaborating with manufacturers such as BASF and DesignTex. These collaborations have led to the development of new carpets and fabrics that are safe, healthy, and ecologically sound. On one project for DesignTex, MBDC joined with Rohner Textil AG, a progressive textile mill in Heerburg, Switzerland, and Ciba-Geigy Ltd., the giant Swiss chemical company, to create Climatex©Lifecycle™. This innovative fabric line was created using only a fraction of the chemicals normally used in the production of fabrics. When ultimately discarded at the end of its useful life, it can be added directly to the compost heap, where it will become mulch that will nourish the soil.[139] In addition, the wastewater coming from the manufacturing plant is actually cleaner than the water going in, resulting in a significant environmental dividend and the elimination of regulatory monitoring of the process.[140]

Value can also be created by the creative use of old materials. The straw-bale building movement that emerged in the 1990s is a good example of this potential. Straw bales were first used as a building material in the Sand Hills of Nebraska. In the late 1800s, following the introduction of modern hay-baling equipment, prairie pioneers used this locally available material to build everything from houses to schoolhouses and churches. After a flurry of construction during the early years of the 20th century, straw-bale construction was largely abandoned by the 1930s in favor of other types of building materials.[141]

A renaissance in straw-bale construction began in the early 1990s. A self-organizing network of builders, designers, owners, and interested parties recognized the ecological value of straw and pioneered the construction of a new generation of straw-bale structures. Straw-bale construction is prized as "an expression of ecological values and as a model for others in ways to 'live lightly on the land.'"[142] Builder-owners also appreciate the potential this material offers for making innovative aesthetic statements, using the bales to form stepped walls, alcoves, benches and window seats, and even vaulted ceilings. Promoted through journals, workshops, and "wall-raisings," the movement has spread across the country, and a self-sustaining network of innovators openly share their knowledge of straw-bale design and construction techniques.

The use of straw-bale construction can create even more value in developing countries. Architect Janet Johnston spent the summer of 1999 in Mongolia supervising the construction of a straw-bale kindergarten for the United Nations Development Program and the Adventist Development and Relief Agency. In areas where schools and clinics often had to close for winter because of a lack of fuel for heating, Johnson's straw-bale structures cut heating costs by 70 percent. These two agencies have also used straw-bale designs to develop new housing projects in eastern China. "Straw bale insulation in 68 houses there will save 6,600 tons of carbon dioxide emissions from the burning of coal for fuel. Funding for the project came from the sale of carbon credits, new on the world market."[143]

New Approaches to Project Delivery

Frank Gehry's ability to implement his vision for the Guggenheim Bilbao Museum depended on his team's invention of ways to translate his sketches into information that engineers could use to detail the building's structure, fabricators could use to cut and shape the titanium panels, and the contractor could use to erect the structure and skin on site. Similar innovations in project delivery processes can yield significant value by saving costs, reducing schedules, and enabling new aesthetic, technical, and functional solutions.

As discussed in chapter 11, architecture and engineering firms can leverage new modeling and prototyping tools to actively involve client and other project stakeholders in design experiences. By deploying their own Decisionariums, firms can create solutions that more successfully address project requirements and facilitate client buy-in and commitment.

Reinventing problem-solving and study processes can yield similar benefits. Michael Baker Engineering realized this potential with a groundbreaking application of personal digital assistants (PDAs) and database technology on a project completed in the mid-1990s for the Department of Defense School System (DODDS). The project required assessing and making recommendations for asbestos and hazardous waste removal for a large number of DODDS facilities scattered around the world. Baker staff, believing there had to be a better way to do this type of work, decided to reinvent the traditional process of field investigation, data recording (paper on clipboards), and analysis (after-the-fact analysis, cost estimating, and report writing). The transformed process required up-front investment in a computer database and handheld computers that could use it. The database combined information about the schools to be surveyed, constructability and cost information, and analysis and report writing capabilities. Using the handheld computers, team members could enter real-time information gathered during field inspections directly into the database. The database software allowed observations to flow directly from the PDA entry into final reports and recommendations. The new process enabled the project to be completed in a fraction of the time typically required for such work, improved the quality and accuracy of reports by eliminating mistakes that inevitably happen when data is moved from notes on paper to spreadsheets to reports, and captured significant cost savings for the client.

Beginning in 1999, the Beck Group, an architecture, construction, and real estate services company based in Dallas, combined three previously separate business segments into a single integrated-services capability, offering clients needing corporate offices, call centers, or light industrial buildings a mass-customized solution structured around the firm's design-build prototypes. This capability is enabled by proprietary software, developed in-house by the Beck Group, that allows real-time parametric analysis of project costs and constructability issues and facilitates exchange of drawings and communications among all members of the project team. Projects completed for clients using this new approach have shown cost savings of almost 15 percent and shortened schedules by nearly 30 percent, while still satisfying the programmatic and image requirements of clients.

Strategies for Technical Innovation

To consistently create value in the three arenas of technical innovation, architecture and engineering firms must transcend existing norms of practice. Four strategies for accomplishing this include the following:

1. Leveraging digital modeling, simulation, and experimentation tools.
2. Changing the social process of design and problem solving.
3. Expanding the established domains of professional practice.
4. Investing in research and development.

Leveraging digital modeling, simulation, and experimentation tools

Chapter 11 outlined the ways in which firms can use digital modeling tools to provide value-rich experiences for their clients. These digital modeling technologies can also provide an infrastructure for technical innovation and creativity.

The adoption, adaptation, and application of CATIA modeling software were essential for turning Frank Gehry's design vision for the Guggenheim Museum Bilbao into reality. CATIA allowed him to express and explore his ideas. The software enabled a virtuous design cycle that began with his imagination and then moved through sketches, digital models, and physical models made with rapid prototyping machines. This iterative design and modeling process also supported ongoing dialogue and creative problem solving among members of the Bilbao design and construction team (architects on his staff, consulting engineers, and contractors), helping them implement Gehry's vision.

"The most important raw material of innovation has always been the interplay between individuals and the expression of their ideas."[144] In his book *Serious Play*, Michael Schrage explores the ways in which many of the world's leading companies are using digital modeling, prototyping, and simulating technologies, as Gehry has, to spark creative synergies. Although architecture and engineering firms have spent heavily to equip their offices with software to automate existing processes, few firms have embraced the transformational potential of digital modeling.

Sketches lie at the heart of the visual culture of architects and engineers. For centuries, the sketch has been the medium preferred by architects and engineers for expressing ideas. Sketches facilitate both individual thinking and interactive communication. Sketches and drawing practices have served as basic building blocks for technical innovation and creativity, and have shaped the way architecture and engineering work is structured. The integration of new computer-graphics and digital modeling software with new generations of staff who speak a new digital sketching language is beginning to transform the way firms think and innovate. Taking full advantage of this trend will require a radical transformation of project delivery processes.

Boeing made this type of radical transformation for the design, manufacture, and assembly of its breakthrough 777 jet. The design and production of the 777 marked Boeing's transition away from a traditional paper-based engineering and design process—plans and drawings on Mylar sheets with disparate departments communicating up and down an established hierarchy—to a new integrated process driven by powerful CAD software. Boeing wanted a process that would encourage experimentation and problem solving well before final assembly, as well as minimize the interferences and conflicts inherent in its existing process. CATIA software tied to a custom-designed add-on to the program called EPIC (electronic preassembly in the computer) provided the infrastructure the company needed to drive this transformation. This infrastructure not only connected Boeing's resident engineering staff but also linked them to suppliers and subcontractors located around the world.[145]

Boeing was able to virtually pre-build the entire airplane on the computer prior to actual manufacturing and assembly. Digital models allowed engineers to try out and modify designs without waiting for time-consuming, expensive physical prototypes to be built. Supported by this modeling capability, engineers were also emboldened to experiment, particularly with design approaches that removed weight from individual parts.[146] The parametric capabilities of the software were used to create a virtual weight marketplace in which designers could exchange pounds and ounces between different subassemblies in an ongoing value engineering process focused on delivering performance with the minimum weight required.

Instead of waiting to find problems, engineers and designers were able to assemble and test digital mock-ups (replacing those costly physical prototypes). The software automatically checked for interferences (two or more parts occupying the same physical space) and notified affected parties of potential problems. For just one 20-piece section of the wing flap, the software performed more than 200,000 checks and caught more than 251 interference problems—errors which, if not caught, would have resulted in significant rework during the final assembly process.[147] This automatic checking process also transformed the way people interacted with each other during the design process. In some cases, engineers temporarily designed interferences into the airplane as a means of seeking out and identifying counterparts with whom they should be meeting to discuss future design issues. Overall, as Schrage notes, "The prototyping medium (CATIA) generated a new genre of interaction between previously segregated disciplines, transcending Boeing's traditional organizational 'silos.'"[148] Using digital modeling technologies, Boeing was able to build the 777 better, faster, and cheaper than any airplane in its history.

New digital modeling, simulation, and prototyping tools offer architects and engineers similar potential. New forms of interaction among members of design teams can spur creative thinking. Increased communication and interaction can, in turn, help build more effective design teams. Digital abundance, made possible by the ever-falling cost of computing, allows design teams to conceive and explore many design alternatives, rather than just a few. New forms of creative conversation and collaboration can be structured around this new medium. And, as Schrage observes, digital models can serve as engines of surprise, challenging underlying design team assumptions and mental models quickly, inexpensively, and with low risk.

Digital modeling technologies also offer designers the prospect of a new virtual master-builder status in the design and construction industry. Modeling software can support real-time value engineering and constructability analysis, facilitating more effective control of cost and time. Modeling can now move beyond 3-D to 4-D simulations that portray the project, step-by-step in time, as it moves through the construction process. Pre-building projects in the computer, before actual construction, allows early identification and elimination of many of the risks inherent in the traditional two-dimensional, paper-based construction document process. In particular, computer models can help eliminate many of the costly coordination errors that continue to cause problems in the field during construction.

Digital modeling tools can be employed to completely reinvent project delivery processes. As discussed in the last chapter, the bridges and structures group at Michael Baker Engineering married its software design ability to its technical knowledge of bridge design and created the company's BRADD software. BRADD replaces the traditional paper-based engineering process for these structures, improves design quality, and significantly reduces design times for clients, such as the Pennsylvania Department of Transportation, who have adopted the software.

Digital modeling tools can also be used to create entirely new services. For example, in partnership with the American Association of State Highway and Transportation Officials (AASHTO), Baker developed software that is now used by more than 30 states to store bridge data, condition information, and load ratings. This capability has now been combined with Geographic Information System (GIS) technology to create yet another new business innovation, a system that allows states automatically to map safe highway travel routes for oversized permit vehicles.[149]

Finally, new computer modeling technologies provide tools that engineers and architects need to respond successfully to increasing project complexity, particularly with respect to building materials and systems. Arup, a global building engineering firm, is leveraging a set of advanced modeling technologies to deal with complex problems and unusual construction problems to create value for its clients. Arup structural engineers have adapted software that was originally developed by the automobile industry for crash-test simulations to help it engineer solutions for complex seismic, vibration, blast, and security problems. The software allows them to virtually test or simulate actual design events (e.g., seismic events) and to study the behavior of structural elements as these events unfold.

The firm's proprietary form-finding program, "FABWIN," is used to develop solutions for extremely complex, organic forms developed by both architects and sculptors. Arup mechanical engineers have adapted fluid dynamics modeling software to simulate the behavior of heating, ventilating, and air-conditioning systems in complex spaces, such as large exhibition greenhouses, building atria, opera houses, and theaters. These models are used not only to improve comfort but also to optimize the sustainable design performance of these structures. Arup's digital modeling tools are part of an iterative design process that helps the firm consistently develop highly innovative engineering solutions that increase performance, save money, reduce client risk, and in many cases make technically feasible concepts that might otherwise have been unbuildable.

Changing the social process of design and problem solving
The second major strategy for enhancing technical innovation and creativity is changing the nature of the social process of design and problem solving. Three major changes could support major advances in firm creativity: enhancing the *diversity* of the people working in the firm and on its projects; increasing the firm's *tolerance* for new types of people, new knowledge, new ways of organizing, and the stresses and conflict that inevitably come with creative work; and *shifting the context* in which the creative work is done.

Diversity. In 1991 an American architect with a background in sustainable design met a German chemical research scientist at a rooftop party in New York City. In their conversation, William McDonough and Dr. Michael Braungart found common ground in their commitment to eliminating the negative outcomes of poor design, whether in the form of consumer products or their packaging, or, on a larger scale, in buildings and their surrounding urban settings.

Three years later, they founded a design firm to do work in chemical research, architecture, urban design, and industrial product and process design with an eye toward transforming industry around ecologically sound principles. Their firm, McDonough Braungart Design Chemistry (MBDC), offers a wide range of services to help clients implement eco-effective projects. Leadership visioning sessions help industrial leaders think outside the box when creating new business strategies. Consulting services help clients create sustainable product designs and manufacturing processes. Education workshops address underlying ecological issues and strategies. Software tools are available to assist in the implementation of projects. A unique co-marketing program using the firm's "optimized by MBDC" label rates manufactured products for their eco-effectiveness. The firm's clients include Herman Miller, Ford Motor Company, Nike, and DesignTex. McDonough maintains his architecture practice, but the creative solutions that emerge from that practice are directly shaped by the collaborative interdisciplinary work at MBDC.

The synergy of McDonough's knowledge of architecture, interior design, and urban planning and Braungart's expertise in the chemistry of materials creates unique value for their clients. Combining their differences in pursuit of a shared goal creates value. Architecture and engineering firms can learn to take advantage of similar synergies by moving away from the monoculture that characterizes most practices and toward the creation of much more diverse workforces.

The lack of racial and gender diversity in architecture and engineering firms is changing, but very slowly. A 2002 industry survey by ZweigWhite indicated that 95 percent of principals in architecture/engineering/planning and environmental consulting firms are Caucasian and 92 percent are male.[150] The mismatch between firm makeup and the reality of national and global demographic and economic trends leaves architecture and engineering firms at a marked disadvantage. They look and act differently from the more representative organizations they work for and with. Further, architecture and engineering firms lose out on the rich potential resources and synergies that could come from diverse staff backgrounds, experiences, and perspectives.

Tom Peters challenged architects at the American Institute of Architects' 2002 convention with his observation that "Women rule!" noting that, according to a special report in *BusinessWeek*, female managers outshine their counterparts in almost every measure. Peters pointed out that, "Women purchase the majority of services and make most of the financial decisions—dominating commercial as well as residential markets. And women buy things differently than men—women create connections when they buy; men simply make transactions." Client organizations populated with high percentages of female employees, managers, and senior executives pose unique challenges for architecture and engineering firms populated mainly by males. Other emerging competitors, such as management consulting firms, which have already figured this out and created this level of diversity, may have a competitive edge in developing relationships and winning work from these clients.

The growth of the Hispanic and Asian populations in Texas, California, and other Southwestern states underscores the increasing racial and ethnic diversity of the United States. This diversity was one of the sources of the creativity that fueled the high-tech boom in Silicon Valley during the 1990s. Computer, software, and Internet companies were populated with engineers and programmers drawn from around the world. Architecture and engineering firms need to more closely reflect the communities they serve and take advantage of the riches that can flow from a similarly diverse workforce.

The lack of diversity in architecture and engineering firms is also reflected in the paucity of people, both technical staff and management, from other disciplines and educational backgrounds. Predictably, engineering firms hire engineers and architecture firms hire architects. The example of IDEO, an industrial design and engineering firm, shows that having a staff with diverse educational and professional backgrounds can pay big dividends for firms. IDEO's pyrotechnically creative environment feeds on the combination of industrial designers and engineers with a rich blend of sociologists, former medical students, biologists, English majors, and others. That population reflects John Kao's Law of Creativity, which posits that creativity increases as the square of the diversity of the people in an organization.[151]

Archeworks, a design school in Chicago, offers another story of the power of cross-disciplinary pollination. Founded in 1994 by interior designer Eva Maddox and architect Stanley Tigerman, Archeworks offers students an alternative, multidisciplinary education that emphasizes learning by doing. The doing occurs on projects for nonprofits and other community organizations that ordinarily lack the financial resources to acquire good design skills. Projects have ranged from furniture for disabled people living in single-room housing to pillboxes to help AIDS patients take prescribed pharmaceutical cocktails and Archeworks's own office and classroom facilities.

One-third of the students are architects and another third are interior or industrial designers. The final third is made up of "what Tigerman calls the 'significant others' or lawyers, art historians, retail-display designers, anthropologists, and experts in others fields."[152] The power of that diversity can be illustrated by the improved design of a wearable head pointer to help cerebral-palsy patients communicate better, which was developed by a team consisting of a nurse, a former shoe designer, and a recent architecture school graduate. "The designer knew of a lightweight metal composite material that the team could use to make the frame and the pointer. The architecture student figured out the optimal construction pattern for the elastic bands that would hold the pointer to the user's head. The nurse helped the team test the product on a patient." The resulting pointer design went into production within a year and rapidly became an industry standard.

Tolerance. Beyond being a fundamental principle of a living company, tolerance is also a major condition for the emergence of innovation and creativity. Tolerant companies are open to new ideas and people and accept the disruptive dynamics that accompany the diversity. They empower staff to experiment with new things, supporting entrepreneurial actions that stretch the organization's sense of what is possible. They nurture and protect innovators from pressures from established operations, until their experiments are ready to succeed or fail on their own merits. Tolerant companies accept the risk of failure as an inherent part of innovation and creativity. A classic story about Thomas Edison and the invention of the light bulb illustrates this point: "When Thomas Edison was intent upon creating incandescent light, he went through more than nine thousand experiments in an attempt to produce the bulb. Finally one of his associates walked up to him and asked, "Why do you persist in this folly? You have failed more than nine thousand times." Edison looked at him incredulously and said, "I haven't even failed once; nine thousand times I've learned what doesn't work."[153]

Tolerance is also expressed when companies create protected organizational space in which innovation and creativity can take root and blossom. The power of this approach was dramatically illustrated by aircraft designer Kelly Johnson's creation of a skunk works at Lockheed Aircraft Company in the early 1950s. To achieve breakthrough aircraft designs for the military, Johnson moved his new group to an airfield hangar, physically isolating the Skunk Works from Lockheed's aircraft division offices and research facilities. In this protected space, Johnson's team turned out a steady stream of the world's most advanced aircraft, including the U-2, the SR-71 Blackbird, and the Stealth fighter, over the next decades.[154]

Throughout its history the discount brokerage firm Charles Schwab has demonstrated a high degree of tolerance for the disruptive effects of creativity and innovation. It has consistently empowered individuals and teams to challenge conventional wisdom and invent new ways of serving customers, even at the risk of sacrificing existing, profitable products and services. The payoff for this tolerance was illustrated by the company's experience in the mid-1990s, when it faced the challenge of figuring out how its retail trading business should respond to the advent of the Internet. Rather than using the conventional strategic management processes of its established organizational structure, Schwab established an independent group charged with shaking up the process, not just developing an add-on to the existing business model. To give the team unrestricted freedom to be innovative, Schwab even moved the group out of its corporate offices in downtown San Francisco into a tilt-up office space south of the city. The team went on to create what is now known as eSchwab, a new way of Internet trading that, when implemented in mid-1996, first challenged and then completely overtook Schwab's core retail trading operations.

Complexity theorists have demonstrated that organizations can also foster creativity and innovation by increasing their exposure to and tolerance for confronting new knowledge and information. This can include networking with other organizations and thought-leaders, creating strategic alliances within the industry, engaging in benchmarking activities, or establishing programs with speakers and guest scholars. Gehry's personal relationships with artists in Southern California and his alliances with collaborating organizations, such as Dassault Systemes (CATIA) and Permasteelisa, the Italian construction company Gehry trusted with the fabrication and erection of his building skin elements, have provided important nutrients for his firm's creative work.

Shifting the context of practice. Technical innovation and creativity can also be spurred by radically shifting the context in which work is done. New ways of thinking can emerge when designers and engineers confront environments with new constraints. The work of architect and academic Samuel Mockbee with his Rural Studio, and the experience of participating architecture students from Auburn University, demonstrated the potential richness of this approach. With Mockbee's guidance, students engaged in hands-on design and construction projects for impoverished clients in rural Alabama's Hale County. They lived, worked, and built in the towns where their projects were located. Budgets were tightly constrained; students were forced to create "simple but inventive structures made of inexpensive, mostly salvaged or donated materials—beat-up railroad ties, old bricks, donated lumber, hay bales, baled corrugated cardboard, rubber tires worn thin, license plates, and road signs."[155]

Beyond this resource constraint, students also confronted the unfamiliar context of the local culture of Hale County, which combines deep poverty with rich cultural traditions. The studio's body of work contributed real value to these communities by re-imagining traditional southern forms and idioms (sheds, barns, porches, and trailers) and creating a unique architectural aesthetic.

Expanding the established domains of professional practice

The third strategy for technical innovation and creativity, expanding the domain of professional practice to take on new challenges facing clients and society, provides opportunities for value creation by opening up the boundaries that often constrain firm action. Architecture and engineering firms are pursuing the potential of this strategy in the fields of sustainable/green design and security design.

Environmental consulting firms are leading efforts to restore brownfields (abandoned or underused industrial, mining, or commercial properties that have real or perceived environmental contamination) to productive status in their communities. These efforts include identification of sites, raising awareness of the public and business communities of the possibilities for redevelopment, and collaboration with community leaders and other industry stakeholders in the process of site remediation and construction of new facilities. These restoration activities not only cure environmental liabilities and rejuvenate decayed infrastructure, but also spur economic growth and expand the tax base in the surrounding communities. Langan Engineering and Environmental Services, a geotechnical and environmental services firm, has demonstrated this proactive approach in staging Urban Core Revitalization Workshops in conjunction with the Urban Land Institute. These workshops brought together land development, legal, municipal, and government decision makers to explore the potential for brownfield restoration in light of New Jersey's smart growth policy for rebuilding its cities.

The new emphasis on homeland security has brought with it a host of opportunities for technical innovation by architecture and engineering firms. Firms are making significant contributions in the design of new types of physical structures and infrastructure that increase security and prevent terrorism for embassies, airports, port facilities, and many other types of facilities. They are also helping to develop new types of security and vulnerability assessment processes as well as risk mitigation strategies.

Architecture and engineering firms are also establishing consulting practices that provide strategic facilities management services for clients. Engineering firms are offering strategic transportation plans for transportation agencies and asset management services for municipalities. Pulling these practice areas into the professional domains of architecture and engineering has the double advantage of recapturing clients that have looked to mainstream management consulting firms for such services.

Investing in research and development

Finally, firms can create value through technical innovation and creativity by investing directly in research ventures to develop products and materials they can use on their own projects and/or sell (license) to others in the manufacturing and construction industry. The principals of Kennedy & Violich Architecture (KVA) established a separate research business called MATx (short for materials) to develop new materials and building products, focusing on the development and commercialization of material innovations that grow out of the firm's architectural work. Results have included a luminous curtain, prototyped for the Opto-Semiconductors division of lighting manufacturer Osram, and a desk that eliminates messy cabling and wiring by embedding ultrathin polymer films in a plywood surface, allowing the wood to carry electricity and data. These efforts create value for the firm in the form of returns on the intellectual property, patents, and proprietary designs that can be licensed to outside manufacturers, and provide economic returns on an ongoing basis (not just selling hours).[156]

Establishing specialized research groups can provide a means to deepen knowledge and understanding of specific building types and construction technologies. Architecture firms involved in hospital work have established research efforts focused on the design, planning, and operation of health-care facilities. Efforts to deepen its knowledge and capabilities in sustainable design led Hellmuth, Obata + Kassabaum (HOK) to develop the *HOK Guidebook to Sustainable Design*, an internal research document later published by John Wiley & Sons.[157] Ratcliff Architects invested in the development of a three-dimensional, Web-based application called the Green Matrix, which provides a guide to designers seeking to implement sustainable design strategies. Ratcliff freely shares the Green Matrix through its Web site, hoping to stimulate further learning through dialogue with architects and clients outside the firm.[158]

Budgeting nonbillable time to support technical innovation is a less obvious but nevertheless significant form of investment. We identified the degrading effect that the tyranny of utilization has on firms, starving them of time needed for reflection, learning, and creative explorations. Firms can look to industrial giant 3M, one of the leading innovators in manufacturing, for a role model. The company allots one day a week for research and development staff members to work on anything they choose, and many of its most important product discoveries have come from these explorations.

Investment can also come in the form of incentives that promote creative behaviors. The Australian real estate development firm Lend Lease uses "culture credits" to reward its innovators. Culture credits work like airline mileage programs, allowing innovators to spend a day working with a senior executive team on a particular problem, or even having lunch with the firm's chairman.[159]

Research activities sponsored by outside agencies and institutions provide another setting for creating value through technical creativity and innovation. In 1999 the Heschong Mahone Group, a Fair Oaks, California, consulting firm specializing in energy efficiency, published the findings of a research project, sponsored by Pacific Gas and Electric, that examined the impact of daylighting on human performance in buildings. Results showed that skylights were positively and significantly correlated to higher sales in retail stores and improved academic achievement in K–6 schools. The study points to significant opportunities for architects to create value for their clients by incorporating daylighting features into their designs.[160] Heschong Mahone has built a practice that specializes in this type of research work for major utilities and government agencies, combining architectural and engineering expertise with computer analysis and professional education training.[161]

Finally, architecture and engineering firms can foster technical innovation and creativity by creating alliances with other industry thought-leaders, organizations, and institutions engaged in research and development for the design and construction industry. Academic institutions, such as Carnegie Mellon's Environmental Institute, Green Design Institute, and Brownfields Center, offer opportunities for collaboration and joint learning to firms that support their efforts. Michael Baker Engineering, along with other engineering firms, collaborated with researchers at Carnegie Mellon to develop wearable computers for bridge inspectors. The computers allow inspectors to speak their notations directly into digital systems, instantly updating databases on a bridge's condition and problem areas, without having to worry about juggling paper and pen while climbing over and under bridge structures.

Stanford University's Center for Integrated Facilities Engineering (CIFE) was a source of inspiration for the Walt Disney Company's development of a simulation technology to design and manage the construction of its new California Adventure theme park in Anaheim, California. Stanford associate professor Martin Fischer helped Disney set up a 4-D CAD application that enabled project managers to see blueprints as an interactive movie with the added fourth dimension of real-time scheduling.[162] CIFE has also joined with other designers to explore the potential of this technology, including a design-build team hired to complete a pilot plant for Sequus Pharmaceuticals. Researchers at CIFE developed a 4-D model, integrating design, cost, and schedule modules, that was used throughout the life of the project by Hathaway-Dinwiddie (general contractors), Flad & Associates (architects), and Mazzetti & Associates (mechanical engineers).[163]

New Social and Leadership Capacities

16 Approaching the New Capacities

The key to full self-expression is understanding one's self and the world, and the key to understanding is learning—from one's own life and experience.

—Warren Bennis[164]

Part III explored key ideas that architecture and engineering firms can use to develop new business models. The new ecology of firms framework enables firms to leverage deep changes being driven by information technology to establish sustainable niches. The experience economy reveals an exciting new form of value professionals have the power to integrate. New approaches to the social process of design and problem solving help create that value and enable firms to turn the social complexity of their projects into a technologically creative force. New approaches to technical innovation show the way to higher margins on the next technology S-curve while retaining focus on technical work. The organization ecosystem model helps firms understand their clients' strategic challenges, how best to position themselves in their clients' ecosystems, and new forms of value they can contribute on an ecosystem level. The value network can help firms assess more accurately—and thus increase—the value of their existing services as well as identify new services to address client needs that have not yet been met. And the three types of work—technical (Type I), collaborative (Type II), and transformative (Type III)—give firms a language for developing and bringing a greater range of leadership capacities to their evolving service offerings.

Part IV will explore the new social and leadership capacities required for the three types of work in today's complex social and technological environment. There are many ways to slice and dice this territory. We have settled on five core capacities for each type of work. Each capacity expresses an overarching concept at the same time that it embraces a broad range of more specific skills and competencies. Our purpose will not be to explain in detail how to acquire these capacities. A vast literature on these subjects already exists, and we will be referring you to some of our preferred sources as we go along. Rather, we intend to provide an understanding of what these capacities entail by describing the disciplines that underlie them in the context of built environment professional work. In essence, we offer a map of the territory professionals can use to pursue their capacity-development efforts.

A dynamic tension at the heart of this material makes it challenging to organize. On the one hand, the capacities for Type II work depend on the capacities for higher value Type I work, and the capacities for Type III work depend on those for Type II. From this perspective, simply starting with Type I and moving sequentially to Types II and III would be the best approach. On the other hand, professionals must focus on the transformation of their own firms if they want to evolve their practice or their business model. Though the Type III work involved in such transformation is the most difficult kind of leadership, it is required in order to experiment meaningfully with any of the ideas proposed here, including those in the arenas of Type I and II work.

For this reason, the first part of this section will include two chapters devoted to the leadership capacities required for Type III work in your firm. Chapter 17 will highlight leading innovation—central to the approaches to innovating in your business model discussed in part III. Chapter 19 will explore the other four Type III work leadership capacities in the context of transforming your firm—work that includes facilitating the shifts of mind for evolving your practice discussed in part II.

Every structure has a trade-off. The trade-off in this one is that we won't get to the Type I and II work capacities that underlie the Type III capacities until later, and we will take them up in the context of client systems rather than the professional firm. To help this structure work for you, keep in mind that you will not have the full picture of all the social and leadership capacities embedded in performing Type III work in your firm until you have read chapters 20 and 21 on Type I and II work in client systems.

With the chapters devoted to leading transformative work in your firm, we include an exploration of a common theme that runs through all three types of work. This theme, developmental awareness, points to a new way of thinking that we believe is as crucial to dealing effectively with project social complexity as the whole-systems thinking embodied in the organization ecosystem and value network models. Though each of the types of work calls for using developmental awareness on a different level and to a different depth, we devote a separate chapter to this new way of thinking because we believe that a general grasp of it is critical to effectiveness in all three types of work.

Social Capacities for Higher Value Technical Work

We believe that most firms must take on Type III work internally in order to thrive in the future and that firm leaders will therefore need to master all three types of work. We don't believe, however, that learning to provide Type II and III work to client systems will be the only way to thrive. Because these service and experience offerings extend significantly beyond the range of most current practice, we offer our map in part to help firms and individual professionals decide whether and how far to pursue these directions. We do believe, however, that firms must develop the new social capacities required to provide higher value Type I work to client systems.

As discussed in part II, a serious gap exists between what clients and stakeholders expect and what professionals are delivering. This gap has caused many clients to view all services provided by architecture, engineering, and related professionals—even those services that have not fully matured on the technology S-curve—as commoditized technical work rather than precious advice from a trusted partner. We have urged that in order to restore the relationship of broken trust and to begin to be viewed as valued partners once again, these professionals must adopt the shifts of mind for catching up presented in chapter 7.

These new mind-sets will not be fully experienced by the client, however, unless the professional brings them alive through the core social capacities required for effective collaboration and partnership. Today's socially and technologically complex environment requires collaboration even for what professionals would consider straightforward technical tasks within their scope of authority. We simply don't believe that any professional can be successful in the 21st century context without greater development of these core social capacities.

Shifts of Mind for Catching Up	**Social Capacities for Technical (Type 1) Work**
Relationship awareness	Deep listening
Whole-project-system awareness	Assertive speaking
Group process awareness and orientation to emergence	Productive conflict
A non-dualistic, inclusive view	Relationship development
Self-aware professional contribution	Learning and self-development

TABLE 16.1. Shifts of Mind and Social Capacities for Technical Work

Table 16.1 lists the five social capacities for higher value technical work alongside the shifts of mind that support them. The capacities may seem simple and perhaps familiar (from training programs or self-improvement books), but in fact each represents a significant challenge for most people, even before considering the new context introduced by the shifts of mind.

Techniques alone will not produce mastery of these complex disciplines. To get a powerful intuitive sense of what we intend by these capacities, first remember the key ideas behind each shift of mind.

Then, imagine the new way of thinking implied by all these shifts as holistically as you can. Finally, place your understanding of these capacities within that different framework, and allow your understanding to expand as a result. Note that the developmental awareness theme appears in both the relationship development and the learning and self-development capacities.

Leadership Capacities for Collaborative Work

Like the social capacities for higher value technical work, the leadership capacities for collaborative work will be effective only if developed in the context of the shifts of mind for catching up. The social capacities enable technical professionals to *participate* effectively in the collaborative processes involved in built environment projects, with strong relationships as a key outcome. In contrast, the leadership capacities for collaborative work enable technical professionals to *lead* these collaborative processes. This leadership involves facilitating individuals, groups, and organizations in collaboratively defining shared project-related goals and in moving through key collaborative steps to achieve those goals. Table 16.2 shows the leadership capacities required for collaborative work alongside the shifts of mind and the core social capacities that support them.

A very different level of capacities is required to lead a group of diverse stakeholders than is required to participate in that group. Technical professionals who have relatively little contact with clients will find these leadership capacities foreign. On the other hand, project managers probably have experience with some level of facilitation. Typically, however, leading project meetings has been seen as a support to technical work and not as a discipline in its own right. Consequently, project managers tend to stay within a narrow, traditional range of meeting management techniques based on hierarchical decision making. A few firms have begun to get more creative in their use of group processes and to use this aspect of their service as a differentiator. Even in these cases, however, these services are treated as an extension of the technical work.

Shifts of Mind for Catching Up	Social Capacities for Technical (Type I) Work	Leadership Capacities for Collaborative (Type II) Work
Relationship awareness	Deep listening	Developing shared vision
Whole-project-system awareness	Assertive speaking	Designing group process
Group process awareness and orientation to emergence	Productive conflict	Facilitating group process
A nondualistic, inclusive view	Relationship development	Group development
Self-aware professional contribution	Learning and self-development	Group learning and self-awareness

TABLE 16.2. Leadership Capacities for Collaborative Work

Some clients with large, complex projects have understood that they have a profound need for sophisticated facilitation and have been willing to pay the much higher rates of management consultants or professional facilitators. They recognize that traditional directive approaches to group management are not viable in the majority of client and stakeholder situations today. Achieving the inclusive participation of stakeholders with differing interests and views while facilitating forward movement is a high art that requires focused practice in its own right. This practice must include continuous learning and self-development and the ability to promote group development and learning. The theme of developmental awareness surfaces again.

Leadership Capacities for Transformative Work

Our definition of Type III, or transformative, work includes Heifetz's adaptive work—the deep internal processes required to change values and beliefs that underlie behaviors—in a larger concept that flows from the notion of adaptation in the largest sense. Given the pace of change in today's social systems, driven both from technological and social sources, organizations and their ecosystems face daunting adaptive challenges. We define transformative work as any work that enables human systems to adapt to significant changes in their environments.

As we've discussed, in these adaptive situations neither the problem nor the solution can be fully understood or defined at the outset. Significant new learning is required for all participants, including those leading the adaptive processes. The changes involved may be technological innovations that in turn drive social change, or they may be social changes. Adaptive social changes may be driven externally by conceptual, process, or behavioral innovations. Or they may be driven more internally—as in Heifetz's examples of groups working through their stresses and conflicts in order to assimilate new complexity. In our definition of transformative work, any or all of these types of change may make a major contribution to a human system's ability to adapt.

If learning has become a premium value generator in the knowledge age, adaptive or transformational learning is even more valuable than the incremental learning involved in collaborative work. It is also at least an order of magnitude more difficult. While we expect a substantial number of architecture and engineering firms will choose to pursue collaborative work as a key element of their new business model, we suspect fewer will venture into the more challenging territory of transformative work. Yet the need and thus the opportunities for Type III are increasing in today's environment. The capacity to include this form of value creation could in turn have a transformational effect on the industry as a whole.

As we've emphasized, the leadership capacities involved in transformative work build on the social capacities for higher value technical work and the leadership capacities for collaborative work, as summarized in table 16.2. Given the need to navigate uncharted territory, the capacities for transformative work also depend directly on the shifts of mind for getting out ahead of clients and stakeholders. Table 16.3 shows the leadership capacities for Type III work alongside these shifts of mind.

Shifts of Mind for Getting Out Ahead	Leadership Capacities for Transformative (Type III) Work
Strategic thinking	Strategic conversation
Systems awareness	Systems thinking
Organization-ecosystem awareness	Organization-ecosystem and organization development
Whole-ecosystem awareness	Leading innovation
Whole-value-network awareness	Leading whole-system learning and adaptive work

TABLE 16.3. Leadership Capacities for Transformative Work

Because technical professionals have focused primarily on technical problems or challenges, they have rarely applied their capacity for strategic or systems thinking beyond the technical realm. Consequently, they have rarely focused on the client system as a whole in the context of its larger environment. Each of these Type III capacities focuses on the client system as a whole. And because each can contribute to the process of successful adaptation to emerging conditions, each might be thought of as a developmental discipline. Organization ecosystem and organization development, however, take the theme of development that we have noted at the level of self, relationship, and group to the level of the client system as a whole.

Another theme runs through the three types of work. Learning is reflected in the social capacity of self-development and learning, the Type II leadership capacity of group learning and self-awareness, and the Type III leadership capacity of leading whole-system learning and adaptive work. We couldn't agree more with Warren Bennis that full self-expression does not hinder but rather is required for becoming a leader. We also couldn't agree more that to express oneself, one must come to "understand one's self and the world," and that such understanding depends on "learning—from one's own life and experience." We will be returning to this theme again and again.

We'll introduce the leadership capacities for Type III work first in the context of transforming the professional service firm. Let's turn now to leading innovation in that context.

17 Leading Innovation

A leader's ability to develop innovative ideas and ask for people's help in implementing them may seem to be obvious keys to success. But the sad fact is that too many of today's leaders resign themselves to the limits imposed on them by flawed systems rather than rethinking those systems.

—Donald T. Phillips[165]

In part III, we broached the subject of innovation from three different angles: (1) the possibilities for innovating in the experience economy by transforming the social process of design and problem solving, in part by leveraging technology; (2) the power of the value network model to reveal directions for business model innovation; and (3) innovating in technical work. Some of these themes will reemerge in this chapter as we look at the underlying leadership capacity that can enable professionals to engage in all these forms of innovation and more.

What Do We Mean by Innovation?

Innovation is a term more often applied to the technological than to the social arena. It can be either or both: any creative change that adds value to products, services, experiences, or other outcomes. Social innovations are designed changes in relationships or processes. While designed social changes may provide new experiences that eventually help to facilitate changes in values or beliefs, such innovation differs from adaptive work in Heifetz's sense. We will discuss that different form of adaptive work in the context of transforming your firm in chapter 19. Innovations, then, can range from materials to design or consulting solutions to technological or social processes, including the value proposition itself.

The mature stage of the architecture and engineering firm business model has produced such a strong, sustained focus on efficient production that developing the capacity to lead any significant form of innovation looks like a very steep climb. Of the many authors who have tackled the challenge of innovation in the last decade, we have found that Clayton Christensen offers some of the most valuable assistance. In addition to building his value network concept into the shifts of mind for getting out ahead (chapter 6), we have used his illumination of technology S-curves, customer expectations, and disruptive technologies to reveal how architects and engineers can innovate in their business models (chapter 14). *The Innovator's Dilemma* provides the first two of nine key disciplines we think make up the leadership capacity required to lead innovation in a firm.[166]

Identify innovation-friendly clients

Christensen emphasizes that in the mature phase of a particular value proposition—the commoditized phase where most professional firms are at this point—the pull toward dedicating a company's efforts to its best existing clients is almost irresistible. Though these clients often need innovation for their own organization-ecosystem renewal, they lack awareness of this need or of what it could have to do with their built environment projects. Instead, they are heavily invested in how the professional firm is currently doing business. Leaders of firms with whom we've worked consistently lament that their clients prevent them from trying new ideas. However, the business model is also at fault. If a firm is optimizing billability in a low-fee scenario, research and development on and for the job is the first thing to go. Research and development outside the project context is out of the question. How can a professional caught in these dynamics lead innovation?

Christensen's most fundamental recommendation is to seek clients outside the firm's existing markets with whom to experiment or offer a different value proposition. Only by finding the means to develop a form of value that initially falls outside of existing clients' stated requirements can firms hope to build a disruptive technology, a new form of value those clients will eventually be willing to buy. In Christensen's approach, the key attribute of the new type of client is different requirements that permit the lower performance characteristic of the early stages of an innovation's life cycle. In the case of David Evans and Associates' bridge practice, however, offering Type II work demonstrated characteristics of a disruptive technology through a different approach. Existing clients' expectations were restructured not by an alternative technology that evolved from lower to higher performance, but by a new process that provided additional forms of value.

Because architecture and engineering project situations are more complex than those addressed by the product-based businesses that Christensen examines, the performance criteria are more complex as well. They can be numerous, and combine some that are clear with some that are fuzzy. Christensen's point about needing to find a different kind of client appears to apply across complex service situations, but the basis for that client's interest and receptivity may vary. In some innovation scenarios, such as that involved in the BRADD software, early clients' receptivity may indeed be a function of lower performance requirements. In scenarios involving more complex services, however, early interest may be more dependent on a client's willingness to step into the unknown. Another well-known curve related to innovation sheds light on the challenge involved in finding such clients.

Like the evolution of an innovation's performance, the adoption of an innovation over time in a linked population reveals an S-curve pattern, proceeding slowly at first, then speeding up, and finally leveling off as individuals who have not adopted become scarcer in the networking dynamic. The curve popularized as "the diffusion of innovations" describes the distribution of adopters in the population based on this pattern, characterizing adopters based on their innovativeness. Articulated by Everett Rogers in 1962 based on research across different types of innovations and populations,[167] the curve exhibits a classic bell shape (Figure 17.1).

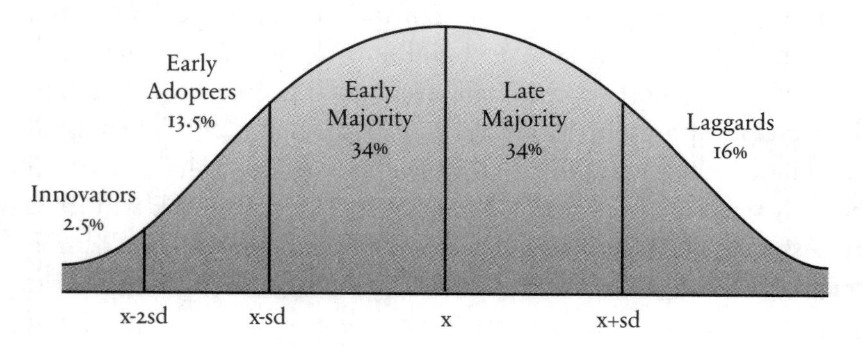

FIGURE 17.1. Adopter Categorization on the Basis of Innovativeness

Note: The innovativeness variable is partitioned into five adopter categories by laying off standard deviations from the average time of adoption (x).

Innovations are initiated by a very small number of people, then depend on a larger, but still small number of Early Adopters to build momentum. People may fall into different innovation categories in different arenas. For example, someone who is avidly interested in new forms of outdoor gear may be highly resistant to adopting new forms of electronic communication. For a given arena, a small number of people will be oriented to early adoption based on some combination of interest, readiness, and character. A much smaller number will be willing or able to be Innovators themselves. The professional seeking to innovate in a project context may need to find a client who is an Early Adopter—or better yet, an Innovator—in the arena of the desired innovation. Willingness to accept lower performance may or may not play a role, depending on whether the innovation entails lower performance on some criteria or whether it constitutes a wholly new form of value.

The diffusion-of-innovations bell curve helps us apply Christensen's imperative to look for a different kind of client across a wider spectrum of built environment situations than architecture and engineering firms usually consider. Though professionals typically have not thought explicitly in terms of these adopter categories, when asked, they have no difficulty identifying individuals likely to be Innovators or Early Adopters relative to a particular arena, either in their firm or among their clients. Though Innovator or Early Adopter clients may be hard to find, professionals are able to recognize them almost immediately. A powerful approach is to break into different communities than those populated by current clients. Publications, leading thinkers, and conferences in the arena of innovation interest are key points of entry.

Gather the right people and provide the necessary degree of isolation for incubation

Rogers's "adopter categories" also help to illuminate Christensen's second major recommendation. Christensen believes that the behavioral patterns of a firm's current system are so strong that it is nearly impossible to innovate without isolating staff working on innovation experiments from all of the firm's existing operations. The skunk works examples cited in chapter 15 reveal this approach. The adopter categories reveal some of the dynamics operating at the individual level. Premature involvement of Late Majority, Laggard, or even Early Majority individuals or groups—who will likely be competing for resources—may kill promising innovations before they have a chance to be fully developed and tested.

While Christensen's primary concern is the isolation factor, the adopter categories also highlight the importance of populating the incubator with the right people. The same criteria for identifying innovation-friendly clients apply to identifying staff, consultants, and other potential participants for innovation efforts. Incubation groups should include a few individuals likely to be Early Adopters as well as demonstrated Innovators in the relevant arena. Early Adopter presence is important partly because the true Innovator population is so small, and we can never know which individuals will be the actual Innovators in a given situation. Equally important, these Early Adopters will already be linked, informed, and ready to perform their natural diffusion function once the innovation begins to gel.

These two powerful recommendations from Christensen are key enablers for innovation, but do not address many other aspects of the leadership challenge. The first of these is the quality of thinking required to form the basis for viable innovations. To generate ideas with this kind of power, professionals must not only think and invest beyond their current business model, but move beyond solving traditionally conceived problems in only incrementally better ways.

Uncover value-rich innovation challenges

Christensen's research reveals that ideas for disruptive technologies are often first developed inside the company with the mature business, but killed for the reasons cited above. These aborted prototypes have typically developed through a tinkering process of experiments that build naturally one upon the next in isolation from customer needs. Such tinkering-based innovations become valuable only if they can find a different market and really valuable only if in a more highly developed form they have the potential to meet a future need of existing customers.

Innovations in the built environment may have the characteristics of disruptive technologies that displace prior ways of meeting the same needs, such as the disk drive and other examples Christensen discusses, or they may address chronic or emerging needs that have not yet been met in any way. In either case, migrating innovation efforts from tinkering-based emergence isolated from customers to intentional pursuit of solutions to currently unmet client needs can take these efforts to a higher level of effectiveness. Recognition and articulation of needs that are driving challenges in client systems—needs of which clients are often unaware—provide the basis for building critical momentum, whether an innovation's seeds are already present in something someone has tried and found to work or whether it must be conceived from scratch.

Understanding such needs and anticipating or noticing possible solutions require mentally disconnecting from the firm's current business and its clients' mind-sets, but not from the clients themselves. These clients must be viewed with new eyes. The first part of this new way of seeing is similar to the initial step in IDEO's famous process for product innovation: intense observation of clients to note what is currently working, what isn't, and what the natural patterns are that are not currently being supported. To achieve this observation in the context of architecture or engineering services requires professionals to free themselves from traditional ways of defining "requirements."

The second part of this new way of seeing involves focusing on the organization ecosystems and larger social environment in which clients are embedded. To uncover needs operating on this larger scale, the professional must cultivate the shifts of mind we have suggested for getting out ahead of clients and stakeholders: strategic thinking, systems awareness, organization-ecosystem awareness, whole-ecosystem awareness, and whole-value-network awareness.

Unmet needs can show up at the detailed heart of technical solutions as well as far upstream in the outer rings of the value network. To be able to perceive across such a wide range of possibility, professionals must cultivate a deep curiosity about the challenges clients and their organization ecosystems face while relentlessly looking for needs that are not currently being met. This means being willing to contemplate arenas where the professional does not possess current capability. Without a willingness to identify and take on new kinds of problems, professionals will be unlikely to innovate in ways that provide significant new value for clients and society.

For a cash-strapped hospital that has been cobbled together through a set of existing but unrelated buildings, the most valuable innovation might involve designing a new kind of physical basis for coordinated information flow to follow the patient. For a community whose drinking water supplies are being depleted by a raft of new golf courses on its fringes, the most valuable innovation might be a next-generation process for using gray or recycled water for their irrigation.

For a town losing manufacturing operations, it might be a master plan for attracting new industries to the former plants with energy-efficient remodeling that enhances worker productivity and reduces life-cycle costs. For a farming community rapidly losing land to the encroaching development of a nearby city, it might involve a new kind of rural-urban planning that keeps land available and affordable for farming, maximizes the city's consumption of local produce, preserves wildlife habitats, and provides for the city's growth by means other than sprawl.

Articulate the innovation challenge in inspiring ways

Once a professional has recognized appropriate innovation challenges for her firm to take on, she must develop compelling articulation of these challenges. A notion from complexity science, the "strange attractor," suggests a powerful process for this articulation. In *Surfing the Edge of Chaos*, Richard Pascale, Mark Millemann, and Linda Gioja explain that an "attractor" is like a "magnet" that "draws a complex adaptive system in a particular direction."[168] A "strange attractor . . . lures a system to the edge of chaos." Here, living systems become intensely dynamic and creative, able to undertake and achieve astonishing outcomes. Arising "from the interaction of an organism and its environment," the process whereby a strange attractor is generated is quite mysterious. From a relatively unorganized state, "multiple nodes . . . align with one another and coalesce into a pattern." If a human system's engagement with its environment is dynamic enough at a particular moment in time, an inspirational idea can emerge that begins to act like a strange attractor.

The authors of *Surfing the Edge of Chaos* cite a compelling example: Robert Shapiro, the CEO who mobilized and presided over Monsanto's explosive innovation in the arena of agricultural biotechnology.[169] We will get to the problematic aspects of this innovation a bit later, but Shapiro's leadership example in some dimensions is simply stunning. When he came on board, Shapiro had absorbed major lessons from complexity science, including the way a strange attractor interacts with a living system to literally pull self-organizing creativity out of it. Pascale, Millemann, and Gioja describe a particular communication event:

> Shapiro points to pieces in the puzzle (life sciences breakthroughs, agriculture, information technology, market knowledge); listeners relate his words to their own experience and fill in the blanks with their detailed knowledge of the business; Shapiro focuses on the unsustainable problems facing humanity—immense challenges that cry out for nontraditional solutions.
>
> Many in the room are moved at the prospect of contributing to the elimination of world hunger and chronic suffering. Those from ... [an acquired company] ... understand the relevance of their expertise in life sciences and bioengineering; those from traditional Monsanto see the usefulness of their knowledge of the food and agriculture industries—the mindset and methods of farmers, and the value of their suppliers and channels of distribution. Others in the audience bring threshold proficiency in the information technology needed to simulate genetic permutations and identify plant applications.

The town hall meetings held by Shapiro in the company had an astonishing galvanizing effect. There were 3,000 proposals out of the first Chicago meeting alone. The share price rose from $16 to $63. Within three years, farmers in the cotton and soybean industries had overwhelmingly adopted seeds genetically modified for disease and herbicide resistance, and "American cotton growers alone had reduced herbicide consumption by $1 billion."

Shapiro's strange attractor was effective in part because he so deeply recognized the innovation challenges that were there for his company to take up. It was also effective because his articulation was inspiring as well as urgent. He invoked current, overwhelming problems that must be addressed, but developed irresistible engagement with the possibility of solution. And he alluded to all the capacities currently in the system and its members as highly relevant and powerful resources for this solution. Professionals seeking to galvanize innovation efforts would benefit greatly from adopting a similar approach. Shapiro's example also shows that in cases of whole-organization scale and urgency, isolation of staff within the firm may not be needed.

Cultivate the conditions for creativity

Once the pull of a compelling innovation challenge has been activated, a leader faces the challenge of optimizing a group's or organization's creative response in the tough early phase of the innovation process. In his seminal book, *Complexity and Creativity in Organizations*, Ralph Stacey forged a brilliant synthesis of psychoanalytic theory and the complexity theory emerging from computer simulation of agent-based complex adaptive systems. Drawing on the simulation research, he was able to summarize the conditions that are present when human systems are most creative in their adaptive responses. These conditions are expressed in terms of five variable characteristics:[170]

- Rate of information flow
- Degree of diversity in schema (mental models)
- Richness of connectivity
- Level of contained anxiety
- Degree of power differentials

Human systems are most creative when each of these variables is operating within a certain range. That is, creativity is less than optimal not only when there is too little of any of these characteristics, but also when there is too much. As we suggested in describing the action of a strange attractor, this space for creativity is not promoted by equilibrium states in the system. Rather, it lies at the edge of chaos—in a realm between comfortable stability on the one hand and intolerable disorder on the other.

The implications for leading innovation are profound. First, understanding that creative adaptation requires a certain degree of instability enables the leader to embrace rather than resist the turbulent conditions of our contemporary environment. And second, the five variables provide the leader with a guide to creating the best possible conditions for innovation. The leader should cultivate a rich flow of information about the system and its larger environment, but not an overwhelming glut of such information. Through hiring, staffing, or alliance building, the leader should ensure diversity in ways of thinking and viewing the world, but not so much of this kind of diversity that system members cannot find common ground. The leader should also build organization or project structure and processes that richly connect system members, though not so much as to produce chaos. Organization or project structure should also provide the optimal degree of power differential: enough equality of position and influence to promote a free flow of ideas while maintaining the minimal degree of hierarchy needed to provide sufficient direction. Finally, the leader should maintain a healthy degree of anxiety in the system by calling attention to issues and problems and maintaining a sense of their urgency and importance. Shapiro's town hall meeting speeches certainly contributed to raising anxiety into the creative range at Monsanto.

The five variables enable the leader to optimize the creativity of a system as a whole because they are focused on cultivating the general environmental conditions for creativity. That system might be a group isolated for incubation experiments or all the members of the firm. While cultivating these conditions, the leader can further support innovation by understanding and promoting the creative processes of individuals and groups.

Promote the disciplines of individual and group creativity

Most studies of creativity since 1926 have referred to British writer Graham Wallas's four-stage model of the process as it unfolds in the individual mind: preparation, incubation, illumination, and verification.[171] The preparation stage is highly disciplined, so-called left-brain in character. It involves immersion in extensive research, inquiry, and exploration of possibilities. The incubation stage involves relaxation of the conscious mind, allowing subliminal processes to participate in the flow of ideas—bringing in the so-called right-brain dimension. This relaxation can be achieved in many ways—from simply focusing on something else to playing to a nap. Following the spontaneous emergence of a creative insight or idea—which characterizes the illumination stage—the verification stage again becomes highly disciplined, rigorously testing the validity and value of the idea. To effectively lead innovation, the leader must understand and nurture this four-stage process among participants, with due regard for the unpredictable and unstructured phase of incubation.

Although Wallas's four stages describe the experience of the individual mind, similar dynamics appear in the group processes for creative problem solving that have been developed over the last three decades or so. The well-known product design firm IDEO, described in chapter 11, has developed these processes to a high art. The firm achieves its extraordinary and relatively consistent yield in part by effectively tuning the five key variables for creativity, such as diversity of thinking among the participants and intervention by an authority figure at key moments. At the same time, its processes include rigorous research and preparation, as well as voluminous brainstorming that leaves the left-brain behind. An elaborate verification stage includes the applied innovation disciplines of rapid prototyping, testing, and combining of possible solutions.[172] A leader can be even more effective in promoting innovation by understanding and supporting these and other group processes that are now known to optimize creative output.

Design for the built environment at the highest levels has often been reserved as an individual activity. Leading design innovation in this context has involved figuring out how to provide enough support and undistracted isolation to "maverick" individuals while fighting inevitable tension between their individual motivation and organizational needs. However, group process has always been present in the charrette, which has certainly made use of short-timeline-induced stress. The role of the highly gifted individual designer is unlikely to disappear. At the same time, with the huge growth of social complexity in built environment processes, we believe that more and more designs—and innovations in general—will be developed in a group context. Each firm, each project, and each professional will need to develop ongoing, unique synergy and balance between individual and group process. Leading innovation must entail being sensitive to and assisting staff and project participants in this effort.

Support prudent risk taking

Because the quality of an innovation cannot be predicted or guaranteed, a further requirement for leading innovation must be support for prudent risk taking and a high tolerance for failure. Built environment professionals whose ethics include core responsibility to preserve public safety are appropriately averse to physical risk. Having been subjected to more than their fair share of the financial risks involved in projects, however, risk aversion has come to dominate their firms' cultures. This more generalized risk aversion has certainly contributed to the suppression of innovation.

Building the capacity to innovate should involve minimizing risk by performing as much small-scale experimentation as possible in order to fail early and often. However, the innovation leader must be willing to allocate resources for failure as well as to cultivate the cultural mind-set that learning from failure is a powerful outcome, well worth the investment. Professionals who have been burned by liability in the industry may not be the best people to take on innovation efforts. Establishing new cultural norms will ultimately depend on the leader's personal ability to respond positively to failures.

For technological innovations, relentlessly integrate social issues and processes

One type of risk that we must all be more sensitive to is the potential unintended consequences of technological innovations for our ecosystem as a whole. Leaders must relentlessly integrate technological with social processes. As we and many others have noted, one of the principal problems faced by our increasingly complex society is the failure of our social adaptation—values, beliefs, processes, and institutions—to keep pace with our technological evolution. Pascale, Millemann, and Gioja's discussion of the problematic outcomes of Monsanto's technological innovations provides a powerful cautionary tale of the importance of attending to this gap.

Following Monsanto's spectacular initial successes with bioengineered crops, European and American objections began to emerge. Eventually, these objections were supported by enough serious unintended consequences to pummel the stock price, reduce revenues, and force the by then over-leveraged company into a merger. Concerns had surfaced early, but Monsanto had treated them as public relations problems or regulatory issues rather than forms of complexity inherent in its innovation arena.[173] How deeply the company has learned the lesson illustrated by its own experience remains to be seen. However, this lesson emerges with great clarity for all who would innovate in any technological arena: never underestimate any social complexity that arises with the process—treat it as integral to the innovation process itself.

The need to ensure that social development and capacity keep pace with technological innovation—and that technological innovation supports the health of our social systems and the sustainability of our planet—must inform innovation in the built environment no less than other arenas of human endeavor. This effort will likely slow the technological innovation process, but the best antidote is simultaneous pursuit of related social innovation and adaptive work on values and beliefs.

Use self-organizing processes to facilitate adoption

Once an innovation is beginning to take shape—through internal firm efforts or a project or two with innovation-friendly clients—the innovation leader must think about the diffusion process. Diffusion provides opportunity for additional development and tests for appeal to a wider market that may eventually include clients of the firm's mature business. Earlier, we used the diffusion-of-innovations bell curve to suggest that for complex services, innovation-friendly clients will be Early Adopters or even Innovators in the arena of a given innovation challenge. This curve also tells us that if just 5 percent of a linked population adopts an innovation, it will not disappear but remain embedded; that is just the front end of the Early Adopter group. A firm could conceivably build a small niche business based on embeddedness if it were willing to cover a broad geographic area.

The exciting part, though, is that if 20 percent of a linked population adopts an innovation—just the front end of the Early Majority—completion of adoption through the whole population is highly likely. This is the so-called tipping point—popularized by Malcolm Gladwell in his book by the same name—that appears in population-related phenomena from epidemics to fashion.[174] Tipping point dynamics can unfold with breathtaking speed, as they did in the adoption of Monsanto's bioengineered agricultural products. Complexity scientists who have used computers to simulate the behavior of complex, adaptive systems observe the same pattern of diffusion that Rogers observed empirically. The natural, self-organizing dynamics of linked populations as living systems do the work of diffusion for us if we can manage to climb up the front-end of the curve. Between 5 percent and 20 percent adoption, then, a firm or client ecosystem is creating an increasingly significant niche that others will likely begin to enter. At 20 percent, market potential really takes off.

The diffusion-of-innovations bell curve offers key adoption-related guidance for leaders right from the beginning of the innovation process. Early Adopters included in incubation groups will be poised to perform their natural diffusion function when the right moment arrives. The leader can optimize this effect by including at least one Early Adopter for whom social networking is natural, the type of person Gladwell calls a "Connector."[175] The leader can further support this effect by linking Early Adopters involved in the incubation to others in the firm's ecosystem. Linking Early Adopters to Innovators and Early Adopters to other Early Adopters is the most efficient and effective way to drive adoption.

Yes, genuine, value-creating innovation is difficult, and few architecture and engineering firm leaders have undertaken it—for both the business-model and the mental-model reasons discussed. However, these leaders have generally not been aware of the stunning power of the diffusion-of-innovations bell curve. There is far more innovation momentum potential here to be tapped than professionals have suspected.

A recent article on tipping point leadership features a public sector leader who is also described in Gladwell's book.[176] Although he has worked almost exclusively in the social innovation arena, his example shows how such innovation has driven at least some of the kind of internal change identified with adaptive work. Bill Bratton has wrought nothing less than transformation in the Boston and New York transit systems and police departments and is credited with being the major force behind the stunning reduction in crime in New York City, now the lowest among the 25 largest U.S. cities. He is clearly a master at building momentum at the front end of the adoption curve for innovations directed at solving social problems.

The article describes a process of focusing on four hurdles that must be gotten over to create this momentum. The leadership for this process is based on sophisticated strategic and systems thinking that provides direction for and most of the elements of the social innovation, such as removing police officers from the paperwork process and holding short-notice, high-profile precinct performance reviews with a panel of top brass. Essentially, the innovation has already been developed, and the hurdle process drives its adoption. Professionals whose transformation efforts include designing social innovations aimed at eliminating significant problems in their firms may find the four hurdles process helpful.

The "cognitive" hurdle involves enlisting those with responsibility for a system in the innovation effort. The leader addresses this hurdle by bringing key managers face to face with their system's current limitations and with the people who are experiencing them. Experience with this kind of direct feedback can enable decision makers to acknowledge issues sufficiently to begin to mobilize their energies for a new approach. The "resource" hurdle is addressed not by increasing resources but by identifying the system's highest leverage points and concentrating existing resources there. The leader approaches the "motivational" hurdle by involving key influencers, who then, as Early Adopters, begin to drive adoption by others. Finally, the leader addresses the "political" hurdle by identifying and isolating internal and external opponents to prevent them from blocking the adoption process through their negative influence.

Because Bratton's process of social innovation has led to some degree of adaptive work in the social systems he has transformed,[177] it forms a good transition from our mapping of nine key disciplines involved in leading innovation to the other Type III work capacities involved in transforming your firm. Let's turn now to an essential building block for these other leadership capacities, and for all three types of work in client systems: developmental awareness.

18 Developmental Awareness

[I]f in the last few hundred years we have succeeded in recognizing a qualitative distinction between the mind of the child and the mind of the adult, it may still remain for us to discover that adulthood itself is not an end state but a vast evolutionary expanse encompassing a variety of capacities of mind.

—Robert Kegan[178]

Development is a theme in the leadership capacities for each of the three types of work, across the human system levels of individual self, relationship, group, organization, and organization ecosystem. The term *development* is often used to refer to any form of learning or growth, and our use of the term does carry this generic breadth. In the leadership capacities and in this chapter, however, it also refers to a particular aspect of learning and growth: the unfolding of human systems through predictable patterns of life-cycle and consciousness development.

All living systems are designed to adapt to changing conditions—incrementally in the context of dynamic equilibrium as well as in the more discontinuous manner that can occur at the edge of chaos. All such adaptation might be considered development in the broad sense. At the same time, all living systems move through stages of development prescribed by an archetypal pattern. A familiar, visible version of such a pattern is the changes that occur in the human body as an individual moves from infancy to childhood to adolescence to adulthood, and then through the various stages of adulthood. To see the whole picture of development, then, we must envision two developmental processes proceeding in parallel. On the one hand, a human system learns and changes in response to its environment. On the other, a human system unfolds through its life cycle. The two processes tend to be synergistic. A human system's creative responses to changes in its environment contribute to its movement through its life cycle. And movement through its life cycle contributes to a human system's capacity to respond adaptively to its environment.

Understanding the developmental stages of client systems enables professionals to match products, services, and experiences to the needs of those systems. And understanding the developmental stages of systems within the professional firm provides critical context for identifying what types of transformation might be appropriate as well as for planning processes for that transformation. In this chapter we will first explore the exterior life-cycle stages that can be observed as system behaviors—at the levels of group, relationship, organization ecosystem, and organization. We will then explore the interior stages of development that characterize subjective awareness—at the individual level in the form of mind-set and at the organizational level in the form of culture.[179]

The Group Development Model

Perhaps the best way to introduce the idea of developmental awareness is through an example that has become familiar to people in the built environment professions, and indeed, business as a whole: the group development model first published by University Associates in 1982. Bruce Tuckman formulated this model based on several hundred studies of small-group dynamics.[180] It identifies four stages in the life cycle of a group, cleverly named Forming, Storming, Norming, and Performing. This model is taught in team skills, team-building, and partnering workshops as an aid to team productivity. Many experts make a formal distinction between a group and a team, but for purposes of the developmental model, we can treat a group and a team as equivalent. According to the model, a team cannot reach the Norming stage without going through Storming, or the Performing stage without going through Norming. Not all teams will achieve the later stages, but the only way there is through the earlier stages. Because teams ultimately return to the early stages when they hit new challenges, the stages are better represented by a cycle than by linear steps, as depicted in figure 18.1.

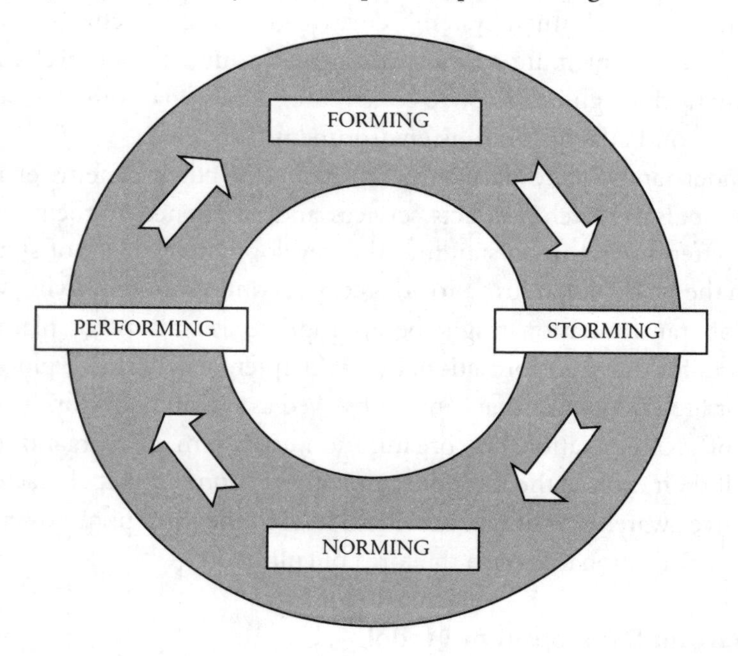

FIGURE 18.1. Stages of Group Development

The major issues of the Forming stage have to do with inclusion and belonging. Individuals ask themselves whether they will in fact fit into a group, whether they will like or be able to work with the other members or vice versa. Members tend to be polite and risk averse. They try to find out what kind of people the other members are and what they really think, but are unwilling to make themselves fully visible in this process. As a result, individuals do not bring their full expertise to the table and productivity is low.

When members are willing to come forward and express themselves more openly, differences emerge. While conflict-averse groups may quickly revert to the Forming stage, healthy development entails moving into Storming. Teams range from overt to covert in their Storming style, but the essence of this stage is conflict among members that threatens the integrity of the group. Members are trying to decide whether they can stick it out, whether they can actually work across these differences. This stage is uncomfortable—sometimes very uncomfortable—but necessary.

Some teams may remain stuck in the Storming stage, never able to resolve their fundamental conflicts or to become highly productive. Healthy development entails breaking through to committing to work with one another, which ushers in the Norming stage. Here, members figure out what it will take to work with each other's differences, developing both group and individual processes for this purpose. Though these processes require conscious effort to do something other than what feels natural, they provide a significant jump in productivity.

If the chemistry of a team never really gels, it may remain in the Norming stage and be reasonably productive. With practice and chemistry, however, team members may move into a natural synergy with each other. Here, high performance begins to emerge without conscious effort, and the team begins to have a lot of fun. However, new members or a major new challenge or change in their organizational environment can take a team back to the Forming stage. Members ask themselves once again, Do I really want to be part of this?

Though this brief summary of the stages lacks much of the helpful nuance of the model, it reveals why so many people have found the model useful. Most professionals, in particular, tend to be conflict averse. They are likely to do everything in their power to smooth over or even suppress significant disagreements, especially early in a project or client relationship. This tendency has dovetailed nicely with professionals' preference for autonomy and prevented many of their teams from moving beyond Forming. This stage can work to coordinate the efforts of individuals, but does not support genuinely synergistic creativity and shared accountability among members, or between the professional and the client. When professionals understand the role of Storming, they are able to suspend their discomfort and move into (rather than away from) the conflict that is essential to team development and to empowering their own contribution in collaboration with the client.

Awareness of the stages of group development is core to professionals doing Type II—and thus also Type III—work. Without the ability to lead a group through the dynamics of Storming, a professional will be unlikely to release the potential of a group to creatively address project challenges. No less significantly, awareness of the group lifecycle stages enables a professional engaged in Type I work to participate fully and creatively on project teams. The different attitude toward collaboration and conflict this awareness engenders may be the single most important factor enabling a professional to move from the role of less trusted, less influential expert to the role of newly embraced collaborative partner.

Tuckman's model of group development sheds light on the concept of developmental awareness. First, this awareness moves our thinking away from the notion that a group is the sum of its members, pure and simple, and likely to feel and act similarly at different points in time based on its makeup. Instead, the group is seen as a dynamic entity, not just coping with or adapting to its environment but also in a state of becoming relative to a deep pattern of unfolding potential. Second, developmental awareness can shift the way we perceive a team that appears not to be working well. Instead of focusing on the fact that it is stuck, we can see where it is in its development, and what might be required for it to move into its next natural state. And third, developmental awareness primes us to observe the difference between a team that regresses to a previous stage in the face of a new challenge and gets stuck there, and a team that under such a challenge cycles back to earlier stages in order to build a foundation for further unfolding. Once we see that teams are not just in motion with purposeful or adaptive activity, but also in motion developmentally, we attain a completely different level of capacity both to participate in them and to lead them.

Relationship Development

A lesser known but strikingly parallel model describing the stages of an individual relationship was articulated in 1980 by Susan Campbell in *The Couple's Journey*.[181] Though her focus was intimate relationships, her stages apply to the life cycle of work relationships as well. As at the group level, individual relationships that don't get stuck tend to cycle more than once through the stages. Figure 18.2 depicts in cyclical form a slightly simplified version of Campbell's stages of a relationship.

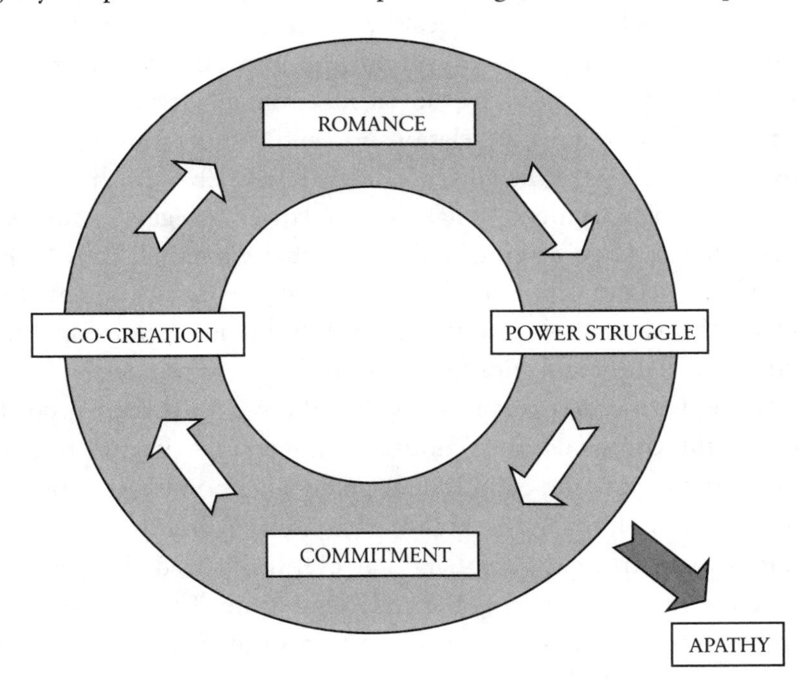

FIGURE 18.2. Stages of Relationship Development

According to Campbell, relationships start in a Romance phase. In work settings, common parlance dubs this the honeymoon period. Here the parties are acting on their initial attraction to one another and the enthusiasm of high expectations. This phase feels terrific, except it is not based on really knowing one another yet. Hopeful projections of one's own imaginings substitute for real exchange and understanding.

As soon as the parties begin to really get to know one another, they discover their natural differences and move into a phase Campbell calls Power Struggle. Here the question becomes, Now that we see how we differ, who really has control in the arena that we are attempting to share? Some individuals take a more covert approach to Power Struggle than others. Either way, this is a challenging and unpleasant phase of the relationship journey, typically characterized by disappointed expectations and bad feeling. From here, the relationship risks ending completely—which can be a healthy outcome for the individuals if differences are too great—or spinning off into a developmental cul-de-sac called Apathy. In Apathy, the parties essentially leave behind all constructive hopes for the relationship and simply put up with the dysfunctionalities of the Power Struggle phase, albeit in a passive form.

If the developmental challenge of Power Struggle is successfully navigated through an intervening step that Campbell calls "stabilization," the relationship can break through to Commitment. In this phase, the parties consciously work at shifting the relationship into productive collaboration. Concerted effort focuses on understanding and working with differences in ways that allow both parties' needs to be met. If this conscious effort continues long enough, it tends to blossom into Co-creation, the culminating phase where synergistic interaction involves less effort and is highly generative. Typically, this rewarding state opens up new potentials that eventually lead the parties into renewed Romance, starting the cycle once again. The relationship thus spirals downward into deeper connection, or upward into higher levels of connection. Either metaphor works to describe ongoing development over a longer term.

The individual-relationship level of developmental awareness is core to the professional performing Type I work. The technical professional who hits a rough spot with a client will tend to attribute it to disagreement about technical issues, or to personality differences, or perhaps to organizational politics. Understanding that the relationship may be going through a necessary stage on the way to becoming truly productive has the potential to completely transform how the technical professional responds to and evolves with the client. Like navigating the Storming stage of group development, navigating the Power Struggle phase can make or break the client relationship. The social capacity for relationship development is thus essential to effective Type I work, and of course, to Type II and III work as well.

Developmental awareness at the level of relationship moves our thinking away from conceiving a relationship as a static constellation of two different personalities who will manifest similar dynamics at all points in time. Instead, we perceive the relationship to be in motion relative to a deep pattern of unfolding potential. It is striking how closely these patterns at the group and relationship levels parallel one another. The role of diversity and conflict in cooperative self-organizing at this interpersonal scale of living systems seems to be fundamental and archetypal.

At a given point in time, the stage of an individual relationship in a team may be very different from the stage of the team itself. Developmental awareness entails being attuned to both. It involves seeing both levels in motion at the same time—not just in activity, but in current or potential developmental unfolding. Let's turn now to a much more complex level in the hierarchy of living systems: the organization ecosystem. As we shall see, the archetypal issues have a new dimension.

Organization Ecosystem Development

We have already alluded to the four stages of business (or organization) ecosystem development introduced by James Moore in *The Death of Competition* in 1996: Pioneering, Expansion, Maturity, and Renewal or Death. As at the lower levels in the living systems hierarchy, organization ecosystems tend eventually to repeat the stages, as depicted in figure 18.3.

Each of Moore's stages is characterized by significant cooperative and competitive challenges. This brief overview will vastly simplify these highly dynamic complexities in order to maintain focus on the larger concept of developmental awareness.

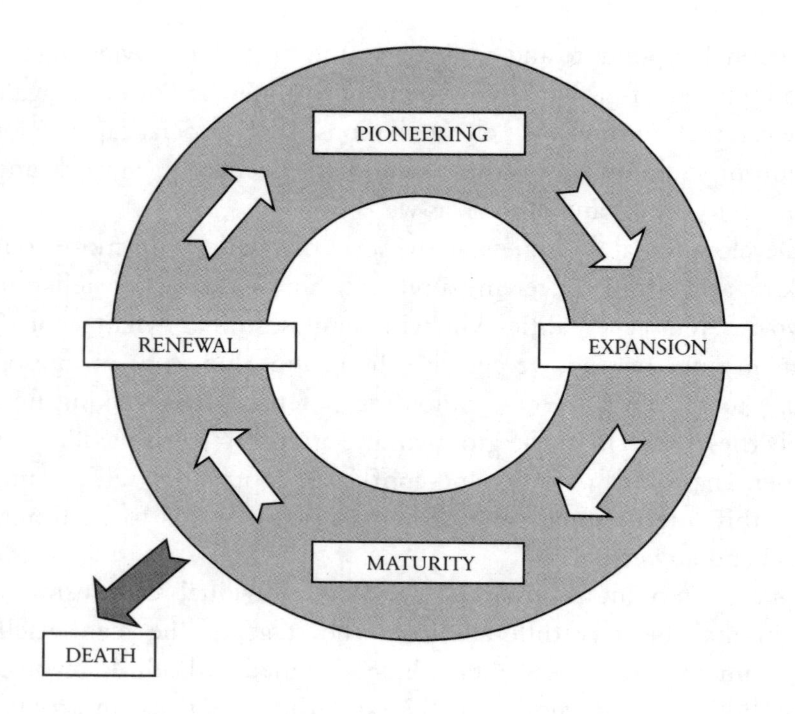

FIGURE 18.3. Stages of Organization Ecosystem Development

In the Pioneering stage, capabilities of different, complementary organizations are linked to create core offers that create value much superior to the status quo. The business or organization ecosystem thus takes form. Once the core offer is sufficiently formulated, members invest in core synergistic relationships to expand scale and scope. The goal of Expansion is to establish critical mass within desired boundaries. Once critical mass is established, the core organization drives the Maturity phase by seeking to achieve and maintain stabilization of its ecosystem. This organization attempts to retain its authority over cooperative relationships and key strategic decisions by embedding its own contributions at the ecosystem's center. As competition for internal niches heats up, the ecosystem pursues continuous improvement in order to hold the line against other ecosystems that are now competing with its value proposition as a whole. Then, as improvement efforts produce diminishing returns, the ecosystem begins to face Renewal or Death. Ways are found to insert new ideas into the old order that enable the ecosystem to stay competitive with other alternatives. Failing this, the ecosystem disintegrates or returns to Pioneering to focus on something entirely different.

Without developmental awareness, all the motion going on in an organization ecosystem can appear chaotic and not terribly relevant to a built environment project. With this awareness, however, deep patterns of unfolding—or arrested unfolding—are visible. Contributing to the organization-ecosystem level of client system development is a core concern of Type III work. However, awareness of the stage of a client's organization ecosystem can contribute greatly to the value of Type I work as well. An engineer presenting a technical solution appropriate to an ecosystem in Expansion will likely miss the mark for one that is struggling toward Renewal. Type II work is also affected by developmental awareness at this level. Even if Type II work does not deal explicitly with organization ecosystem stage issues, the value it delivers will depend significantly on alignment of facilitation processes with the issues currently dominating the relationships among ecosystem members.

Dynamics of cooperation and competition inside the ecosystem or between the ecosystem and other ecosystems are present in all stages, though they play out in different ways. These dynamics reveal the issues of diversity, conflict, and cooperative self-organizing that we observed in the patterns underlying group and relationship development. However, the stages themselves reveal a pattern linked to the life cycle of a way of producing value for the larger system: the sequence of states that create a technology S-curve, thinking of technology in the broadest sense. Let's look now at life-cycle stages one level down from organization ecosystems in the hierarchy of living systems: the organization itself.

Organization Development

In the early 1990s, a group of organization development consultants felt the need to articulate a model of development that would enable them not only to facilitate the development of their client organizations, but also to achieve appropriate fit for any services their clients requested. One of these consultants, Herman Gyr, published the four-stage model for the life cycle of an organization that resulted from their efforts.[182] Though Gyr included the potential to cycle through the stages more than once, he represented them as a linear progression. Figure 18.4 shows his model in cyclical format.

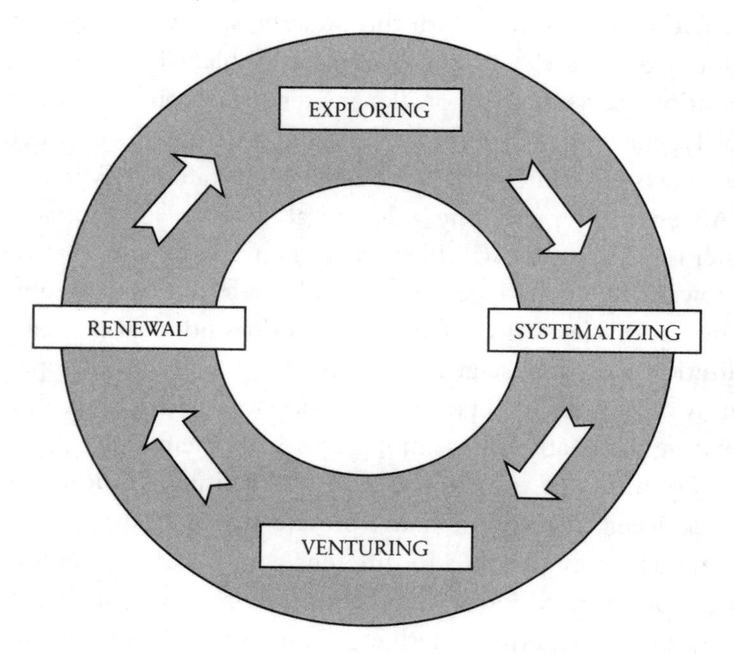

FIGURE 18.4. Organizational Life-Cycle Model

Gyr summarized the stages as follows:

> In the Exploring Stage everything about the organization needs to be defined, from the organization's mission to the specific jobs, relationships, and tasks through which the mission is accomplished; organizational expectations and arrangements are necessarily loose and fluid. In the Systematizing Stage order begins to emerge, routines and controls are established, and organizational life becomes

> predictable. In the Venturing Stage power and authority are delegated within clearly defined parameters and monitored accountabilities, freeing people to exercise their growing competence with greater autonomy. In the Integrating Stage a compelling shared vision unifies and offers priorities for the diverse endeavors on which the organization is embarked."[183]

As in all developmental models, each new stage does not simply replace the prior stage. Instead, each builds on the previous, incorporating its strengths and adding new ones that now receive primary focus.

Like the organization ecosystem life-cycle stages, the stages of the organization life cycle are a core focus for Type III work. These stages are equally important for the other types of work, however. The need to align Type II processes with life-cycle stages becomes even clearer at the organization than at the organization ecosystem level. An organization in the Systematizing stage will tend to require facilitative leadership that dovetails with its need for clear, linear, and well-managed processes, while an organization in the Venturing stage will need processes that give individuals room to exercise their own initiative.

Similarly, the format of a Type I work product would need to be different for an organization in Systematizing than one in Venturing. A standardized deliverable for a particular kind of technical problem may fit the needs of some client organizations, but not all. Much better to understand where a client organization is in its development, and match your product and process to what it is attempting to achieve in that stage of its life cycle.

Gyr's organization life-cycle stages reveal strong functional correlations to Moore's ecosystem stages and its underlying technology S-curve. Exploring at the organization level parallels Pioneering at the ecosystem level: experimenting with an initial value proposition. Organizational Systematizing parallels ecosystem Expansion: clarifying and replicating the value proposition. Organizational Venturing parallels Authority at the ecosystem level: improving on the original value proposition through creative initiatives. Organizational Integrating parallels ecosystem Renewal: driving significant innovation through co-creation of parties who are deeply aligned. At any point in time, however, stages at the organization and organization ecosystem levels will not necessarily line up. Developmental awareness entails sensitivity to the deep patterns operating at both of these levels. And it means being attuned at the same time to stages at the relationship and group levels.

Regardless of the level of social system, developmental awareness enables us to see beyond surface activity and motion to deeper, inherent patterns of unfolding potential. Just as the organization ecosystem and value network models enable professionals to see the bigger picture of client systems *in space*, developmental awareness enables them to see the bigger picture of client systems *over time*. With the understanding that social systems are not static but evolving, and with the ability to see where they are in their developmental process at a given time, professionals can forge more effective relationships with and exercise more effective leadership in those systems. Professionals can thus achieve a better fit with developmental stages present at different levels of a client system as well as promote healthy development at those levels. They can also more effectively lead appropriate evolution or transformation in their firms.

For developmental awareness to fully yield these benefits, however, professionals must also factor in the interior dimensions of individual mind-set and organization culture.

Interior Dimensions of Human Systems

Our examination of social system life-cycle stages so far has focused primarily on observable behaviors, or what we have called their exterior dimension. All four levels—relationship, group, organization, and organization ecosystem—also have an "interior" dimension. This dimension consists of subjective experience, of consciousness itself in its full spectrum from thinking to feeling states. Understanding subjective experience of the individual is probably more familiar than considering it at the more complex levels of social systems. However, students of group experience note the presence of a group mind, which includes an unconscious as well as a conscious dimension, emotional and spiritual as well as mental states. Group consciousness may also be understood to be present at the small scale of a one-on-one relationship, and we are all familiar with attempts to understand the consciousness of larger collectives. Anthropologists study the cultures of tribes and nations as well as organizations, while historians and philosophers have alluded to the notion of large-scale collective consciousness evolving over time through the concept of zeitgeist, or spirit of the age.

The ability to grasp unfolding patterns in the exterior behavioral dimensions is immensely useful to the professional, but not without awareness of the interior dimensions. The professional works with and through the mind-set of the individual client in forming the client relationship, and with and through the culture of the client organization in the processes of design and problem solving. He experiences the client organization culture not only in the organization as a whole, but in groups within the organization and in individuals who have absorbed and are working within it. On the other hand, the professional works through his own mind-set and through the culture of his own organization as it manifests in project systems as well as in himself and other involved colleagues.

The meshing of these interior dimensions at the individual and organizational levels is the context in which professional work is performed. Above all, or underneath all, the professional must be able to navigate this consciousness context in order to create value for a client system in any of the other ways described here. In fact, misalignments between clients' and professionals' individual and organizational mind-sets can compromise value-creating capacity even more significantly than life-cycle-stage mismatches. Misalignments in mind-sets between leaders and other members of a professional firm can produce equally debilitating dynamics in the context of internal transformation efforts. Developmental awareness that includes the interior dimensions is the best defense against these problems.

Navigating the interior dimensions is more challenging than the exterior dimensions because we are often unconscious of the assumptions, beliefs, and behavioral norms that drive our consciousness. In an organization, members are often not aware of the deepest aspects of their shared culture. To use a common metaphor, their system of values, beliefs, and ways of working is like the water in which fish swim. It is all-pervasive, yet it is not the focus of their attention. Similarly, individuals are often unaware of the values, beliefs, and ways of working that characterize their own mind-sets. These elements typically function like invisible foundations beneath the thoughts, feelings, and actions to which our conscious minds attend.

Like the life cycles of relationships, groups, organizations, and organization ecosystems, individual mind-set and organization culture reveal underlying patterns of unfolding. These patterns are related to exterior life-cycle stages at the individual and organization levels but cannot be equated with them. The consciousness stages often unfold more slowly than life-cycle stages, yet like the life-cycle stages, they are best understood and worked with not as static states but in their potential-laden developmental context.

Individual Mind-Set Development

Perhaps the simplest formulation of the stages of individual mind-set development is one popularized by Stephen Covey in *The 7 Habits of Highly Effective People*. Covey defines the "maturity continuum" as unfolding from dependence—"the paradigm of *you*"—to independence—"the paradigm of *I*"—to interdependence—"the paradigm of *we*."[184] Children are naturally dependent in their mind-sets and behaviors, while adolescents strive for greater independence as they learn to function apart from their parents. Ideally, young adults begin to develop the capacity for interdependence as they become enmeshed in more complex social responsibilities. However, we all know individuals who have remained to some degree oriented to dependence throughout adulthood. In the United States, where we are so heavily grounded in a culture of individualism, moving beyond independence into the interdependence stage—in which relationships are considered to be as important as oneself—is a significant achievement. The three stages are predictable, but moving through them entails confronting and successfully navigating developmental challenges. Consequently, a given individual's current mind-set may bear little relationship to his life-cycle stage.

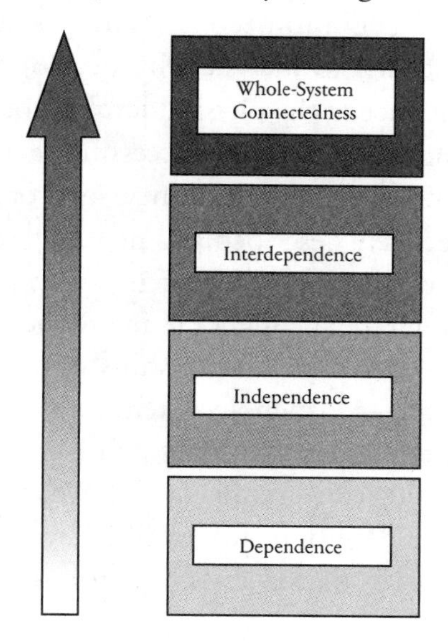

FIGURE 18.5. Individual Mind-Set Developmental Stages

Covey's three-stage model does a good job of describing mind-set corollaries for life-cycle challenges up through the beginning of what we now think of as midlife. In midlife, individual concerns tend to shift from the interdependence issues of relationship, family, and immediate community to issues of larger community and the human family as a whole.[185] A desire to engage with and contribute to issues at a larger system level is the natural next step as an individual's children grow up and leave home—if that individual successfully navigates the developmental challenges involved. While this next step includes an orientation to interdependence, it defines a new level: the orientation is not just to relationship with others, but to relationship with large systems in which others are embedded. If we were to name this new mind-set *whole-system connectedness* and add it to Covey's three stages, we would have the developmental model depicted in figure 18.5.

We depict these stages as a linear progression rather than a cycle. Though consciousness may regress, healthy unfolding continues forward motion. We can see the contrast by observing that individuals reaching very old age begin to become more physically dependent as their bodies break down. Under healthy developmental conditions, however, this dependence at the physical level is not accompanied by a return to the dependence mind-set. Successful navigation of the challenge of mortality entails passing into a new level of awareness and wisdom that is not effectively described as a new form of dependence.

The traditional domains of the professions mirror a mind-set that reflects at least as much independence as interdependence. Our new set of values and ethics—founded on the shifts of mind for catching up and getting out ahead—adds whole-system awareness to interdependence, encouraging professionals to develop fully into the third and then into the fourth stage described in this model. We will see that the stages of individual mind-set development described by Covey closely parallel the mind-sets inherent in the stages of organization culture.

Organization Culture Development

About the same time that Gyr published his model based on the organization life cycle, another organization development consultant, Roger Harrison, presented a model of four fundamentally different kinds of cultures that he believed also revealed a developmental progression.[186] Figure 18.6 depicts his model.

In observing these cultures across a large number of organizations, Harrison noted that an organization's culture is rarely unitary or pure. Across individual and subgroup differences, however, one culture tends to dominate. We might think of this as the organization's cultural center of gravity.

Harrison labeled the first type of culture Transactional and described it as "oriented to satisfactions and rewards controlled by others" with the qualities of "exchange, hierarchy, dependency, control and compliance, [and] bureaucracy." The second he called Self-Expression and described it as "oriented to the self as source of satisfactions" with the qualities of "creativity, pleasure in doing, competition, self-development, and empowerment." The third, the Alignment culture, is "oriented to the achievement of tasks, goals and purposes" with the qualities of "idealism, dedication, subordination of personal needs to task, voluntarism, commitment, [and] focus." Finally, the Mutuality culture is "oriented to relationship, caring and connection" with the qualities of "reciprocity and cooperation, seeking consensus, giving, sharing, empathizing, considering the context, and looking after stakeholders."

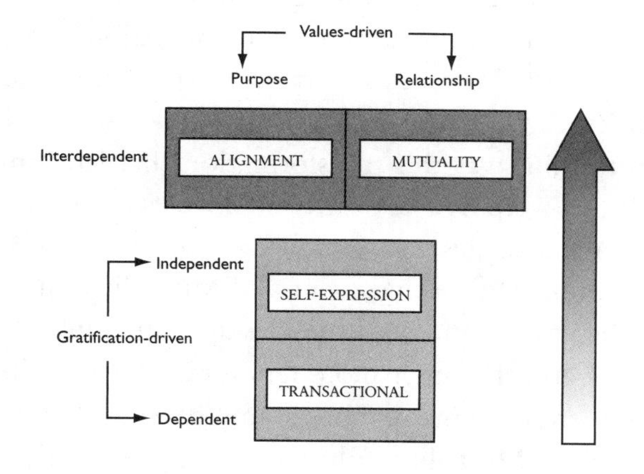

FIGURE 18.6. Model of Organization Culture Development

Harrison characterizes the Transactional culture as "dependent," the Self-Expression culture as "independent," and the Alignment and Mutuality cultures as "interdependent." This resonance with Covey's descriptors of the stages of individual mind-set development suggests that organization cultures are grounded in individual mind-sets. The developmental leap to a culture of interdependence depends on replacing gratification with values as the primary motivator. The Alignment and Mutuality cultures differ in their central value, with Alignment focusing on purpose and Mutuality on relationship.

Harrison did not set out initially to define organization culture development. Rather, he simply observed the different cultures present in the organizational landscape in which he was working. His observations led him to note a striking shift from a dominance of Transactional cultures in the 1970s to a dominance of Self-Expression cultures in the early 1990s. This shift appears to be related to the transition de Geus describes during that same time from a capital-based to a knowledge-based economy—to an economy centered on people rather than physical resources as the core asset. The significant adaptive challenges faced by U.S. corporations during that period of declining competitiveness likely played a key role in bringing about this shift. Harrison implies that though some organizations with Transaction cultures died and some new organizations were founded with Self-expression cultures, this shift occurred mainly as a result of Transactional cultures being successfully transformed into Self-Expression cultures. It was this transformation that led him to think that these cultures were not just different species, as it were, but different developmental stages.

Harrison represents the Alignment and Mutuality cultures as two possible branches to the interdependent stage, one embodying more masculine values and the other more feminine. However, he observes that in the 90s there were many more examples of Alignment cultures in our society than Mutuality cultures. He suggested then that we might be on the way to seeing the development of more Alignment cultures into Mutuality cultures, just as we once experienced the emergence of more Self-Expression cultures out of Transactional cultures. He notes: "Mutuality cultures offer strong support for high order learning, particularly in dealing with several fundamental issues with which organizations currently struggle. These are 1) dealing with complexity and chaos through organizational learning; 2) improving quality and service through appreciation; 3) healing the trauma of rapid change, and 4) working with diversity."[187]

This description is consistent with our formulation of the social and leadership capacities required to catch up with current trends in the social context of the built environment. Anchored in relationship awareness, these capacities parallel a Mutuality culture that has not simply replaced but transcended and included the prior three cultures

Harrison identifies. These capacities incorporate appropriate hierarchy and systems, empowerment and self-expression, and purpose-based alignment into a relationship-centered worldview.

Harrison's cultural stages partially mirror Gyr's life-cycle stages. The Transactional culture appears to be the social counterpart of the Systematizing stage and the Self-Expression culture of the Venturing stage. The Alignment and Mutuality cultures seem to reflect different aspects of the Integrating stage. A cultural corollary to Gyr's Exploring stage is missing from Harrison's model, however. Perhaps few organizations at this early stage had sufficient scale and resources to hire Harrison as a consultant, so they did not appear in his sample. If we were to fill in a cultural corollary for the Exploring stage, it would be even more dependent than the Transactional culture. Because that dependence is based upon the more personal relationship to founders, an appropriate label might be *Familial*. A simplified version of cultural stages corresponding to life-cycle stages is depicted in figure 18.7. Table 18.1 summarizes the characteristics of these cultures, including the first, Familial stage.

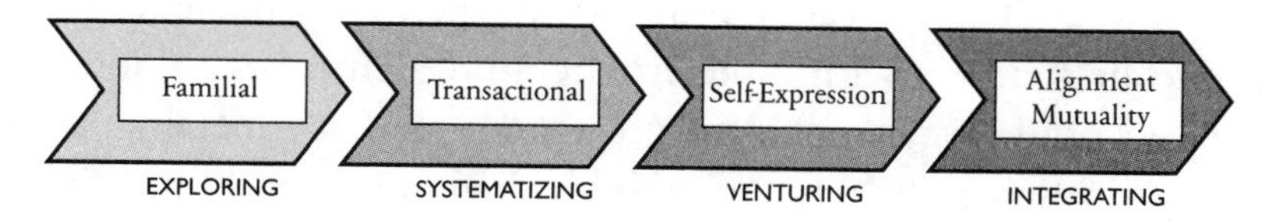

FIGURE 18.7. Life-Cycle-Linked Organization Culture Stages

If an organization allowed its culture to evolve with its life cycle, the cultural stages would likely evolve in parallel—although not necessarily reaching the Mutuality stage. While organization leaders can respond to the need for a new life-cycle stage by initiating new structures and processes, such an effort must work against cultural drag. The culture of the organization, like the values and beliefs of the individual, must change organically based on new experience—from the inside out. The prior cultural stage would tend to persist for at least some period into the current organization life-cycle stage.

Culture	Source of Satisfaction and Rewards	Values
Familial	Leader/founder	Approval, dependency, control
Transactional	Others	Exchange, hierarchy, dependency control, and compliance
Self-expression	Self	Creativity, pleasure in doing, competition, self-development, empowerment
Alignment	Achievement of tasks, goals, and shared purposes	Idealism, dedication, voluntarism, focus
Mutuality	Relationship, caring, and connection	Reciprocity and cooperation, consensus, giving and sharing, empathizing, considering context and stakeholders

TABLE 18.1. Key Characteristics of Organization Cultures

Such natural evolution is a rare event. Though originally tied to the organization life cycle, all five cultures have now been established by broad societal experience and are available in the larger social environment as what Harrison calls cultural archetypes. This enables organizations to be founded with a preferred culture that tends to dominate regardless of life-cycle stage. This preferred culture is generally the one that most closely reflects founders' individual mind-sets. In practice, an organization's culture may differ from its current life-cycle stage in a variety of ways. Such misalignments produce two kinds of challenges.

Recently founded organizations born as Self-Expression, Alignment, or even Mutuality cultures cannot return to cultures that appear earlier in the developmental progression. However, they must pass through all of Gyr's organization life-cycle stages. These organizations tend to resist devoting time and attention to organization life-cycle stages that correspond to earlier stages of cultural development. They may fail at these earlier stages or move through them too quickly to build sufficient foundation for success at later stages.

Older or more traditional organizations founded with a Transactional or Self-Expression culture face a different challenge. They must move through the organization life cycle with a culture that tends to obstruct the structure and process changes needed for Venturing or Integrating. Culturally resonating with earlier life-cycle stages, these organizations risk stalling developmentally. Cultural challenges can also occur for the client organization when other organizations in its ecosystem do not align with its culture. The professional focusing on Type III work can offer assistance to client systems in meeting the challenges involved in the tensions between culture and life-cycle needs.

Consultant-Client Misalignments

Just as culture may present the knottiest developmental challenges for an organization, it may present the greatest potential blind spot for the professional attempting to provide value to a client system. Even if a professional recognizes that the client organization's culture calls for a different approach than his own firm's culture would dictate, he may not recognize all the aspects of his process, service, and product that express his firm's culture.

Many professional service firms have a Self-Expression culture. Can you think of times when, in the context of delivering technical work, professionals in your firm have encountered great discomfort from clients who felt that they just weren't getting enough information and documentation—that things were out of control? These may have been situations in which a client organization's Transactional culture went unrecognized. Conversely, can you think of times when professionals in your firm were seen as not sufficiently sharing the goals of the client—not really being the team players the client wanted? The client organization culture in these cases may have been in the Alignment stage while the professional remained oriented to Self-Expression.

Pitfalls for professionals providing Type II or Type III work are just as serious. Imagine a professional from an Alignment culture designing project processes that subject clients in a Self-Expression culture to more group process than they can tolerate. Or a Mutuality culture professional leading processes for a client organization struggling into the Venturing stage with a still Transactional culture. How would clients in this organization feel if this professional could not help telegraphing her impatience with their high need for linear process and control in the face of the anxiety they are experiencing?

To provide optimal value for a client system, a professional must understand the culture of the client organization in the context of its life-cycle stage. She must also recognize the culture of her own firm and how that culture manifests itself in all that she and her colleagues do. And she must integrate this cultural awareness into her developmental understanding of specific relationships and groups operating in the project context. We believe that the majority of consultant-client relationship casualties occur in situations where the consultant is unaware of or does not know how to respond to misalignments between her firm's and the client's organization cultures.

At the same time, the mind-sets of individuals may diverge significantly from the one embedded in their organization's culture. A professional also needs to be aware of her own mind-set and that of the individual clients with whom she works. Such divergence—on either the consultant or client side—can produce misalignment even when the organization cultures are aligned. On the other hand, professionals can use a workable degree of diversity of individual mind-sets in their own firms to provide staffing that can be effective across a broader range of client mind-sets and organization cultures.

Organization culture and individual mind-set alignment are not always required for client satisfaction, however. A professional operating from an Interdependent mind-set can work effectively with a client in an Independent or Dependent mind-set or culture, though he must take great care in honoring where the client is. He may derive less enjoyment in these cases than in working with clients with an Interdependent mind-set or culture, but he can be quite effective. Working in the other developmental direction across the mind-set stages, however, is unlikely to produce good results. A professional operating from an Independent mind-set is likely to struggle mightily in understanding and meeting the needs of a client operating from an Interdependent mind-set or culture.

Let's summarize our exploration of this essential way of thinking about human systems over time that we are calling developmental awareness. The three types of work require cultivating developmental awareness at different levels and to different depths, with Type I work emphasizing relationships, Type II emphasizing groups, and Type III organizations and organization ecosystems. From the point of view of fit for relationships, services, and products, however, all three types of work benefit from developmental awareness at all these levels.

As we shall see, the Type III work of transforming the architecture or engineering firm can also be profoundly enhanced by developmental awareness. Let's turn now to the other Type III leadership capacities that must be coupled with leading innovation to successfully transform your firm.

19 Transforming Your Firm

The stakes for us are raised considerably by . . . the traumatic nature of the adaptive process. An old order falters, dissolves, and (if fortune smiles) a new order emerges, more complex and better able to cope with the changing environment. If we were simply rocks, without knowledge and feelings, the process might be negotiated with equanimity. But we do have knowledge (awareness) of our past and present, and typically we have very strong feelings about both. Truthfully, we do not like losing what is old and familiar, and typically will resist such change to our dying breath, in some cases quite literally. Yet for the adaptive process to succeed we must let go of the old in order that the new may appear.

—Harrison Owen[188]

If you have attempted significant change in your firm, you know that the process is challenging. When you push in the direction of the desired change, the system pushes back! Living systems reveal an amazing paradox. On the one hand, they are always changing in response to their environment—they are designed to be responsive and adaptive. On the other hand, they are oriented to homeostasis. Except under special conditions, they seek return to an equilibrium state. Human systems share this paradox with other living systems, while the human capacity for self-awareness adds conscious intervention potential to less conscious layers of adaptive process.

One of the more useful insights into change in human systems is the notion that people don't resist change, they resist being changed. Many change-related hurdles have been successfully crossed by involving members of an organization wishing to change in determining both the nature of that change and how it will be implemented. In fact, this insight is one of the many streams feeding the movement toward participation and collaboration in all things to do with organizations and communities. At the same time, studies of the implementation of major programs that require some degree of culture change—and involve at least some participation—have all indicated a failure rate of 70–75 percent.[189] Participation may be a necessary but not sufficient condition for effective change.

One of the reasons participation tends not to be sufficient is wide individual variance in comfort with change. As several personality-typing instruments have highlighted, orientation to change varies across populations in much the same way that innovativeness varies across the adopter categories.[190] Even when individuals participate in planning and implementing change, they will exhibit a wide range of motivation and capacity to implement it at a personal level.

A second reason participation tends not to be sufficient has to do with the nature of the change being attempted. If that change entails shifts in fundamental values, beliefs, or attitudes, even a change-friendly individual may balk—perhaps not on the surface, but below self-awareness. Some individuals will both choose and be capable of such change, and some won't. When it can occur, such change will likely take time.

A third reason participation may not be sufficient is that organizations are complex systems with emergent properties—characteristics present at higher levels of system that are not present at lower levels. Even if we populate an organization with individuals who have a preference for proactive change, we may find ourselves meeting significant resistance. Equilibrium-seeking may be operating at the larger system level despite the best intentions of individuals desiring to drive change.

The good news is that 25–30 percent of significant organizational change efforts do succeed! At least one study has indicated that 70 percent of these successes depend upon "a wake-up call."[191] This data corresponds to another characteristic of living systems that we have touched on both in Robert Shapiro's "strange attractor" approach at Monsanto and in Ralph Stacey's discussion of the five key parameters involved in optimal creativity. When environmental and internal conditions pull a human system out of equilibrium toward the "edge of chaos," that system becomes capable of astonishing change that can establish a completely new basis for equilibrium. While leaders can make proactive use of these conditions, this data highlights the key role that reaction to externally-induced stress tends to play in successful change efforts.

This same study indicated that over 90 percent of successful organizational changes were characterized by the following:

- Passionate commitment: conviction that the change is both desirable and possible.
- Clarity about goals and first steps to take.
- Disciplines, structures, and mechanisms to ensure repetitions.
- An unconditional support base.
- Integrity: truth to organizational values and furthering larger purpose.

These characteristics are by no means easy to achieve—particularly the unconditional support base! Note that these characteristics do not imply that these organizations have begun with understanding of the problems they have to solve. They have envisioned goals and identified initial steps. They have also identified disciplines and structures that will be practiced on a consistent basis. But they do not yet understand how they will achieve their goals other than by the practice of those disciplines, with integrity. These features of successful efforts reveal some of the proactive aspects of leading Type III work.

While some forms of business model innovation will require less transformation than others, we believe that in addition to leading innovation all will require at least some use of the other four leadership capacities for Type III work. You will need strategic conversation, systems thinking, and organization development to help determine which innovations to undertake and how best to undertake them. You will need leading whole-system learning and adaptive work for ongoing course correction and to evolve the core of professional practice in your firm—to lead the deep internal aspects of change that will be required to make your innovations successful. Building on our discussion of leading innovation in chapter 17 and developmental awareness in chapter 18, this chapter explores the other four leadership capacities as applied to successful Type III work in firms.

Strategic Conversation

Strategic conversation is consistent, informal discussion inquiring into and exploring strategic issues facing a human system. The capacity to lead ongoing strategic conversation is grounded in the mind-set of strategic thinking and draws on two core disciplines: strategic planning and scenario planning. Scenario planning is an adjunct to strategic planning that addresses environmental conditions of medium-to-high turbulence and uncertainty—essentially, the conditions that most firms find themselves in today.

In characterizing strategic conversation as one of the five key leadership capacities for Type III work, we are reflecting the sea change that has occurred in the strategic planning and scenario planning disciplines over the last 15 years or so. Both have migrated away from purely formal, event-based processes producing detailed plans that are often little used. In the face of rapid change and high uncertainty, business leaders have realized that such plans quickly become obsolete. The more powerful approach is to treat strategy as a discovery process[192] in which all members of an organization participate on an ongoing basis.

In this context, formal strategic planning events have not disappeared, but they have changed their focus. These events now seek to achieve deep alignment among organization members that prepares them to discover appropriately cohesive strategy through the organic processes of the organization's work. The questions that used to be posed when detailed plans were developed are still relevant, but the process of finding ever-emerging answers occurs through ongoing, informal conversation among staff and between staff and customers or clients.

The strategic planning discipline concerns itself with three fundamental sets of questions. The first set focuses on an organization's core identity and long-term aspirations. For a firm seeking to become or sustain itself as a living company, we suggest the model presented by Collins and Porras in *Built to Last* as guide to this first set of questions.[193] Vision in their model is based on the yin-yang relationship between those things about a firm that don't change—or change very slowly—and those that do change very dynamically. The unchanging yin is a firm's *core ideology*: its core purpose and core values. Core purpose is a firm's reason for being. This purpose should be general enough to be seen as underlying past and current businesses as well as having the potential to guide development of future businesses that may be needed in order to adapt. Core values should express the enduring DNA of the organization. The yang, a firm's *envisioned future*, also consists of two parts. A "BHAG" (big, hairy, audacious goal) addresses the very long term, often as long as 30 years. A vivid description envisions the state a firm will be in when its BHAG has been achieved. Vision as these authors conceive it thus encompasses both who we are and what we aspire to as an organization.

If a firm has not already done strategic planning that addresses these questions of core identity and aspiration, it should consider doing so—and strategic conversation should be used to plow the ground for that effort. If a firm has already formally answered these questions, strategic conversation should constantly refer to core purpose and core values, while relating current issues, challenges, and possibilities to the envisioned future. Strategic conversation thus provides ongoing interpretation and animation of your firm's vision as a whole.

The second fundamental set of strategic planning questions centers on what is going on in an organization's environment at a given time in order to identify opportunities and threats. Trends impacting client organizations and involving the firm's competitors must be surfaced through continual streams of market intelligence from people in the firm as well as other sources. While formal strategic planning relied on formal intelligence gathering, strategic conversation assumes that members of a firm already possess quite a bit of market intelligence. The conversation provides a venue for sharing, consolidating, and making sense of that intelligence, and for motivating firm members to pursue more of it, more proactively. Integrating the organization ecosystem model into reflecting on what is going on in the firm's environment can provide a powerful new context for understanding opportunities and threats. In a recent *Harvard Business Review* article, "Strategy as Ecology," Marco Iansiti and Roy Levien elaborate a helpful framework for understanding whether a firm is currently positioned as a commodity, niche, or keystone player in its ecosystem and what kinds of opportunities and threats tend to go with each position.[194]

The third set of strategic planning questions focuses on what modified or new directions the firm should take based on its aspirations and current environmental conditions. Some of these directions might be aimed at mitigating perceived threats, but most should seek to take advantage of perceived opportunities. As traditional SWOT (strengths, weaknesses, opportunities, and threats) analysis has always guided, current strengths and weaknesses should also play into the consideration of a firm's mission or current business model: what businesses it wants to be in at a given time, in what markets, with what kind of clients, providing what forms of value, and with what kind of approach.

Integrating the value network model into strategic conversation, together with accurate feedback about the organization from its marketplace, can sharpen understanding of both the firm's current state and new opportunities. Inquiring into the true value criteria that clients are currently using to buy—or not buy—a firm's services can expose significant gaps that should be factored into possible modifications of the firm's current value proposition. Mapping the value networks of desired clients can reveal opportunities to meet currently unmet needs that might not be seen any other way.

Before conditions were quite so turbulent, firms could extrapolate in fairly straightforward ways from current trends to future probabilities. In today's conditions of rapid change and unpredictability, incorporating the discipline of scenario planning into strategic conversation is almost a necessity. Scenario planning adds two core questions to the strategic planning discipline. The first of these is, given the critical uncertainties in our environment today, what alternative futures can we imagine? Arie de Geus, one of the pioneers of scenario planning, characterized its key value as the capacity to create "memory of the future."[195] His point was that human perception is constructed in such a way that we can only perceive based on patterns already established in the brain. We literally cannot see new phenomena coming at us if we have not prepared our brain to see these phenomena.

Peter Schwartz's *The Art of the Long View* is still the best primer on the scenario planning process.[196] Rigorous analysis of local and global trends related to the strategic issue you are considering forms the basis for identifying driving forces, predetermined elements, and critical uncertainties. You then combine the critical uncertainties in different ways to produce the foundation for very different alternative futures. Imagination assists you in weaving these foundations together with other driving forces and predetermined elements to create compelling narratives of these alternative futures. The narratives are critical, because without them, you cannot see or communicate the unanticipated directions that might unfold or consider their potential impacts on your firm. The narratives also help you identify leading indicators to monitor whether a given scenario may be emerging.

The second scenario-planning-based question is, given those alternative futures, what strategy could your firm pursue that would enable it to thrive in all or the majority of these possible futures? At this point, you examine all the modified or new strategic directions you are considering in the light of these imagined futures. How many are they viable in? If only one or two, what alternative or contingent strategies could you consider that could cover you in the nonviable futures?

Because fewer professionals are familiar with scenarios than with strategic planning, you will probably need to provide some formal learning experiences. Once people become familiar with both the rigorous trend analysis and the imaginative part of the process, they can more easily integrate these dimensions into the informal processes of strategic conversation.[197]

The most difficult challenge you are likely to face in leading strategic conversation in your firm is battling preference for tactical thinking alongside relentless demands of project work. Staff may understand the importance of strategy at strategic planning time, but may not see the profound necessity of ongoing strategic conversation to promote continuous adjustment to changes in the firm's environment. One approach is to identify key moments and contexts when staff are more likely to see the value. Deciding when to invest in developing a new relationship without yet having a project or whether to go after a particular project provides one type of strategic conversation opportunity. Attempting to learn from your failure to win a project or from a client's decision not to offer your firm repeat business can lead to a different type of strategically rich conversation. Another good approach is to insist that management or other teams dedicate separate meetings to strategic conversation. When such conversations coexist on an agenda with operational concerns, they will tend to be crowded out.

The professional developing the capacity to lead strategic conversation must become well versed in the strategic- and scenario-planning disciplines—as able to make provocative contributions to strategic thinking as to articulate the endless stream of questions required for the process to bear its richest fruit. As in all leadership, setting a good example is critical.

Systems Thinking

The organization ecosystem and value network models are good examples of the mind-set of systems awareness and of the first level of systems thinking required for leading transformation in your firm. They provide a means of seeing how separate entities are connected in a larger picture of interdependent relationships and of seeing how the firm fits into that system of relationships. When considering changes inside their firms, professionals must take into account how these changes will ripple out into external relationships. They must also consider how changes in one part of a firm's system will affect changes in other parts.

For this purpose, a working model of the firm's subsystems can be very helpful. Using organizational structure as the sole organizing principle for such a model produces a familiar set of considerations: how will a change in one office affect another, or a change in corporate departments affect business units? More helpful is a multidimensional model such as the one created by Chuck Schaefer at Chevron in the early 1990s. He defined the organization's subsystems as depicted in figure 19.1.

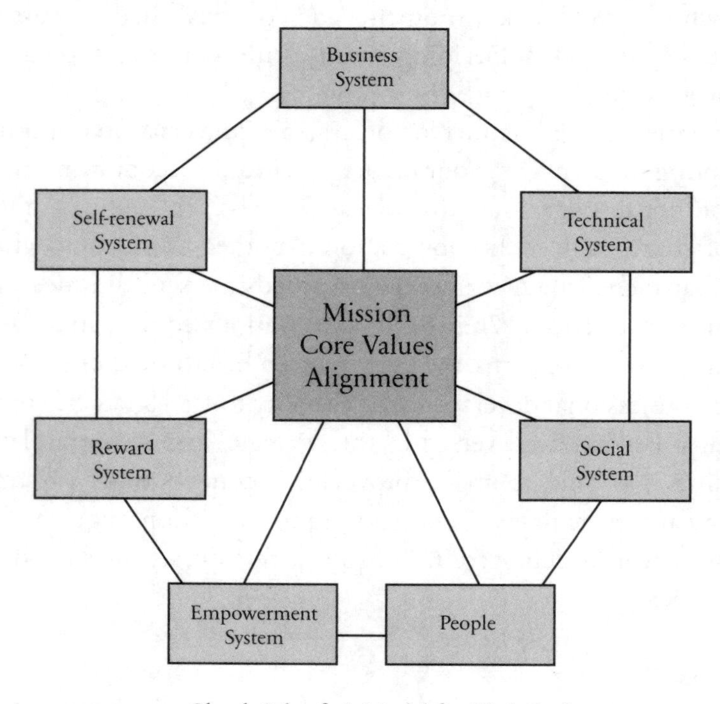

FIGURE 19.1. Chuck Schaefer's Model for High Performance Work System Planning and Development[198]

Consistent use of this kind of systems model for considering the impacts of decisions related to transforming your firm can help reduce unintended consequences. This first level of systems thinking can be further enhanced by considering the "Laws of the 5th Discipline" introduced by Peter Senge in *The Fifth Discipline.*[199] These include such counterintuitive insights as: "Today's problems come from yesterday's 'solutions,'" "Faster is slower," "Small changes can produce big results—but the areas of highest leverage are often the least obvious," and "There is no blame."

To take full advantage of these insights, however, a professional must move to the second level of systems thinking, based on the causal loop diagramming that Senge introduced as the heart of the discipline he called "the cornerstone of the learning organization." Well over a decade after Senge popularized it, practice of this discipline is building, but adoption remains slow. His other four disciplines—personal mastery, mental models, shared vision, and team learning—have made more rapid inroads into organizational leadership and management culture. While we have made significant use of causal loop diagramming in our own work, we have invested a number of years learning and practicing it. Most of our clients have found this effort too daunting.

We believe, however, that causal loop diagramming is one of the few disciplines available that provides a means of understanding the complexity we are confronting in our emerging world. As many scientists have observed in different ways, searching for simplicity in our understanding of how things work is essential—but not too much simplicity, and not forms of simplicity that fail to reflect the complexity we observe. Causal loop diagramming has proven its power in part by identifying generic "patterns of system behavior," called archetypes,[200] that have appeared across time and across all kinds of systems. Taking the first step into the second level of systems thinking may involve understanding archetypes such as "Limits to Growth," "Shifting the Burden," or "Success to the Successful" and learning to sense when they are operating in your firm.

Causal loop diagramming involves looking beneath our awareness of interactions—which this context calls the level of events—to uncover the underlying dynamics of system behavior. The elements of these patterns are aspects of the system that vary over time—say, profitability or number of employees—and the ways in which these variables influence one another. Since everything is ultimately connected in systems through mutual influence, feedback loops rather than linear flows are used to describe these patterns. Two kinds of loops operate everywhere in living systems: reinforcing loops that drive growth or decline, and balancing loops that hold these reinforcing dynamics within limits. Reinforcing and balancing loops form the basic language of causal loop diagramming.

Anyone who has experienced a price war has felt the effects of a classic reinforcing loop. Though the short-term effect of a firm lowering its fees may be to win the competition for a piece of work, the long-term effect of lowering fees is to make it difficult for all firms to charge fees that provide reasonable margins. In other words, lowering your fee actually produces more pressure to lower your fee. Choosing the short-term benefit of winning a particular job without understanding how the variables in the underlying dynamic drive each other causes the participants in a price war to shoot themselves in their collective foot. One balancing loop in this kind of system includes the variables involved in firms' inability to sustain lack of profitability over time. Eventually, the effect of this balancing loop will kick in, or at least some firms will die.

We all have grasped such dynamics intuitively on occasion. However, few of us regularly think this way about important problems or decisions. A professional seeking to lead transformation in her firm can gain a clearer understanding of the dynamics driving the current state of the firm and any problems that must be addressed by beginning a personal practice of this second, more challenging level of the systems thinking discipline. This practice must involve identifying key variables that bear on an issue, constructing reinforcing and balancing loops from these variables, and uncovering mental models that drive relationships between the variables. Only when we become aware of these mental models can we see how our prior solutions have created today's problems—and how current thinking will only perpetuate those problems. Only through this kind of reflective process can we identify ways to do things differently that will actually make the difference we seek.

At the same time, no individual has enough information and experience to understand the system as a whole. Because taking on firmwide issues requires many different points of view, a professional must bring others along in the practice of the discipline as well. This leadership requires developing the ability to teach others to think in the new ways required, and enough mastery of the team learning and personal mastery disciplines to effectively engage viewpoints very different from her own in potentially charged contexts of group problem solving.

In addition to helping professionals understand the problems and issues driving the form that their innovations and evolutions need to take, the second level of systems thinking provides a powerful way of working with the dynamics of transformation itself. Nine years after the publication of *The Fifth Discipline* and five years after *The Fifth Discipline Fieldbook*, Peter Senge published *The Dance of Change*. This third book was a comprehensive attempt to learn from the experience of achieving no greater success rate than other methodologies attempting profound change since the 1970s—the 25–30 percent already mentioned.

At the heart of *The Dance of Change* is a systems-thinking-based approach useful to anyone attempting to lead transformation. The approach involves recognizing that core reinforcing loops drive change and that balancing loops constrain these drivers. The key to freeing up the reinforcing dynamics is to mitigate the limiting processes inherent in the balancing loops. *The Dance of Change* provides extensive advice for applying this strategy to the limiting processes that arise at all stages of transformation.

Organization Development

The leadership capacity of organization development helps a professional lead transformation that achieves sustainable success because it honors and flows with the natural dynamics of the firm as a living system. Two main thrusts are involved. The first entails choosing strategic directions appropriate not only for the core identity of the organization, but also for its life-cycle and culture stages. The second entails pursuit of those directions in ways that promote healthy movement of the firm through its life cycle and, in the spirit of the living company, support the realization of the firm's full potential in reaching its aspirations. This second thrust depends upon effectively facilitating development of the firm's culture.

As mentioned in chapter 18, the dominant culture of a given organization may not correspond to its life-cycle stage. Because cultural preference is so tied to individual mind-set, the first time an organization passes through the life cycle, the mind-sets of the founders will tend to override the natural culture for the stage unless the founders evolve their mind-sets with the stages. Culture stages—like individual mind-set stages—don't cycle but form a linear progression. On subsequent passes through the life cycle, an organization will tend to retain the culture stage to which it had developed in the prior pass—unless it is overridden by the mind-sets of the current leadership group. Figure 19.2 unfurls the life-cycle stages into a linear progression and uses the Self-Expression culture as an example to show the variety of relationships that can exist between organization culture and life-cycle stage.

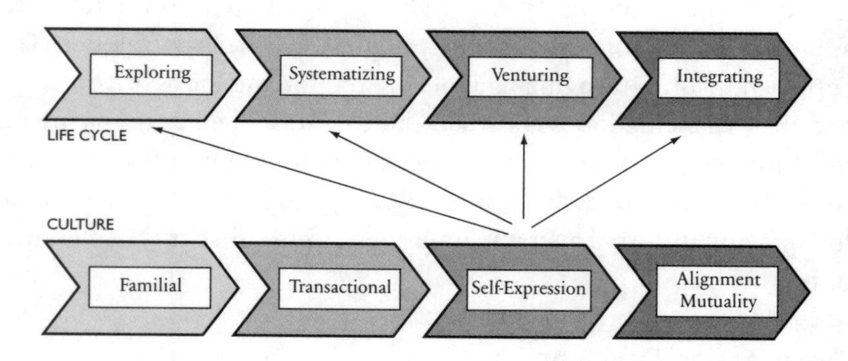

FIGURE 19.2. Relationships Between Organizational Life-Cycle and Organization Culture Stages

Every firm is unique and unfolds through the life-cycle and culture stages in a unique way. At the same time, firms exhibit a great deal of commonality in the deep archetypal patterns of the developmental stages. If you are reading this book, there is a good chance that your firm is somewhere between the Venturing and Integrating stages in the organization life cycle. If so, you are likely struggling with services that have reached the later stages of maturity on the technology S-curve of the industry as a whole. If so, you know you are struggling because you are losing clients or market share of which you were formerly confident. Given what your clients are paying you, they may feel that you are failing to

- Achieve what they would consider sufficient creativity and innovation in your solutions,
- Provide the collaborative co-creative process that they are looking for given the social complexity of their projects or their orientation to experienced-based value,
- Deliver product and service quality that are minimum expectations—simply assumed—in this market.

Product or service quality problems may stem from a staff top-heavy with senior people pressured to market and do damage control and with too few intermediate people to oversee and coach the inexperienced juniors. To deal with the aging maturity of your current services in general, you may have incubated a new business or two that you likely pursue through separate entities that are currently in the Exploring stage of the organization life cycle.

If this is your scenario, your cultural center of gravity is likely still in Self-Expression, even as you are trying to foster greater Alignment in order to move effectively into the Integrating life-cycle stage. This culture minimizes the legacy of the Systematizing stage, which it may have caused to be weak in the first place. It may thus run substantial interference on any attempts to address product and service quality problems with standardization-based approaches. This culture may at the same time make it very difficult to move to more collaborative service delivery that gives clients a greater role in creative outcomes.

While the Self-Expression culture may help you run out the tail end of the S-curve for your mature business, it is unlikely to help you achieve renewal because it is so steeped in the mind-sets that must shift in order to catch up with most clients and stakeholders. From this perspective, you would best serve your firm by beginning to shift this culture based on the Independent mind-set to an Alignment culture based in the Interdependent mind-set. Yes, we are recommending the long-term work of culture change—profound change in Senge's terms.

One of the great challenges we have seen among firms attempting to make this shift is that individuals in the Independent mind-set see the Alignment culture as harmfully opposed to their own self-expression. They believe they must give up their individual creativity in order to team effectively. When these individuals participate on teams, team activity is often ineffectual because they believe they must exercise their individual creativity separately—and they continue to do so in maintaining their former activities. When this transitional phenomenon occurs, teams take up time and add little value, having the sole effect of making the former way of doing business less effective.

Contrary to this tendency of initial thinking, teams are only effective when they fully leverage everybody's individual talents. Professionals must come to understand that moving to the next cultural stage does not mean giving up the prior one, but incorporating it into a stronger and more powerful synthesis. Your initial and ongoing communications regarding your vision and strategic plan set the stage for strategic conversations throughout your firm that can help develop this awareness. Coaching by Early Adopters or individuals who are already operating from the Interdependent mind-set can show how that mind-set can add value to existing work situations. However, the challenge we have described reveals how difficult it is for many individuals to go through the deep internal change of mind-set shift. Achieving this deeper basis for culture change requires more than communications and coaching. We'll explore these additional dimensions under the remaining Type III work leadership capacity, leading whole-system learning and adaptive work.

The change from Self-Expression to Alignment culture can take quite a while, and of course, not everybody comes along at the same pace. In the meantime, if you have a vibrant Self-Expression culture that still has lots of life left in it, you might want to start with technical innovation as your core innovation strategy. The use of collaborative processes for technical innovation, such as those employed by IDEO, could act as a bridge to an Alignment culture by demonstrating the value that the group adds to individual creativity, and vice versa.

The professional firm trying to move from Venturing to Integrating with a still dominant Self-Expression culture is only one scenario, however. If your firm is in Exploring, you have the opportunity to build in a new value proposition from the outset. If your firm is in Systematizing, you may want to reevaluate how much life is left in your value proposition. Is it low or high on its technology S-curve from an industry perspective? You might also want to check how much geographic expansion looks possible and desirable. If the life of your current value proposition seems limited, you might want to build another one in parallel.

If your firm is in Integrating, you are in a strong position to undertake a wide range of innovation efforts. The form of innovation you choose in these situations, however, should also depend on your dominant culture. With a Transactional culture, you might want to explore the mass customization arena as a starting point. With an Alignment culture, you might choose innovating in the social process of design or problem solving, or moving into Type II work. And with a Mutuality culture, you might want to consider Type III work, realizing that you would likely need some interim steps.

As you choose clients with whom to innovate, consider the stage of their organization ecosystem and its appropriateness for your strategy as well as for your organization life-cycle stage and culture. As you experiment, you will probably need to iterate back and forth between your aspirations and your current state, adjusting for what is possible at a given time. At the same time, experimentation will likely take you in the direction of multiple efforts. Stewarding an organization that consists of significant centers of more than one culture is a challenge and a risk. Depending on your size and regional economy, however, you may want to consider more than one business with more than one business model in more than one market—resulting in more than one culture—especially as a transition strategy.

As you move forward, you need to attend not only to where the organization is developmentally, but to how fast it is moving or needs to move. The pace at which life-cycle stages unfold depends on conditions in the organization's environment. The more quickly the S-curve of a firm's value proposition develops, the more rapidly it must pass through its stages. At moments of environmental stress especially, firm leaders may tend to back away from the edge of chaos where the most creative evolutionary leaps can be made—leaps that respond adaptively to environmental conditions while taking the organization forward in its development.

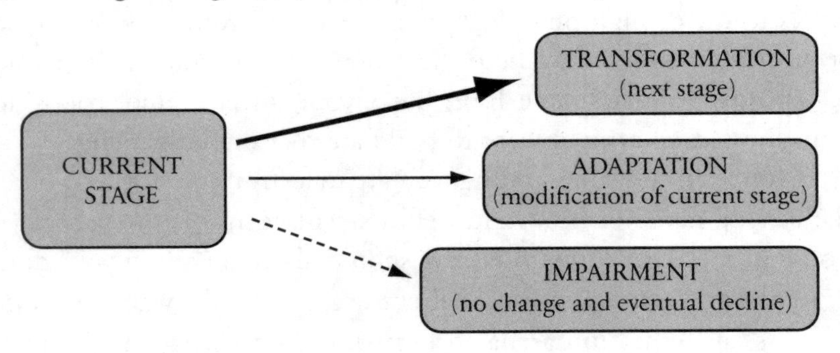

FIGURE 19.3. Developmental Turning Points in the Organizational Life Cycle[201]

In addition to defining four stages, Gyr's organization life-cycle model addresses the dynamics by which a firm moves from one stage to another—or does not. Toward the end of each stage, its limits begin to emerge and eventually produce a developmental turning point when "the organization's leadership can choose among three options: ignore the emerging need and change nothing (impairment), address the need but remain basically the same kind of organization (adaptation); or develop a new way of functioning (transformation)." Figure 19.3 shows these three possible paths, and table 19.1 identifies the three versions of these paths that open up at the end of each stage.

Stage	Impairment Path	Adaptation Path	Transformation Path
Exploring	*Chaotic Organization:* Cumulative lack of coordination produces burnout	*Responsive Organization:* Flexible case-by-case approach provides more reliability	*Systematizing Stage:* More consistent processes, roles, and responsibilities
Systematizing	*Rigid Organization:* Bureaucracy suppresses individual creativity	*Operative Organization:* Employees are engaged in continuous improvement efforts	*Venturing Stage:* Individuals make unique contributions in context of solid systems
Venturing	*Reckless Organization:* No one is responsible for unproductive investments and efforts	*Empowered Organization:* Diverse efforts are coordinated through reporting structures and processes	*Integrating Stage:* Overarching purposes and goals unify efforts
Integrating	*Overzealous Organization:* Insider focus cuts off from clients, stakeholders, and community	*Aligned Organization:* Key outside parties are involved in core functions	*Exploring Stage:* Difficulty establishing close external partnerships leads to experimentation with new organization functions

TABLE 19.1. Developmental Paths from the Organizational Life-Cycle Stages

All change efforts must start with where an organization is and be undertaken with the utmost respect. However frustrated you may be with your firm's challenges, it is where it is for good reason. With and from its unique identity, it has evolved over time in response to its external and internal environments. Whatever forms of transformation a firm may be pursuing, it is unfolding through its stages, experiencing their predictable limits, and facing developmental turning points. The leadership capacity of organization development enables you to understand your firm's current stage and to recognize the external and internal dynamics of developmental turning points. It enables you to help your firm avoid impairment and choose appropriately between adaptation and transformation—especially at times of environmental stress—to optimize your firm's developmental health. A developmentally healthy firm is creative, resilient, adaptive, and primed for longevity.

Leading Whole-System Learning and Adaptive Work

Business model innovation and practice evolution through culture change will likely be inextricably bound up with one another in your transformation effort. Both are Type III work: the leader fully understands neither the problem nor how a solution can be reached. Such work can proceed only through a process of experimentation. Effective experimentation depends on the capacities for ongoing learning and quick course-correction. We might envision this process as the learning loop shown in figure 19.4.

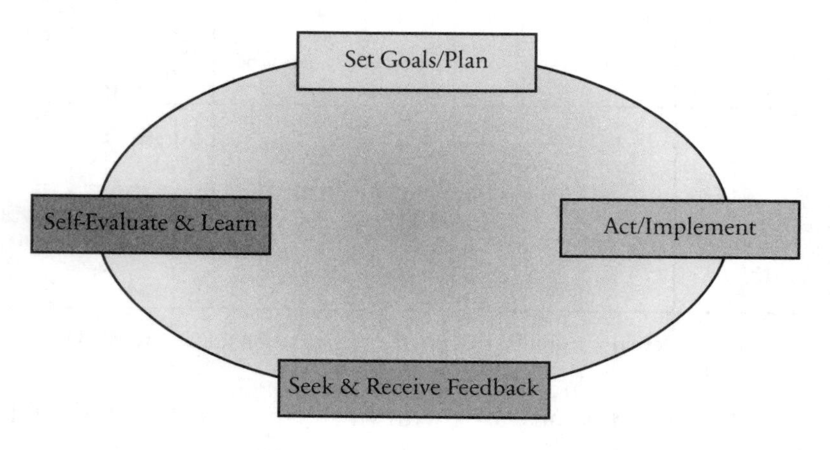

FIGURE 19.4. The Learning Loop

To be effective, this loop must be closed consistently. While individuals may manage to incorporate what they have learned into their goal-setting and planning with relative ease, enabling learning to pass from one individual or group to others is a much greater challenge. Whole-system learning requires processes for sharing learning across the organization. An even greater challenge is to enroll people in the change efforts to get the loop rolling in the first place.

Commitment to change spreads through your firm through reinforcing processes. A key strategy for initiation is to use what we know about the diffusion of innovations to drive these processes. Once you have involved as many people as practical in developing your initial steps and communicated those steps to everyone, encourage those who are excited about the change effort—the Early Adopters—to take leadership roles. As we discussed relative to the systems thinking capacity, however, the reinforcing processes of change are always coupled with balancing loops that limit them. *The Dance of Change* identifies archetypal limiters that arise in most organizations—such as "we don't have enough time" to address the change initiatives—and suggests multiple strategies for dampening their effects. We highly recommend this approach to professionals leading transformation in their firms.

Some of the learning that needs to take place relative to your change efforts will be of the single-loop variety: how to do better something that you are already doing. But a great deal will likely be of the double-loop variety: understanding that what you are already doing is not going to get you where you want to go—that you need to do something entirely different. This kind of understanding involves becoming aware of and being willing to shift your current mental models. While these mental models reside in your organization's culture, they ultimately operate in the minds of individuals. To transform your firm, you will need to be effective at leading individuals through changing their mental models as well as at leading the whole system through culture change. As your change effort unfolds, you may find dynamics at the individual mind-set and organization culture levels inhibiting one another at times, building on one another synergistically at other times.

Promoting individual development of the personal mastery and mental models disciplines can lay the groundwork for deep individual learning and change. Practicing the team learning discipline through dialogue can provide contexts in which individual learning can be deepened. We will have more to say about personal mastery and team learning in chapters 20 and 21, respectively. Peter Senge explores the conceptual foundations of these disciplines in *The Fifth Discipline* and, with a number of colleagues, provides much guidance for their development in *The Fifth Discipline Fieldbook*.[202]

While we encourage professionals to use Senge's learning organization methodology as much as possible, two inherent limitations may make it insufficient to transform your firm. First, the methodology can only be effective to the degree that members of your firm actually practice it. The disciplines are challenging. To individuals whose past, usually more reactive ways of responding to evolutionary challenges have seemed sufficient to them, the disciplines are likely to seem more trouble than they're worth. For the reasons Roger Harrison articulates, Alignment and Mutuality organizational cultures will tend to support practice of learning disciplines, while Transactional and Self-Expression cultures will likely have difficulty. In essence, these disciplines focus on relationships between and mutual effects among individuals and groups. This focus expresses and requires an Interdependent mind-set.

The second limiting aspect of learning organization methodology is its heavy reliance on reflective processes. Reactions to individual, group, and whole-system feedback are charged with emotions tied to current underlying beliefs and values. Ideally, reflective processes access these emotions. However, when emotions are threatening enough to remain unconscious, this access cannot be assured. Immersion in real-time experience may be needed to surface such emotions for reflection. As Heifetz says in his second book, "To meet adaptive challenges, people must change their hearts as well as their behaviors."[203] If people are not emotionally ready for a particular awareness, they may be able to articulate that awareness and even engage in new behavior, but will not be able to dwell in and act from that awareness on a fundamental level.

"Adaptive work" as Heifetz defines it can effectively address many situations that learning organization methodology cannot, particularly in the context of Transactional and Self-Expression cultures. Though Heifetz's approach requires the leader to engage in disciplined practice of key principles, it requires no formal disciplines to be practiced by other system participants. And though its messy and conflictual processes can produce reflective awareness, they do not rely on it. Heifetz's approach is certainly not infallible, and some firms may simply not be ready for profound change. However, by insisting that people work through to their own resolutions without the help of defined practices, Heifetz's adaptive work tends to engage people exactly where they are. In so doing, it has a good chance of accessing the underlying stream of emotional experience required for change at the deepest level.

Four key insights form the core of Heifetz's model. The first is the role that formal authority plays in human systems, no matter how participatory and nonhierarchical they have become. Critical functions of formal authority include choosing the direction of group movement, protecting the group from dangers at the boundary, orienting members to their status and place, controlling conflict, and maintaining norms (including resource allocation). Internalized norms enable members to deal with familiar issues and work, but when stress brings new challenges for which members are not currently equipped, members look to the leader to provide guidance and reduce the stress. This behavior of members endows the leader with authority to guide the system in responding to the stress.

The second insight is the central role that navigating conflict plays in successful adaptation. Stress produces conflict, and conflict's appearance is often an indicator of the need for adaptive work as well as of the issues that work needs to address. Consequently, conflict is both the context within which and the means whereby the adaptive work must be done. The third insight is the tendency of human systems to avoid conflict and the stressful effort of working through it. Heifetz labels this tendency of groups facing Type III situations "work avoidance."[204] The fourth insight is that in order for adaptive work to be effective, change must occur inside the members of the human system and not be imposed upon them, by the leader or anyone else.

The inescapable implication of these four insights is that the leader must respond to two contradictory needs. On the one hand, he must respond to the members' expectation that he will reduce the stress on the system. To the extent that he serves at the consent of the members, he will be vulnerable to deposition by them if he fails to meet their expectations. On the other hand, he must resist the full thrust of the members' dependence on him to fix the problem and insist that members do the adaptive work. No wonder Heifetz invokes the metaphor of "the razor's edge" to describe the work of leadership. The key is the leader's ability to use his authority to create a container for the work that is safer for people in the system than being fully exposed to the challenge. Using the boundary of the container, the leader moderates the stress to a bearable level while keeping the pressure on the members of the system to do the required work.

Heifetz's central notion of regulating distress—keeping anxiety high enough, but not too high—corresponds to Stacey's presentation of how "level of contained of anxiety" affects the creativity of a human system, which we discussed relative to leading innovation. Similarly, Heifetz's premise of the role of authority in human systems reflects Stacey's conclusion that a human system cannot be optimally creative without a significant degree of power differential or hierarchy of authority. In addition, Heifetz's affirmation of the value of diverse views to adaptive work parallels Stacey's assertion that too little diversity in fundamental values and beliefs diminishes creative capacity. However, Heifetz does not refer to the other end of the complexity theory range: the case in which a very high degree of diversity in fundamental values and beliefs might make creative adaptive work virtually impossible. In transforming your firm, you may occasionally discover such a case.

According to Heifetz, adaptive work can be led from a position of either formal or informal authority, though leading with informal authority presumes working in relation to a formal authority holding the container for the system-as-a-whole. We will discuss leading adaptive work without formal authority in the context of Type III work in client systems in chapter 22. For purposes of transforming your firm, we will assume that you hold and can leverage some formal authority. Table 19.2 summarizes the principles of this discipline.

Group experiences are as essential to Heifetz's adaptive work as they are to learning organization methodology. Ultimately, however, adaptive work occurs one person at a time. Working with individuals in this approach differs from traditional coaching in two important ways. First, advice plays virtually no part. And second, though the leader may have a general idea of how an individual needs to develop—say from an Independent to an Interdependent mind-set—he has no idea of how this will look in this individual. The leader's work is to articulate the adaptive challenge, sharing information about problems, highlighting conflicts, and asking tough questions. Some individuals will choose and be able to respond creatively to this kind of regulated distress from their own personal edge of chaos. And others won't.

Get on the balcony
- Move back and forth between field of action and a larger view
- Look for patterns

Identify the adaptive challenge
- Listen to concerns inside and outside the organization
- See conflicts as clues
- Frame key questions and issues

Regulate distress
- Create a holding environment, including appropriate sequence and pacing of work
- Manage the rate of change
- Clarify business realities, key values, and new roles and responsibilities
- Expose conflict or let it emerge
- Challenge unproductive norms while preserving productive ones

Maintain disciplined attention
- Minimize work avoidance by counteracting distractions from the conflicts at hand
- Deepen the debate by asking the tough questions

Give the work back to the people
- Help managers let go of control and move to supporting employee responsibility
- Help employees take initiative in defining and solving problems
- Develop collective self-confidence

Protect voices of leadership from below
- Delay the impulse to resist voices of uncomfortable dissent
- Legitimize voices who may speak awkwardly, but have a key perspective
- Actively inquire into dissenting perspectives from subordinates and front-line workers

TABLE 19.2. Leading Adaptive Work With Formal Authority[205]

Like organizations, individuals move through their developmental stages in part in response to stress from their environment. In part II, we identified increased social and technological complexity as the driver for the shifts of mind we believe are necessary for catching up and for getting out ahead of clients and stakeholders. Harvard psychologist Robert Kegan has studied the mental stresses produced by the increased complexity of contemporary life and the stages of cognitive development that unfold to cope with this complexity.[206] Maureen O'Hara of the Saybrook Institute in San Francisco agrees with Kegan's view that we have experienced rapid emergence of mental challenges more complex than we have been used to—mental challenges that have profound consequences for our being as a whole.[207] Building on and helpfully extending the framework of general psychology, she summarizes three possible responses to specific challenges individuals may face.

One response is to deny the existence of the new complexity, to disconnect from it in order to remain in one's existing reality. Taken to an extreme, this represents a "psychotic" response, where psychosis is defined as living in an internal world that is disconnected from the collective perception of the community around us. A second response is to develop a coping mechanism that acknowledges the existence of the challenge, but does not really engage it. O'Hara refers to coping mechanisms that avoid full engagement with new complexity as the "neurotic" response. A third response, which she calls "transformational," involves assimilating and integrating the new challenge in such a way that it empowers rather than diminishes the individual's capacity to fully engage his more complex world.

O'Hara's three responses strikingly parallel the three developmental paths that Gyr believes an organization may take at the end of each life-cycle stage. In the context of natural progression to the inherent limits of a mind-set stage, increased complexity in an individual's environment catalyzes a developmental turning point. From the Independent mind-set, for example, an individual could simply deny the emerging complexity (a psychotic response), figure out a way to cope with the more complex environment while remaining in the Independent mind-set (a neurotic response), or make the shift to the Interdependent mind-set (a transformational response). A leader doing adaptive work in his firm can use awareness of these potential responses to provide optimal developmental support to individuals as they find their way through the firm's transformation.

Professionals leading transformation in their firm should consider using both learning organization and adaptive work approaches to complement one another. In an Alignment culture and an Integrating stage, an overall learning organization framework for the transformational work will likely make sense. Heifetz's approach may be helpful in dealing with specific emotionally charged situations. In a Self-Expression culture and a Systematizing or Venturing stage, Heifetz's approach may provide a more effective overall framework. Within this framework, learning organization disciplines may begin to be selectively practiced as an additional spur to developmental progress.

Now that we have explored the new leadership capacities required for transforming your firm, we turn to the social capacities for providing higher value Type I work to your clients.

20 Social Capacities for Higher Value Technical Work

Building emotional intelligence happens only with sincere desire and concerted effort. A brief seminar won't help, and it can't be learned through a how-to manual. Because the limbic [emotional] brain learns more slowly—and requires much more practice—than the neocortex [thinking brain], it takes more effort to strengthen an ability such as empathy than, say, to become adept at risk analysis. But it can be done.

—Daniel Goleman, Richard Boyatzis, and Annie McKee[208]

Whatever new value-creation strategies a firm decides to pursue in transforming its business model, the capacity to provide technical work is likely to remain in the mix. As we mentioned in chapter 16, for those architects and engineers who will continue to provide Type 1 work, we don't believe that developing social capacities for delivering higher value is optional. This capacity-development effort is required to catch up with current expectations of clients and stakeholders in built environment projects. The social capacities we define correspond to a great degree to the social skills dimension of the emotional intelligence model developed by Daniel

Goleman and colleagues. Our first four capacities of *deep listening, assertive speaking, productive conflict,* and *relationship development* all depend on the self-management and social awareness that make social skills possible in the emotional intelligence model. Our fifth capacity of *learning and self-development* entails the most fundamental dimension of emotional intelligence: self-awareness.

In addition to drawing on new emotional capabilities, however, our social capacities are best grounded in a new cognitive orientation—the shifts of mind for catching up discussed in chapter 7. These shifts require individual development from an Independent to an Interdependent mind-set and flourish in a professional firm that has moved beyond a Self-Expression culture to one centered in Alignment or Mutuality. You may want to refer to table 16.1 to review the shifts of mind for catching up in the context of the social capacities for higher value technical work.

Some of our readers may feel that they have already been sufficiently exposed to at least some of these capacities, or that they are too basic to be discussed in a book about new models of practice. We start with these basics for three reasons. First, we believe that the current marginalized position of built environment professionals indicates that these capacities are not sufficiently present. Second, we believe that all of these capacities are life-long disciplines that no one ever fully masters. And third, we believe that the leadership capacities required for Type II and Type III work require a high degree of mastery of these fundamentals. We want to make sure we render the landscape comprehensively for those who are new to this material, and we hope to inspire those who have already spent some time developing these capacities to take their practice to the next level.

Deep Listening

The most fundamental social capacity is the capacity to listen deeply and well. It is probably the capacity we most take for granted, and yet it is probably the most challenging of all. Part of its challenge is its sheer complexity. First we must physically hear the sounds another utters.

Then we must assign meaning to the words to establish the literal meaning—a process that is interrupted if we don't speak the same language or dialect as the other person, or if the person speaks with a thick accent from another tongue. Then we must interpret that literal meaning in the context of the situation.

That interpretation depends on our giving broader meanings—to the words and to the shared context—that are similar to those the speaker gives to them. It also depends on our being able to integrate the layers of meaning—often the emotional context—provided by the tone of the speaker's voice, use of silence, and emphasis, as well as facial expressions, hand gestures, and body posture. That interpretation depends too on the assumptions we ourselves bring to it. These assumptions reflect values, beliefs, and reactive patterns (including emotions) that we are often not conscious of in the moment.

All of this presupposes that the speaker has in fact said what he means, which is its own complex process. That's a lot to navigate, and more than enough justification for developing the discipline of constantly checking for mutual understanding and for appropriate recognition of feelings as an aspect of the listening function.

Understanding the challenge of deep listening involves much more, however. Studies in human perception have shown that our primary perceptual process is pattern recognition. The brain matches patterns registering in the perceptual mechanism—ear, eye, nose, tongue, skin—with previously stored patterns. As we noted in the context of scenario planning to develop "memory of the future," we are wired to recognize what we already know a lot more effectively than to be able to take in something new. Some have estimated that 80 percent of what we perceive at a given time actually represents material projected from our brain rather than information coming in.

If information from the outside is too remote from our existing pattern recognition library, we will likely be unable to take it in at all. If such information is close to a pattern we recognize, however, we are at risk for collapsing it into what we already know rather than using it to add nuance and new pattern. This distinguishing is exactly what we are capable of, however, if we are willing to work at the conscious discipline of active listening. While we often think of listening as relatively passive, to do it well, we must transform it into a very active process demanding enormous presence, attention, awareness, and, yes, learning.

A key challenge to the effort of active listening is what Chris Argyris referred to as "defensive routines." In his seminal article, "Teaching Smart People How to Learn,"[209] Argyris argued that professionals actually have a greater learning disability than people who have less education. Because professionals are so invested in their expertise, they tend to be unwilling to act in ways that would suggest they do not know something, or are not on top of something they should be on top of. Defensive routines are habitual behaviors individuals use to demonstrate their knowledge and avoid taking in new information that might be inconsistent with it. Such behaviors prevent professionals from engaging in dialogue that would lead to learning. Defensive routines seriously undermine the potential of conscious, active listening.

In *The 7 Habits of Highly Effective People*, Stephen Covey wrote extensively about the discipline of what he called "empathic listening," based on the principle, "seek first to understand, then to be understood."[210] Empathy is certainly at the heart of deep listening. The most effective way to get beyond the ego's interference with our ability to tune in and connect with someone else is to put ourselves in the place of the other. The imaginative act of empathy literally lifts us out of ourselves and into the other person's shoes. From here, we can listen at a whole different level, one that resonates with the other person emotionally and thus honors him more deeply.

The other common foe of deep listening in the professional setting is the primacy of task. Of course, there are times when urgency requires us to pursue task at the expense of process or relationship. But to understand fully the nature of this core social capacity, we must recognize that the thrust to closure or the attachment to an outcome diminishes the professional's ability to hear anything else. With the capacity to detach from the desired outcome, as with genuine empathy, comes the opportunity for the emotional person, not just the thinking person, to enter the exchange. It is this emotional person who forms the relationship, and through whom the task-related exchanges flow.

Beyond overcoming defensiveness, maintaining empathy, and the willingness to temporarily detach from a desired outcome, we find perhaps the most powerful core of active listening: deep curiosity. The engine of active listening is inquiry: noticing all input that is not understood or that challenges our current understandings at any level, then formulating questions that engage the other person in helping us understand new or dissonant information more fully. Without the pursuit of artful questioning, the capacity for full exchange between the parties to a relationship remains only partially realized.

Our capacity for inquiry is best served by an ability to be aware of and move up and down what Peter Senge calls "the ladder of inference."[211] The ladder diagrams the process whereby we select data from our observations, add meanings to that data, make assumptions based on the meanings we add, draw conclusions based on our assumptions, adopt beliefs based on our conclusions, and take action based on our beliefs. We must be able to observe this process going on in ourselves—in effect, listen deeply to ourselves—at the same time that we listen deeply to another person. Only then are we able to inquire into that other person's process in ways that enable us to fully take that person in, and thus to fully connect in robust relationship.

Assertive Speaking

If deep listening is the cornerstone social capacity, assertive speaking is its equally essential cornerstone on the other side of the house. Why assertive? We use the term *assertive* as distinguished from *aggressive*, which we most assuredly do not mean. But why do we need a term of this ilk at all? If self-protection is a primary driver in professional settings, then the second most potent tool we have for self-protection after Argyris's defensive routines—which protect us from hearing challenging information—is holding back anything that might cause someone to challenge us. Thus, as Argyris's famous "Left-Hand Column" exercise[212] demonstrates, professionals generally go to great lengths to avoid sharing challenging information with others.

The most common kind of challenging information that we withhold is performance or behavioral feedback—in fact, the most important kind of information that individuals or groups need in order to learn. Leaving this kind of information uncommunicated to the parties who need to hear it leads to disastrous results—for relationships as well as projects. Professionals must be able to muster a good deal of courage as well as artfulness to be able to get over their natural tendencies to avoid this kind of communication. Hence our use of the term *assertive*.

Much of what makes giving performance or behavioral feedback so difficult is our knowledge of the negative emotions people typically experience when they perceive that they are being judged or blamed. At the same time, we ourselves may be experiencing a negative emotional charge based on whatever problem has emerged. The art of assertive speaking centers on stepping back from an emotionally charged impulse. In order to step back, we must understand that our thoughts are the product of our own perceptions and ladder of inference rather than the "truth." Because we are always operating on partial information, our conclusions are often based on erroneous assumptions. Assuming that our own analysis is correct is thus rarely appropriate. The assertive speaking process is founded on the need to respectfully compare perceptions, data, interpretations, and assumptions with the other parties. And it depends for its success on deep listening in this phase.

When we take the blame and judgment out and offer instead the so-called I statement—putting forward our perceptions for the purpose of being checked out and compared with others' perceptions—the need for a defensive response diminishes. At the same time, the potential for a deepening of the relationship, based on authentic information and exchange, increases. Viewed more broadly, assertive speaking can be understood to include not only challenging communications, but a holistic process of full self-expression in as close to real time as possible. We know that we are mastering this capacity if we rarely find ourselves holding back something that needs to be said, and if we find ourselves able to discover a way to say it that others can productively hear. That level of authenticity enables the professional to hold his ground amid project relationship dynamics and complexities that can feel disempowering. And that level of authenticity is the stuff of which robust relationships are made.

The other common assertive speaking challenge is delivering bad news that at least initially is not of a personal nature. Needed scope expansion or additional services, missed deadlines, or exceeded budgets are typical examples. Communicating these uncomfortable facts requires the professional to be accountable for what has occurred and take responsibility for what is to come. Assertive speaking in such situations also requires staying engaged as a full partner in the relationship, ready to work with what the other parties are willing and able to do in response. The more robust the relationship—significantly enhanced by assertive speaking and deep listening—the more likely bad news will be followed by healthy recovery. In its fullest sense, assertive speaking involves taking up appropriate self-responsibility, both at the level of communications and at the level of project work.

Productive Conflict

If parties to a relationship are using assertive speaking, disagreements are inevitable. When speaking assertively, we are not protecting each other from our natural differences, but bringing them out into the open. The ability to engage in conflict that is productive is thus a core social capacity. Though some cultures and families incorporate this capacity better than others, just about all of us enter our professional fields with some level of conflict aversion intact. As the recent work on emotional intelligence has shown, our emotions derive from more primitive parts of our brain than our cognition.[213] Emotional reactions precede cognitive function, and must be managed by it. For men, in particular, the natural response to conflictual situations is fight or flight. Other recent research shows that women also have a "tend and befriend" response,[214] but conditioned to perform in work contexts defined primarily by men, women, too, often experience fight or flight.

This instinctual response—leading to avoidance, defensiveness, and occasionally attack—undermines the flow of exchange and in many cases damages the relationship. Being able to channel conflict productively requires not only being able to manage our emotional responses to differences, but being able to reframe conflict as potentially highly creative. As Ralph Stacey's work has highlighted, living systems made up of individual agents require a sufficient degree of diversity in ways of thinking to be optimally creative.[215] Common sense confirms that groups who think too much alike tend to reproduce what they already know rather than create new things. The stimulation and challenge of difference is needed to open up new challenges, and hence new opportunities and solutions.

If we really take this in, we must conclude that conflict is a necessary condition for creativity. Defensive reactions to assertively expressed difference probably cannot be eliminated entirely. However, cultivating a welcoming attitude toward conflict can help us disarm those reactions internally so that we can quickly turn to bringing out the creative potential in the interaction.

As Roger Fisher and William Ury suggested in their now classic *Getting to Yes*, a fundamental step in productive conflict is to "separate the people from the problem."[216] As long as a person feels attacked, there is no relationship through which a productive exchange about the issue can flow. The key is to focus the disagreement on the issue, while staying connected to the person in the context of the relationship. If the problem is the relationship, it should be dealt with as such (see Relationship Development).

From a connected state, collaboration is possible. In collaboration—consistent with the mind-set of a nondualistic, inclusive view—all parties are interested in getting others' needs met as well as their own. The conversation organizes itself around everyone's concerns being addressed. Though passionate intensity is hopefully still present, it is now flowing beneficially in the service of this more inclusive goal. Collaboration thus creates the opportunity for a win-win outcome and the possible emergence of a new, co-creative solution. Along with *Getting to Yes*, Covey's 7

Habits is a good source for practicing this discipline, as are many books on teamwork that address processes for reaching consensus.[217] Two recent books entirely devoted to this subject are helpful as well: *Difficult Conversations: How to Discuss What Matters Most*[218] and *Crucial Conversations: Tools for Talking When Stakes Are High*.[219]

A win-win conversation benefits enormously from balancing impassioned advocacy with the reflective inquiry of deep listening. In addition to assertively stating their positions, participants listen deeply, moving up and down the ladder of inference to understand what is underlying others' positions as well as to reveal what is underlying their own. Very often, misunderstandings are exposed and common ground uncovered. From this new ground, new possibilities appear.

Peter Senge calls the art of balancing advocacy and inquiry to reach a resolution or outcome "skillful discussion." The quite different art of subsuming advocacy into inquiry for the purpose of exploration only is "dialogue." Often, if skillful discussion doesn't produce a win-win outcome, spending time in the less structured format of dialogue can be a very productive step. Letting go of urgency and making space for pure reflection can sometimes shift a group into alignment without having to move through a fully articulated resolution. We will explore dialogue as part of the group learning and self-awareness leadership capacity in chapter 21.

In relationship, as in investing, the higher the risk, the higher the potential reward. The more acute and frightening a conflict, the more power it has to build a resilient relationship—but only if it is productively navigated.

Relationship Development

In today's socially complex and experience-oriented environment, relationship has at least equal value with other project outcomes. To create and leverage this value for their firms as well as their clients, professionals must develop, not just maintain relationships. Developing relationships involves bringing to them a new level of awareness—both of oneself and of the other person, investing in them more and differently, and taking a proactive role in shepherding them through their predictable life-cycle stages.

A relationship begins with a bit of a chicken-and-egg situation. To develop a relationship, a professional has to have enough information about the client to know how to develop it. To get that information, a professional must jump into developing the relationship without benefit of that information. The processes of getting to know a client and developing a relationship with a client unfold in parallel and are inseparable.

Given that the Power Struggle stage is required for a relationship to evolve into co-creative capacity, and that this stage will pose a significant threat to the health of the relationship, laying a strong foundation from the beginning is essential. Deep listening is central here. Inquiring deeply into and empathizing with your client's goals, concerns, and needs will lead to her feeling heard and valued. In this process, you may find that her goals, concerns, and needs are similar to your own. If so, the relationship will tend to develop quickly and feel like a lot of fun, but may tend to present greater challenges to surfacing crucial differences later on. If you instead discover that your differences outweigh the similarities, the relationship will tend to develop more slowly and feel like work. It may, however, hold greater potential for complementarity and creativity born of diversity.

Some of the differences may have to do with your client's organization culture and individual mind-set versus your own. As discussed in chapter 18, being attuned to these values-based differences enables you to shape your process, service, and product to the expectations of the client and her organization. If the consultant-client relationship involves differences in these domains, you will likely need to make an extra effort on an ongoing basis to be guided more by the client's needs and preferences than by your own.

You may notice other values-based differences that appear to derive from your respective national, ethnic, or religious backgrounds. Given the diversity of the U.S. population and its likelihood of increasing as globalization proceeds, as well as the likelihood of international work, demonstrating sensitivity to such differences is another essential ingredient for early relationship building. All professionals should take the time to become familiar with the key differences appearing across domestic and world cultures.[220] One important variation involves the point at which people become comfortable sharing information about their personal lives. While some cultures are more reserved in this respect, others presume it will be an early part of any interaction. At some point, every work relationship can be strengthened by the care and concern implied by sharing life beyond work to some degree. When it becomes comfortable to do so, inquire about your client's family and other activities as well as notice and respond to her state of being. Or start the process by volunteering this kind of information about yourself. This attentiveness to the whole person will deepen your bond over time.

Beyond cultural differences, personality differences present significant challenges for consultant-client relationships. Every professional should also take some time to understand himself in terms of a well-known personality style instrument, such as the Myers-Briggs Type Indicator (MBTI), discussed in chapter 8. Many professional firms and partnering consultants have used this instrument to improve teamwork, so you may already have been exposed to it. If not, resources are available both in print and online.[221] Becoming familiar with a system such as the MBTI can enable a professional to recognize a client's style even if the client has not used the instrument. Most such systems provide guidance on how to use awareness of style differences to optimize communication. For example, recognizing that your client is a "J" while you are a "P" on the MBTI's Judging-Perceiving continuum can help you compensate for the discomfort your closure-oriented client will likely feel when you must leave a possibility open at the end of a meeting.

Conceivably, if you are an "E" and your client is an "I" on the MBTI Extravert-Introvert continuum, you could err by communicating too much. Our observations indicate that this is rarely the case. Instead, clients generally complain about—and problems arise from—too little communication. In part, this may indicate that too many relationships go the other way; clients are the "E's" and consultants are "I's" who don't work hard enough at overcoming their natural reserve. This problem seems to have additional dimensions, however, such as all parties having too much task-related work to do—all the time.

The role of communication in developing relationships, especially in the early stage, can hardly be overemphasized. Sharing responsibility for communication in a working relationship as a 50–50 proposition feels natural, but taking 100 percent responsibility is a more effective approach. A few clients will take a more than 50 percent share, in which case you will be well covered. But most will take less than 50 percent, and in that case if you don't take 100 percent you can get into trouble, from a task as well as a relationship perspective.

Communication is both a cause and an effect of relationship, forming a basic reinforcing loop. Building a relationship in the early stage is actually all about reinforcing dynamics. While deep listening implies that you value the client and her concerns, making appreciation explicit can add substantial momentum to the virtuous cycle. Most of us tend to go on to the next task-related topic rather than stop to recognize someone for a contribution or interaction. If we do say thank-you, we rarely take the time to describe specifically what it was that we appreciated. Appreciating your client in the context of your relationship inspires the client to invest more in the relationship, and the relationship grows as a result.

The reinforcing dynamic of appreciation can also be used to build specific aspects of a client's contribution to a relationship. Positive reinforcement has long been acknowledged as one of the most powerful coaching and development tools for managers and leaders. Aubrey Daniels thinks it is also the most powerful performance-generating tool, as he explains at length in *Bringing Out the Best in People*.[222] In *The Dance of Change*, Peter Senge and his co-authors show that we get what we want less by resisting what is not consistent with it, than by nurturing and thus reinforcing what is consistent with it.[223] When a colleague does something we find valuable and we express our appreciation, the person will be more likely to do more of that kind of thing, which in turn calls forth more of our appreciation. In this way, appreciation (as in investing!) amplifies value.

Another reinforcing dynamic that contributes to early relationship-building is the process of modeling what it is that we desire to create. When we honor and keep our commitments to the other person, our integrity encourages the other person to act with integrity toward us. We are thus encouraged to continue to act with integrity, and so on. When we show flexibility in give-and-take, that flexibility will tend to be returned to us, which will in turn inspire us to show more of it in future situations. When we feel compassion and spontaneously express generosity, these precious human gifts will tend to come back to us and give rise to more. And when we trust people and give them the benefit of the doubt, they will tend to rise to action consistent with that trust. Of course, this dynamic is not infallible. We are all highly sensitive to broken trust or lack of reciprocity. However, those who start from a position of mistrust, demanding that trust be earned, will find themselves at a disadvantage. Relationship does not thrive on suspicion and doubt.

Greater awareness of both self and other, along with more investment in these reinforcing processes, can help you lay optimal groundwork for client relationships in the early stages. However, take care not to let this great work suppress the emergence of conflict required for the Power Struggle stage. Though some relationships will have a rockier time than others, all relationships that seek to reach a high level of functioning must, at least to some degree, pass through this stage. The key social capacity here is to understand, recognize, and effectively navigate this dynamic of the relationship life cycle.

If you have done your early relationship-building well, you will have plenty of deposits in your "emotional bank account," as Covey describes it.[224] If a relationship appears not to be reaching a sufficient level of exchange to be healthy, these deposits will enable you to engage in and to prompt in the client the assertive speaking needed to ensure that the relationship is progressing into the Power Struggle stage. If conflict does surface on its own, you can respond by engaging it productively, rather than avoiding it or responding defensively, knowing that this is an essential phase of the relationship's development.

If the conflict turns out to be about the relationship itself, rather than an isolated issue, the best initial approach is to treat the problem as a function of behaviors that can be changed rather than of less malleable aspects of personality. If the problem turns out to involve deeper aspects of identity, resolution may still be navigated with caring and mutual respect that keep both parties' self-esteem intact. Tuning in to the rhythms of the cycle enables the professional to recognize when it might be most helpful to shift to a more systematic approach to recommitting on a new basis. Essentially, with this model in mind, the professional can monitor the health of the relationship and intervene in ways that promote rather than suppress development. Rather than feel at the mercy of the unnerving dynamics of the stages, he can assume a proactive role in working through them.

So what if our best relationship-development efforts fail? What if we get stuck in Power Struggle—overcome by the disappointment of a perceived breach of trust or lack of reciprocity? Or what if we get stuck in Apathy, or never even get to Power Struggle at all? As complexity theory teaches us, creativity emerges from an optimal range of diversity in ways of thinking: one can have too little difference, or one can have too much. If either of these scenarios is present with a client, it may be an indicator of lack of sufficient fit. If either is present with another stakeholder, the professional may need to fall back on an approach to the relationship that is based more on appropriate political processes than on personal ones.[225]

Learning and Self-Development

In the shift of mind from assumed authority to self-aware professional contribution, a professional should focus as much on what he doesn't know as what he does know. This focus on not-knowing—or beginner's mind, as it is called in learning organization theory—makes room for others in design and problem-solving processes and prompts lifelong learning, for optimizing professional contribution, not just continued certification. The mind-set of self-aware professional contribution also relies on the most fundamental arena of emotional intelligence—self-awareness. Self-awareness tends to produce self-development, which in turn tends to produce self-awareness. The social capacity of learning and self-development enables the professional to bring the mind-set of self-aware professional contribution to fruition in action. All learning is a feedback loop with at least the basic elements shown in figure 20.1.

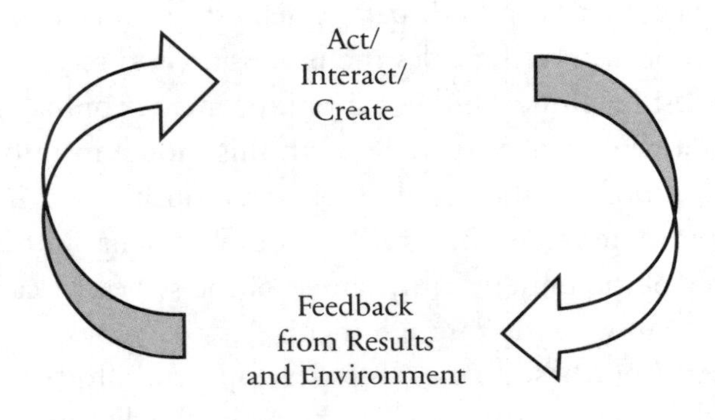

FIGURE 20.1. Basic Learning Loop

All living systems learn; that is, adapt and behave differently in response to feedback from their environment. In human systems, not all learning is conscious. Our unconscious mind and body will process feedback and redirect our actions to some extent. The more complex and mental an activity, however, the more participation by the conscious mind is required for learning to be effective. Conscious learning requires the discipline of the learning loop, depicted in figure 19.4, to be operating at all levels of the firm to achieve transformation. In that loop, action is preceded by the highly conscious acts of goal-setting and planning. Feedback is consciously sought and received. And learning emerges from the also highly conscious processes of self-evaluation and feedback interpretation. While closing the loop and achieving the agility of ongoing course correction is easier at the individual level than at the group or organization level, it is by no means easy.

Senge defines the learning organization discipline of personal mastery in terms of the goal-setting step in conscious learning.[226] In his definition, the aspiration we cultivate contrasts with awareness of our current state to produce creative tension that literally powers our learning. If the gap between our desired future and our current reality is too great, the tension that generates the energy of our pursuit cannot be sustained. If the gap is too small—either because we lack sufficient vision or ambition, or because we make an overly optimistic assessment of our current reality—no creative tension is present. As long as the gap is in the generative range, it continuously motivates our learning and development.

Core to the social capacity of learning and self-development, then, are three enablers of personal mastery. First, we must cultivate the aspiration to grow professionally and personally. As Daniel Goleman and his colleagues stress in the context of emotional intelligence, an individual can only grow if he is genuinely motivated to do so.[227] Second, we must develop the ability to accurately assess our current state. And third, we must develop and maintain sufficient self-esteem to maintain our motivation to pursue self-development efforts—and our psychological health—in the face of ongoing experience of the gap. These three enablers inform the "theory of self-directed learning" developed by one of Goleman's colleagues, Richard Boyatzis. In his model, self-development occurs through what he terms five discoveries:[228]

1. My ideal self—Who do I want to be?
2. My real self—Who am I? What are my strengths and gaps?
3. My learning agenda—How can I build on my strengths while reducing my gaps?
4. Experimenting with and practicing new behaviors, thoughts, and feelings to the point of mastery.
5. Developing supportive and trusting relationships that make change possible.

According to Goleman and his colleagues, emotional intelligence requires more practice than rational intelligence, because long-established neuro-pathways in the brain must be reprogrammed. However, all forms of learning require a good measure of experimenting with new behaviors, thoughts, and feelings.

In addition to the shifts of mind closely related to emotional intelligence—relationship awareness; a nondualistic, inclusive view; and self-aware professional contribution—the individual learning loop for higher value technical work should activate the shifts of mind to whole-project-system awareness and group process awareness and orientation to emergence. The professional should focus on his impacts on the project as a whole and on groups in which he is involved, as well as on individuals. That means seeking and receiving feedback from all three of those levels, and then learning from feedback at these levels to modify goals and plans.

While everything—including no feedback—may ultimately be viewed as feedback, in practice, professionals receive insufficient feedback to do the learning they need to do on a timely basis. On the whole-project-system and group levels, feedback that would drive course-correction usually doesn't come until a problem has reached the point of pain. The best way to overcome this problem is to develop the ability to sense emerging problems holistically through a combination of observing people and monitoring key project dimensions. Timeline and budget are obvious indicators, but issues like scope, focus, number or effectiveness of meetings, the form of solutions, or unresolved differences involving technical issues in your arena can be equally or more telling. The time to begin asking probing questions is very soon after you begin to sense things going in an unsettling direction.

On the individual level, feedback that implies the need for course-correction is often withheld for the reasons we discussed in the context of assertive speaking: project participants are avoiding the challenge or conflict they imagine will result. Especially in relationships that have not reached Power Struggle or teams that have not reached Storming, individual feedback can be very difficult to get. Developing a practice of regular inquiry in a safe setting is the best approach to individual feedback. You are likely to get valuable positive feedback that has been lacking simply because so many people don't stop to focus on that. Take this in as deeply as you can, but don't conclude the conversation too soon. Keep asking probing questions linked specifically to project activities and developments and your contributions until you are satisfied that the person has shared all that they are thinking.

Effectively interpreting feedback from all these sources requires you to balance what you are hearing from others and observing in the project evolution with what you have been observing and concluding yourself. The learning you ultimately glean from these more proactive feedback-gathering efforts will be as much about the whole system, specific groups, and other individuals as it is about you. At the same time, learning about yourself is always the most difficult—partly because receiving constructive feedback is so emotionally challenging and partly because having confidence in your own self-evaluation is so mentally challenging. If our self-perception is out of line with others' perceptions of us, our interpretive responses to feedback will tend to be off the mark from what could strengthen our contribution to the project.

An excellent process for comparing your self-perceptions with how others perceive you is 360-degree feedback. If your firm has not used this process yet, you might suggest it, or simply decide to do the process yourself through a self-development course or consultant-based coaching.[229] In this process, supervisors/managers, peers, and people who report to you all answer questions about aspects of your performance that are important to success. Including client feedback can add a powerful dimension. Because the process is so emotionally difficult, we don't recommend you pursue it without some kind of support structure. As Boyatzis's model of self-directed learning emphasizes, "trusting relationships that help, support, and encourage" you all along the way are essential.

The purpose of 360-degree feedback is not to take others' perceptions as the real truth about you. Rather, knowledge of others' perceptions makes you aware of areas in which you judge yourself more harshly than others do, perhaps because your standards may be higher. That knowledge also makes you aware of areas in which others hold you to a higher or different standard than you hold yourself. Though these perceptions from others are as much about them as they are about you, they form key aspects of your work environment. Whether or not you agree with those perceptions, they point to key blind spots that may be hampering your ability to be effective in that environment.

Through the feedback you gather, you may learn that you need to strengthen aspects of your technical performance or that you have gaps in social skills, social awareness, or self-management. Some of the learning you need to do may be of the single-loop variety. Other self-development you need to undertake may involve double-loop learning. Senge calls the change resulting from double-loop learning—change that combines deep internal shifts with external ones—profound rather than transformational.[230] We have been using *transformative* in this book. Whatever the term, such change involves a leap to a new level of being and capability. The core social capacity of learning and self-development includes both incremental change and the ability to experience and take advantage of developmental leaps. The shifts of mind for catching up, or the further shifts for exercising leadership, are examples of the most fundamental kind of developmental leaps: moving to the next stage in the stages of mind-set development.

How can a professional prepare for this more profound kind of learning? Perhaps the most important practice is self-reflection: regularly making quiet time for focusing solely on your own being and thoughts. Use at least part of this time to pay attention to anything that remains unresolved in you, whether feelings about yourself or others, or thoughts that won't go away because they haven't been dealt with. Noticing your physical state can often help. Sometimes upset manifests itself as pain somewhere in your body. Allowing yourself to explore what you are feeling physically often surfaces the emotional or belief-centered core of what is going on.

Once you have brought the issue to your awareness, do your best to inquire internally about it. Why are you having these feelings or thoughts? What part of their source resides inside you? Are you holding a belief that is currently not serving you? What do you have at stake in holding on to that belief? What could you do differently internally? Externally? The process of sticking with such self-inquiry often brings at least partial resolution, usually with at least a minor shift in your own being.

Remind yourself that your own creative adaptation is likely similar to that of other human systems: most generative at the edge of chaos. A certain amount of quite uncomfortable anxiety—usually provoked by challenges coming from your environment—will befriend you in the process of self-development. Remember too that we all face greater degrees of complexity than we have had to navigate in the past and that with the help of O'Hara's model, we can be aware of three potential responses in ourselves. With this awareness, we can ward off any tendency to deny the complexity confronting us that might lead to a psychotic response. And with this awareness, we can accept rather than suppress the internal turbulence that will likely be required to help us find a transformational rather than merely neurotic response.

The Enneagram, a personality typing system, may also be a powerful aid to transformative self-development. This system has evolved from ancient roots and resonates both with modern psychology and with the great spiritual traditions. While this system resembles others in that it can be used to understand and work with other people better, it differs from other systems in that it focuses deeply on internal states and emphasizes development over time. Many books and online resources are available for exploring the Enneagram.[231]

Professionals who begin to practice the core social disciplines of listening deeply, speaking assertively, engaging in productive conflict, and developing relationships in high-stakes contexts will likely receive more feedback than before, even without proactive feedback-gathering efforts. Such efforts will produce even more. For a professional who already feels undervalued, navigating the passage to effectively taking in all that feedback will tend to be a significant challenge. A more accurate picture of the gap between desired state and current reality may deal a further blow to self-esteem. As we touched on in chapter 7, the professional must understand that the problem is personal while not taking it personally. Self-compassion plays a critical role in the social capacity of learning and self-development.

If a professional steps out of self-protective behaviors into the rigorous social capacities we have described, her need for self-care will increase rather than decrease. We monitor the degree to which we are getting our physical, emotional, mental, and spiritual needs met through internal feedback. As we interact with others, we have a stream of internal experience that will likely include all of these aspects of ourselves. For example, we have a wave of exhaustion as a meeting goes overtime, or a gut feeling that something is going very wrong with a project as we listen to a client. Or we have a flash of an idea that solves a problem we have been laboring over for weeks. Or we experience the emotional gratification of a relationship that is going well, noting that a particular interaction even gives us a glimpse of the reason we got into this work in the first place.

To factor appropriate self-care into learning and self-development, we must add feedback from our insides to all the other sources being monitored in the learning loop. While some of this internal feedback will lead us to modify our behavior to enhance relationships or project impacts, some of it will teach us about our own desires, needs, and limits. This increased awareness of ourselves will expand our ability to take responsibility for the impact of our own behaviors on our well-being. And it will help us balance our own well-being with the needs of others and the project.

We can all probably recognize times when our energy and attention have been externally focused in the extreme. Many members of the U.S. baby-boom generation have spent time overextending themselves with too much work or too much care of others and then flipped into periods of self-concern that others often viewed as narcissistic. The flip was often catalyzed by a wake-up call like a health problem. According to Frederic Hudson—who has developed a four-stage model of renewal process that propels us through the stages of the individual life cycle—occasional periods of withdrawing focus from our external environment are natural and essential.[232] However, taking ourselves to the extreme of requiring a wake-up call of a life-threatening nature indicates that the personal mastery challenges involved with our own internal processes can be every bit as challenging as those involved in relationships with others.

Learning and self-development that arrives at a healthy Interdependent mind-set naturally balances care of self with care of others and the efforts in which they are involved. It forms the basis for the continued learning and self-development required to pursue Type II work—leading collaborative work in client systems. And a healthy Interdependent mind-set provides the basis for the transformative learning and self-development that is required to leap to a whole-system-connected mind-set, the best state from which to undertake Type III work in client systems. As we have suggested here, a healthy Interdependent mind-set also provides the best means for a professional to provide higher value technical work in today's socially complex environment.

As Warren Bennis has said, all leadership depends on knowing oneself and knowing the world.[233] Though technical work in architecture, engineering, and the related professions has not in recent times involved much leadership of social processes, it still involves exercising technical leadership in the context of social processes led by others. Every one of the basic social capacities we have discussed in this chapter, including learning and self-development, is essential for building the strong relationships required to regain the trust of clients and stakeholders. Each of the capacities is also essential for meeting client and stakeholder expectations for effective participation in collaborative process. Only when professionals demonstrate these capacities will clients begin to value technical work at the high level that will enable professionals to exercise effective technical leadership—and to express themselves fully and creatively on behalf of clients and society.

21 Leadership Capacities for Collaborative Work

Think of organizational data as a wave, moving through space developing more and more potential explanations. If this wave meets up with one observer, it will collapse into one interpretation. All other potentialities are lost by that act of observation. An organization swimming in many interpretations can discuss, combine, and build on them. The outcome has to be a much richer sense of what needs to be done. The more participants we engage in this participative universe, the wiser we become.

—Margaret Wheatley[234]

Collaborative, or Type II, work—as we presented in chapter 12 based on Ronald Heifetz's definition—requires clients and stakeholders to participate in developing solutions with the technical professional. In these situations, the professional understands and can define both the problem and a path to solution, but a viable solution cannot be achieved without client and stakeholder participation and learning. The learning that occurs in the context of Type II work is significant, but incremental, as distinguished from the deeper learning that occurs in the context of transformative, or Type III, work.

Recall the panel of client representatives who resoundingly asserted that the most important thing they were looking for from their engineering and architecture consultants was leadership. The socially complex situations these clients faced provide examples of the kind of learning required in collaborative work. The citizens who were obstructing additions to the municipal water system in their neighborhood needed to learn what was at stake, the design alternatives available, and the advantages and disadvantages of the proposed addition, as well as have an opportunity to have their input heard and responded to. The faculty at the university needed to understand the issues and trade-offs involved in facility decisions under current budget constraints and have an opportunity to help make some of those decisions. Solutions in these situations could not be reached without the participation and agreement of these groups. While the learning resulting from their participation would likely help them be more informed community and organization members in the future, it would probably not require fundamental change in their values, beliefs, or behaviors.

Collaborative work addresses the technological and social complexity involved in these examples. Projects requiring integration of many technical disciplines—and different firms—can be as challenging as those that involve diverse nontechnical stakeholders. Management consultants and program and construction managers are often hired to provide the collaborative leadership required for such projects. But few of them have mastered all five leadership capacities for collaborative work. With this mastery, we believe architects and engineers could out-compete them.

As we have made clear, providing social process leadership is not the only way for technical professionals to escape commoditization. However, the demand for collaborative work is increasing, and clients who expect leadership from technical professionals represent a significant segment of those with whom the professions must catch up.

A number of architecture and engineering firms that focus strategically on service and relationship have already ventured into Type II work. Whether the work has involved gathering design input from different groups within the client organization or discussing solutions in public meetings, it typically has been treated as part of the Type I work. For example, a number of firms have used collaborative approaches to public school design to differentiate themselves from their competitors. These firms have generally sold these services as hours valued on the same basis as their technical work.

One of the business model innovations we want to suggest is offering Type II work as a service that is distinguished from and given a different kind of value from Type I work. To support this innovation, however, a firm must broaden and deepen its collaborative leadership capacities. Its professionals must be able to deliver the forms of value associated with participation, alignment, co-creation, and learning in ways that are visible to clients. And professionals must be able to provide these forms of value across a range of settings and in the face of challenges that clients feel unprepared to address through their own leadership capacity.

You may want to refer to table 16.2, which shows the leadership capacities for collaborative work alongside the shifts of mind for catching up and the social capacities for technical work on which these leadership capacities build and depend. At first glance, these capacities—developing shared vision, group process design, group process facilitation, group development, and group learning and self-awareness—strike the technical professional as not his job. Yet anyone who has participated in complex built environment projects understands that such projects cannot succeed without someone performing these functions. They represent a significant opportunity for engineers and architects.

Again, our aim is not to teach technical professionals *how* in detail to develop these leadership capacities. Instead, we seek to explore *what* these capacities entail in the built environment context. We hope this exploration will help firms decide whether they want to offer collaborative work as a differentiated service, and if so, get them started on the effort. Throughout, we will point to key resources that will enable professionals to take their next steps. Let's turn now to the first of the leadership capacities required for collaborative work.

Developing Shared Vision

It is a rare situation where one individual can determine the direction for a group or organization of any size. Stakeholders expect to participate in setting direction. Leaders prefer to involve stakeholders to ensure that they are truly aligned with the direction, and thus willing to invest the necessary time, effort, and creativity to pursue that direction. Built environment projects—even relatively simple ones—are no exception. The larger and more complex the project, the larger and more complex the process of developing shared vision, and the greater the risk of failing to achieve alignment. When the clock is ticking on hundreds of people's time, the costs of unaligned efforts become immense.

Virtually every stakeholder has a personal view of what is at stake in the project and what its potential holds. When a client organization includes groups with diverse and potentially competing needs—such as doctors vs. nurses vs. administration vs. patients in a hospital, or faculty vs. students vs. administration vs. service groups in a university—the range of personal and subgroup views can seem unmanageably broad. When members of a client organization ecosystem have still different interests, the challenge becomes even greater. The art of developing shared vision involves teasing out and honoring the various views while moving toward a meaningful unity that can define direction for the project. Allowing more time at the outset for this process will likely pay for itself many times over in both monetary and nonmonetary benefits for the project.

Peter Senge urges that the process of developing shared vision for organizations be ongoing. He believes the vision should always be changing, at least subtly, to reflect the ways the organization is evolving, both internally and externally. The vision for a complex project will surely evolve over its long cycle-time. In some projects, the vision must be substantially revised as major constraints are discovered in the programming and early planning phases. However, projects involving physical construction differ from organizations in that they must proceed based on a clear shared vision at a certain point in time. The strange-attractor phenomenon we discussed in the context of leading innovation can assist with this requirement. If a human system's engagement with its environment is dynamic enough at a particular moment, a strong vision can emerge that not only aligns diverse participants but also elicits creativity beyond what might have been imagined.

Does developing shared vision mean the professional should not be developing her own vision for the project? The answer is no on two counts. First, if a shared vision is so beneficial in aligning and releasing the creativity of the diverse participants, no project can afford to be without one. Where it originates is less important than that it be present. It can come from a client or stakeholder, or it can come from the professional. As long as it is not imposed, then, a strong vision developed by the professional is a valuable project resource. Second, the gelling of shared vision often occurs through the professional's ability to sift, synthesize, and articulate the visions of the other stakeholders. Invariably, such a synthesis is informed not only by the professional's nondualistic, inclusive view, but by her own visionary capacity.

This subtlety of interaction begins to reveal the essence of this leadership capacity. On the one hand, a professional's ability to provide strong vision can be pivotal for a project. On the other, a professional's ability to adopt and amplify someone else's vision or to synthesize everyone else's vision can be pivotal. The collaborative visioning process requires the professional to honor the visions of others as much or more than her own—to be very careful not to dominate or over-direct the process. At the same time, the professional may be most effective in helping to bring forth others' visions by trying out her own ideas, especially if stakeholders are having difficulty clearly articulating the challenges present in their environment or forming a compelling response to those challenges. In this process, the professional must be sensitive to when her ideas take with stakeholders and when they don't. Backing off when they don't is critical. The work is to evoke, not impose, a vision through a process that—like the arising of a strange attractor—remains a little mysterious.

This leadership capacity thus involves both a subtle form of leading and a complex form of following what the other stakeholders are expressing. With this capacity, the professional begins to put the mindset of group process awareness and orientation to emergence into action. A successful shared vision effectively expresses both the identity of the human system being served and an appropriate response to the challenges the system faces. The professional's self-expression must occur in this context.

Once a professional has succeeded in aiding and abetting a shared vision into coherence and articulation, developing evolves into sustaining over the remaining life of the project. Enabling the vision to inform all participants' decisions and work entails a continuous process of keeping the vision alive so that its potential can be fully explored and unfolded. This unfolding may require some evolution, and the professional leading collaborative work must be attuned to facilitating this process as well. To achieve ongoing inspiring communication as well as the capacity to honor the core through any evolutions, the professional must be fully engaged with the vision herself, whatever the means have been for bringing it forth.

Group Process Design

Developing shared vision is only one of many collaborative processes that will need to be undertaken by groups in the project context. Though the ultimate goal of these processes may be to create opportunities for groups to self-organize, the best means to this end is to provide structure for generative flow. The ability to design processes that enable groups to be productive together is a core leadership capacity required for all projects except those of very small scale.

While skillful discussion in Peter Senge's sense is a mainstay of collaborative project work, even it needs to be structured. What is the nature and form of the outcome that is needed from the discussion? What is the context for this outcome? What has happened to date? Are there any likely obstacles to this outcome? Given the desired outcome and likely obstacles, logical interdependencies and project critical path, what is the appropriate sequence of issues or topics? What is the best way to state or form a question around an issue, or to frame a needed decision?

These questions are just the beginning, however, of thinking about the process in a situation where skillful discussion appears to be the obvious choice. Who will or needs to be present to have needed constituencies and disciplines covered? How well do they know one another? Are there any problematic relationship issues? How can participants best be prepared to start working effectively together, given the need to balance task and relationship? How much should you involve participants in developing the design? And how much of the design detail should you provide to them in advance?

How large is the group? If it is between 5 and 20 or so, how will discussion moderation be handled? How will recording and timekeeping be accomplished? Who will fill these roles and how will this be decided? What ground rules would be helpful and what processes will be used to hold participants accountable to those ground rules? If more than 20 people will be present, is skillful discussion the best process? What alternatives might be more effective? How can the design enable appropriate participation of all attendees? How can the design help to ensure a quality outcome? How can it engage participants where they are in the process of their involvement with the project and enable them to generate or move to the next step?

Stepping outside the realm of skillful discussion, a vast lexicon of processes is available for groups of all sizes. Facilitators have developed processes for getting to know one another, building teams, visioning, gathering input, sharing information, processing information, generating ideas, prioritizing, fleshing out options, testing ideas under different scenarios, simulating processes, analyzing workflow, diagramming systems, action planning, reflecting on experience, and, of course, making decisions. In recent years, so-called large-group technologies, designed to bring together and work with "the whole system in the room," have made it possible to handle anywhere from 60 to several thousand people simultaneously.

Architects and engineers interested in offering Type II work must at least be familiar with the major forms of group process design that have developed over the last 30–40 years. With this knowledge, they can develop processes customized to fit the needs of their projects, based on the roles they are playing and the challenges they are facing. Over time, a set of core processes that work especially well for a person's style and context begin to take shape. As that occurs, most designs become variations on those basic forms.

The literature on group process design is vast. *The Fifth Discipline Fieldbook* introduces the process of skillful discussion and its uses. An excellent overview of structured process designs appears in *The Change Handbook: Group Methods for Shaping the Future*, edited by Peggy Holman and Tom Devane (1999). It includes chapters on major methodologies written by their originators. One of the best resources is Laura J. Spencer's *Winning Through Participation: Meeting the Challenge of Corporate Change with the Technology of Participation* (1996). For whole-system and large-group processes, consider Marvin R. Weisbord and Sandra Janoff's *Future Search: An Action Guide to Finding Common Ground in Organizations and Communities* (1995) and Harrison Owen's *Open Space Technology: A User's Guide* (1998). Jane Magruder Watkins and Bernard J. Mohr's *Appreciative Inquiry: Change at the Speed of Imagination* (2001) explores a multitude of applications for a positive-energy-generating group process technology that can be used at any scale.

Group Process Facilitation

In the early years of organization development, group process facilitation meant process only. The facilitator's job was to rigorously manage himself to stay completely out of content. The purposes here were twofold. First, practitioners doubted that a facilitator could effectively observe and work with process if his perception was clouded by involvement in and attachment to content issues. And second, practitioners felt that since the facilitator was not a member of the system being served, he should not have a say in the decision making for that system. Neutrality was felt to be the only legitimate ethical stance.

This view of process facilitation has evolved considerably in recent years, mainly because clients want to benefit from all the forms of knowledge a consultant possesses. To limit this to the consultant's process knowledge would deprive clients of the entire arena of his content knowledge. Two other factors have significantly contributed to this evolution. First, the living systems orientation has revealed that although the consultant is not a member of the organization that will go on after he has finished his assignment, he is a member of the temporary system working on a particular project or problem. This temporary role in turn makes him part of this organization's ongoing web of relationships, or organizational ecosystem, as we have called it. Neutrality becomes less relevant in the context of this view. Second, sufficient mastery of process awareness and process design enable a facilitator to be an effective contributor both to process and to content, just as some members of the organizational system also may have reached this more complex capacity. More participants aware of both process and content generally make a group process more effective.

This evolution of the process facilitation role is especially important for engineers and architects, given the essential part their expert knowledge plays in the expectations of clients and stakeholders. The technical professional's challenge is to evolve from content contribution to process facilitation, which includes, but is not dominated by, content contribution. This is a tall order, but the professional who has made the underlying shifts of mind we have discussed and developed some mastery of the core social capacities will have a good head start.

In experimenting with different types of process designs, the professional should be alert to the different types of facilitation roles these designs call for. Because the discussion context is so common in professional work, discussion moderation is the cornerstone of the facilitator's role, except when other participants may be moderating. In this role, a facilitator must not only model deep listening, assertive speaking, productive conflict, and balancing advocacy with inquiry. He must also inspire and guide others to practice these disciplines. Agreed-upon ground rules are an important element here—a basis for empowering the group to monitor and hold itself accountable. But the facilitator's role in skillful discussion entails much more than helping the group agree on and use its ground rules.

The facilitator should expand deep listening by inviting, honoring, and balancing participation by all parties. An essential aspect of honoring is recording individual contributions in a form of group memory that is visible to all, such as a flipchart or a projected computer screen. This technique represents a significant shift from traditional documentation where an individual takes notes independently and without visibility to the group. The traditional method provides neither recognition for contributions nor real-time opportunity to check for accuracy and shared understanding of issues, agreements, and action items captured. In certain circumstances the essential role of recorder can be performed by the facilitator, but in most cases the group benefits when it is performed by someone else. This frees the facilitator to engage in the active inquiry associated with deep listening and to prompt others in the group to do so as well. Group monitoring of visible recording significantly promotes the deep listening process.

When discussion heats up with the energy of disagreement, the facilitator must override personal emotional responses that drive toward either personal attack or avoidance. The discomfort of seemingly out-of-control divergent energy produces an almost overwhelming temptation to resolve the conflict for the group. This is the right thing to do in some circumstances, as we indicate in the group development section. However, in most circumstances, this use of authority to avoid working through conflict suppresses the group members' ability to take ownership of their own process as well as their creative capacity to resolve the conflict. Instead, the facilitator should use self-management and productive conflict skills to "hold the container" for the group's conflictual energy, such that the group feels safe to work through it.

Further, the facilitator should support the group in "working through" conflict by using assertive speaking to feed back what is going on in the group dynamics—including any violation of ground rules—and by balancing advocacy with deep listening and inquiry. Summarizing various positions—including their strengths and limitations—and posing key questions that arise from this reflection pave the way for checking for consensus. Ground rules may require that decisions be made by consensus, or may specify other processes, such as majority vote or authority-based decision by a senior client representative. Generally, however, checking for consensus will be productive and will lead to the opportunity to hear out dissenting views. Facilitator decisions regarding when and when not to make a contribution are especially delicate in conflictual situations.

The facilitator's larger challenge is to be able to modify the process design in real-time if it is not achieving desired outcomes. No matter how good the design, unfolding group dynamics will challenge it to some degree. The facilitator must let go of attachment to the carefully constructed design when better options emerge, either explicitly from the group or as a result of the facilitator's own observation of what isn't working. Meeting this significant challenge depends on suspending notions about where the group should be at a given moment—at least enough to be able to perceive and honor where it actually is. The facilitator must also be able to recognize or design a new process in the moment. Returning to desired outcomes either to reaffirm or to revise them, and then keeping focus on those outcomes so that they guide the newly emergent process can help ensure a good result. Like developing shared vision, being effective in group process facilitation involves improvisational capacity to both lead and follow in the service of releasing the creative potential of the group.

Facilitation, like design, is an aspect of the now highly developed group process discipline. Practice is the ultimate means of mastering this capacity, but reading some good books on the subject is a great way to begin. Roger Schwarz's *The Skilled Facilitator* (2nd edition, 2002) is an excellent starting point. *Facilitator's Guide to Participatory Decision-Making* (1996), by Sam Kaner and others, is also a helpful basic text. David Straus's *How to Make Collaboration Work: Powerful Ways to Build Consensus, Solve Problems, and Make Decisions* (2002) offers the now long-honed wisdom of one of the pioneers in the field. This book has the additional benefit for built environment professionals of building on Straus's early experiences as an architect. *How to Make Meetings Work* (1976), by Straus and co-author Michael Doyle, is a classic in the field.

Group Development

Though built environment projects typically involve many people flowing in and out of the complex process, generally one or more core groups are located at its center. Some of these groups are teams in the formal sense, though many are not. Membership in these core groups tends to change with the stages of the project, creating a significant challenge for group effectiveness. However, many are stable enough for long enough to benefit from developing a group identity. Group identity entails members understanding themselves not just as a collection of individuals, but as a whole that is more than the sum of its parts. Groups, like individuals and relationships, benefit from developmental attention.

The professional who is moving from the social capacity of relationship development to the leadership capacity of group development learns to expand the elements of relationship development to this more complex entity. The leadership capacity involves bringing more awareness to group interaction and to oneself in the group setting, making greater investment in the reinforcing dynamics of group building, and effectively navigating the stages of group development.

While formal roles are present in all project relationships, the positions occupied by individuals—both in their organizations and in the project—take on greater importance in the group context. Knowing the roles of participants from the client system and something about the politics of those roles is critical. Who has formal authority? And who has powerful, informal influence? Knowing the relationships that exist among the client and other organizations involved in the project and a bit of their history is helpful as well. These roles and relationships form the backdrop for group member interactions, determining at least initially the kinds of things that can be said.

If you initially have little clout within these dynamics, you will have an uphill climb to establish your credibility in the process leadership role. At the same time, to the extent that you are able to facilitate effective outcomes, you will be able to establish a level of informal authority greater than you have experienced in technical work roles. Being aware of your informal as well as formal authority in the context of the power relationships will enable you to design and facilitate processes that work better with the realities of the situation.

The culture and mind-set differences we have discussed in the context of relationship development will of course also be in play. In groups, you will have an opportunity to gain fuller awareness of the client organization culture and life-cycle stage. You will also have a bigger opportunity to fail to align with these characteristics if your mind-set or the mind-set of other group members does not match that of the key clients. Though it may be frustrating for a professional in the Interdependent mind-set to work with the Dependent or Independent mind-sets likely to be encountered in Transactional or Self-Expression cultures, the Interdependent mind-set best supports a professional in seeing and responding to the needs and preferences of individuals with these mind-sets.

Personality differences can exert even greater impact in groups than in individual relationships. In MBTI terms, you will need to know whether the dynamics between Extraverts and Introverts and between big picture Intuiting types and detail-oriented Sensing types are blocking full communication. Are the dynamics between Thinking and Feeling types and Judging and Perceiving types resulting in less than optimal decision making? As you attempt to compensate for such imbalances, you must be careful to avoid simply driving the preferences of your own type.

In addition to cultivating these forms of awareness, group development requires the professional to focus on team building. The partnering movement appeared to acknowledge the importance of team building in the early stages of a project and returning to team process periodically. In practice, however, not all partnering efforts have achieved team building at a significant level. The pull toward task remains strong in the built environment context. Persuading many players to invest the necessary time, dollars, and sincere participation is often a challenge. At the same time, professionals consistently report that projects that have gone well have included significant team-building efforts.

As a collaborative process leader, a professional can help a group invest in itself by insisting on adequate time to develop a shared vision and to agree on roles, processes, success criteria, and milestones in the early stages. The leader can also help a group develop its identity by observing the emergence of its unique capabilities. Recognition and appreciation are as key to group development as they are to relationship development. Noticing when group synergy leads to a significant outcome and feeding this back build the group's ability to take hold of its unique capacity *as a group*. Celebrating achievements supports this process, as does providing resources and time for the group's work, to the extent that this lies within your authority. All groups need help understanding, working effectively with, and sometimes recovering from difficulties met along the project path. An effective collaborative process leader can transform a group's adversity into the potential for new creativity by balancing the challenge of high performance with trust in the group's capacity and support for its heart and spirit.

Common wisdom acknowledges that projects—and the groups that steer them—undergo stages of development. In *Design. Market. Grow!* Craig Park outlined the "Six Phases of a Project," which will be familiar to readers:[235]

Phase I	Enthusiasm
Phase II	Disillusionment
Phase III	Panic
Phase IV	Search for the guilty
Phase V	Punishment of the innocent
Phase VI	Praise and honor for the non-participants

This cultural archetype reflects the unhappy ending so often experienced by groups that shoulder responsibility for built environment projects. Though this view inevitably produces laughter and camaraderie among those who have had the experience, it sadly points to the process failures that have led to broken trust with clients and stakeholders.

The model of group development presented in chapter 18 comprises an alternative that provides at least the potential for happy endings. While there are a number of variants on numbering and naming, virtually all of the studies of small group process include some version of Tuckman's four stages: Forming, Storming, Norming, and Performing. The team-building processes we have discussed—including developing shared vision—occur in the Forming stage and are essential to providing initial momentum and laying the groundwork for the rigors of the Storming stage. However, participants are not yet expressing themselves freely. Outcomes of this phase are limited by its inherent lack of synergistic creativity and usually must be revisited to some extent in the Norming stage.

As differences inevitably emerge, most groups will be tempted to avoid conflict by staying stuck in the relatively unproductive Forming stage. When hierarchical power relationships don't keep the lid on and members have productive conflict skills, the passage into Storming will happen more easily. Even when rank is operative and these skills are not present, however, the leader can help a group make this passage with two complementary actions. Acknowledging and valuing disagreement when it appears, or eliciting it when it doesn't, gives group members permission to express themselves more fully. At the same time, communicating to the group that disagreements can be productive rather than inevitably destructive enables the leader to create a sense of safety. Sharing the stages of group development and emphasizing the importance of the Storming stage can also be helpful.

Once the group has moved into Storming, the dynamics change fundamentally. Even if the Storming process is more covert than overt, it is challenging and uncomfortable for the participants. In the covert form, members will withhold support from each other or the leader. In the overt version, members challenge one another and the leader—often at the level of personality as well as content. In either case, the struggle is about who will control the group's work. At this point, the modest productivity that was being achieved in the Forming stage can grind to a halt. The struggle for control suppresses the ability to perform at the task level.

Researchers such as Wilfred Bion have studied the ways in which the group unconscious operates in a realm apart from task.[236] One common dysfunctional pattern is attributing unrealistic power to and becoming excessively dependent on the leader, then attempting to topple him in the wake of inevitable disappointment. Another is forming factions and fostering divisions outside of the group context and refusing to bring them up to the group as a whole. Storming is usually the most difficult stage for the leader because he becomes the focal point for so much of this unconscious activity. To remain intact in this process, the leader must work hard at keeping his natural defensive reactions under control. He must be prepared to deal with attacks that have little to do with how he has been leading the process. Unconditional regard for members, combined with dispassionate negotiation and mediation, is the best approach to navigating this emotionally perilous stage. The leader will need all his emotional intelligence at the ready. Remember above all, that individuals will not be able to reach consensus or move forward in development unless they feel respected and included and fully heard exactly where they are.

Many groups get stuck in Storming, particularly of the covert type, which looks a lot like the Apathy cul-de-sac in Campbell's relationship life cycle. In O'Hara's terms, this might be called a neurotic response to the challenging complexity presented by differences among group members. A transformational response occurs, however, when members have sufficient social capacities and the leader sufficient leadership capacities to find common ground from which to build the shared commitment needed to break through to the Norming stage. Here, members arrive at respect for what each can offer and are willing to work with differences in order to access and blend those offerings. They commit to the effort required to work consciously with those differences. Though group process may feel somewhat self-conscious and awkward in this stage, productivity begins to flow.

If the group persists in this practice, increasing fluidity evolves into high performance essentially unobstructed by group process issues. This Performing level is not free of conflict, however. On the contrary, it is characterized by highly productive conflict focused on key issues. Deep listening, assertive speaking, and a balance between inquiry and advocacy consistently produce unanticipated, creative solutions. Members are highly challenged and having great fun as they take pleasure and pride in their accomplishments. Significant changes in the group's membership, goals, resources, or organizational response to its work can send a group back to any of the prior stages. But with the help of good leadership, a team that has reached the Performing stage will be likely to recover that stage relatively quickly.

The approach to leading group development outlined here can be helpfully supplemented by a core leadership discipline that highlights the way in which a leader's style must change as the four stages unfold. Situational Leadership, developed by Paul Hersey and Ken Blanchard,[237] addresses leading individuals but can be easily extrapolated to the context of groups. Situational Leadership holds that while leaders naturally favor one or two of four basic styles, they must be able to use all four well to be effective across the full range of situations they are likely to encounter. The current version of the model presents the four styles of Directing, Coaching, Supporting, and Delegating as combinations of supportive versus directive behavior.

In working with individuals, four developmental levels correspond to these four leadership styles. Individuals with low competence and high commitment—those who are new to a responsibility and still learning—respond best to the Directing style, which provides required task-related guidance. Individuals with low to some competence and low commitment respond best to the Coaching style, which builds their competence and ownership through supportive questioning and collaboration. Individuals possessing moderate to high competence and variable commitment become most productive with the less involved Supporting style, which builds their confidence and provides back-up guidance if needed. Individuals with high competence and high commitment respond best to the mostly hands-off Delegating style, which allows them to optimize their own productivity. Progress through the four leadership styles thus supports an individual in moving through these natural developmental levels. Effective use of the Coaching and Supporting styles enables a leader to facilitate an individual's development from a dependent state to a fully empowered one.

Relating this progression to group development, we can see that in the Forming stage, group commitment tends to be high, but competence is low because of the cautious, exploratory nature of the group's process. While maintaining the peer-based stance appropriate to project groups, the leader can often be most effective during the Forming stage by doing more leading than following. Even when individuals of higher rank are group members, the Directing style—providing hands-on guidance, more in the arena of process than of expert content—is what the group needs to progress.

When the group moves into the Storming stage, competence as a group has begun to increase with the expression of differences, but commitment moves into dangerously low territory. To build a strong foundation for working together, the group must take responsibility for finding its own solutions at the same time that discomfort and risks are higher. As we have already suggested, the leader facilitates progress through this stage by encouraging and supporting members' assertive speaking and productive conflict at the same time that she remains an active participant in guiding the process to keep it safe. This guidance may involve mediation or even directive resolution if the group is unable to resolve a conflict that is escalating to a destructive level. This blend of supportive and directive behavior is the essence of the Coaching leadership style.

In the Norming stage, the group moves into greater competence and commitment, but is still finding its way to some extent. The Supporting style encourages and reinforces emerging strengths, shifting the balance from leading to following in the arena of process. The professional will likely have more opportunity to provide expert content at this point, but should be careful that it does not spill over into too much direction in the process leadership. In the Performing stage, the leader can rely on the group's high competence and high commitment to essentially lead itself in the process arena. The Delegating style is thus most effective at this stage for process leadership, enabling the professional to concentrate on optimizing her content contribution in collaboration with that of other participants.

Effective group leadership requires a professional to focus energy and attention on nurturing the group as a living system as well as on the quality of the group's output. Knowing the stages of the group life cycle and having a good grasp of the Situational Leadership styles enable her to recognize the group's developmental unfolding and to respond in ways that release its creative potential.

Group Learning and Self-Awareness

So much is riding on the results produced by core groups involved in long-cycle-time projects that the capacity for immediate learning and course correction is critical. As with individuals, learning in groups depends on continuous movement through the Learning Loop depicted in figure 19.4. Like individuals, groups must seek, interpret, and learn from an ongoing stream of feedback on their actions and interactions—from the individual, group, and whole-project levels—and then use that learning to redirect their actions and interactions.

The larger and more complex the project, the longer the delay is likely to be before whole-project-system results will be visible. The relationship of more immediate kinds of feedback to whole-project results remains to some degree uncertain, and care must be taken not to be inappropriately reactive. Nevertheless, a project group must learn to navigate with all feedback that is available. Senior decision makers or key stakeholder groups in the client organization or its ecosystem will likely begin responding to their interactions with the group or to their observations of its efforts early on. Even before this occurs, groups will confront feedback from their own internal process. The behaviors of individual members and the group dynamic as a whole provide a basis for improving the group's effectiveness and promoting its development.

As at the individual level, effective learning from feedback followed by effective implementation of learning is by no means guaranteed. The challenges groups face are typically more complex than those faced by individuals and require more conscious learning to be effectively addressed. However, groups are no more likely than individuals to move from unconscious to conscious learning without concerted effort to develop self-awareness—in this case, at the group level.

The stage of the group's development has a major impact on its ability to make this concerted effort. Prior to the Norming stage, members are unlikely to be able to do the assertive speaking and deep listening required to make internal feedback explicit and to reflect rigorously on it along with any external feedback received. The group cannot afford to wait until then, however, and reaching the Norming stage in fact depends on developing these capacities—not just as individuals, but as a group. The best way to promote the capacity of a group to learn is for the leader to establish structures and model behaviors that support the ongoing discipline and practice of group self-reflection from the beginning.

Structured discussions that support reflective practice should be consistent elements of group process designs. Agendas for regular meetings or one-time events should include time to address external feedback. These agendas should also include a segment at the end for internal feedback. What did we do well today as a group? Where could we have done better, and how? The regular practice provided by these activities can establish the discipline of learning in all areas of group performance. To make these exercises as meaningful as possible in the early stages of group development, the leader must do more than design and insist on these exercises. He must fearlessly bring forth problematic issues and guide the deep listening and inquiry needed for learning in the context of these exercises.

Well-facilitated reflective discussions of this kind will typically lead to the single-loop learning that characterizes continuous improvement. To help groups move from getting better at what they are already doing to the possibility of double-loop learning and doing something entirely different, Peter Senge and others recommend the process of dialogue as the most powerful means.[238]

In contrast with discussion, dialogue has no goal or structure other than exploration and is based more on inquiry than on advocacy. Instead of remaining attached to their own points of view, participants in a dialogue explore why they hold those points of view and become curious about why others hold the views they do. They set aside the rush to disagreement or to conclusion, and become comfortable with reflective silences. Focused dialogue poses an initial, extremely open-ended question to prompt a fruitful area of inquiry; unstructured dialogue may pose no initial question at all. While we recommend focused dialogue for project settings, groups who practice dialogue frequently will often begin the more unstructured kind quite naturally.

In dialogue, participants do their best to surface and "suspend" their assumptions in front of the group so everyone can reflect on them. Because our patterns of behavior are determined by the assumptions that make up our mental models, surfacing and examining those assumptions creates the possibility of new patterns of behavior. While this deeper learning occurs in the individuals practicing the deep listening and self-awareness disciplines that make the dialogue possible, what seems to emerge at the same time—perhaps because of the degree to which our mental models are shared—is deeper learning by the group as a whole. When individuals let go of as much defensiveness as possible, shared thinking in a shared context promotes emergence of a "group mind."

If group dialogue depends on the participants' individual ability to practice deep listening and self-awareness, how can a professional lead dialogue with clients and stakeholders who have had no prior exposure to reflective disciplines? The burden of this question is somewhat lessened by remembering that dialogue is not a science, but a self-organizing process in which people learn by doing. The group leader should explain this at the outset, also noting how dialogue is different from discussion. She should also explain that in facilitating the dialogue, she will refrain from traditional discussion moderation tactics—that she will instead hold the context of dialogue while participating along with everyone else.

In addition, the process leader should provide some simple guidelines to help participants focus their attention differently. Perhaps the most important of these is to set aside all considerations of rank or authority. Other helpful guidelines include

- Speak truthfully and only for yourself,
- Build on what others have said,
- Seek to expand the inquiry.

Facilitators sometimes also suggest that participants speak to the center of the room rather than to another individual in the group, and especially to avoid sustained cross-talk between individuals. This practice seems to encourage the emergence of the group mind.

The professional should not enforce these guidelines in the manner of ground rules the group uses for other processes. Rather, she should use her participation in the dialogue to model the rules. For example, when advocacy is predominating, she might pose a question rather than make a statement. If the dialogue seems to be stuck on the surface of an issue, she might share a deeply felt personal experience related to it. Or if the dialogue seems to be moving too fast for real reflection, she might say that she is having difficulty reflecting at this pace.

Although skillful discussion may be adequate for most project situations, groups that consistently practice dialogue begin to notice additional benefits. The capacity to think together with deeply embedded group self-awareness leads to greater efficiencies in discussion processes as well. In cases of knotty conflict, it may be possible to bypass cumbersome consensus processes. Exploration in dialogue, leaving a conflict uncomfortably unresolved for the moment, can lead to spontaneous emergence of alignment later. Well-honed inquiry skills can nearly eliminate the gyrations of defensive routines. Learning flows freely as feedback pours through such groups, enabling them to respond quickly and flexibly.

Professionals who develop the five leadership capacities discussed in this chapter will have done far more than create the optimal environment for their own technical contributions. They will have met and perhaps exceeded client and stakeholder expectations for leadership in addressing the social and technological complexity that surrounds their built environment projects.

Firms interested in taking on the challenges of offering collaborative work to clients should recognize that this form of value is fundamentally different from technical work, and that it involves a substantial effort to develop. These firms must factor this different form of value and this substantial effort into their new business model. Firms already offering it to some degree, but not differentiating it from technical work, should consider whether they want to get more serious about it and how to begin the process of differentiation and revaluation.

22 Transformative Work in Client Systems

No one can force change on anyone else. It has to be experienced. Unless we invent ways where paradigm shifts can be experienced by large numbers of people, then change will remain a myth.

—Eric Trist[239]

According to Richard Foster and Sarah Kaplan, the "success" life span of major U.S. businesses has been declining since about 1932. "Extrapolating from past patterns . . . by the end of the year 2020, the average lifetime of a corporation on the S&P will have been shortened to about ten years, as fewer and fewer companies fall into the category of 'survivors.'"[240] Like Christensen in *The Innovator's Dilemma*, Foster and Kaplan attribute this increasing failure rate to companies being too attached to existing customers, overlooking competition from unfamiliar quarters, and failing to innovate.

In *The Death of Competition*, published in 1996, James Moore pointed to the dearth of business leadership and strategy at the ecosystem level. Michael Porter and Mark Kramer implied a similar assessment six years later in their critique of corporate philanthropy.[241] We believe a lack of ecosystem-level leadership characterizes other kinds of organizations and institutions as well. Though organization failure takes different forms across different sectors, it appears to be increasing across the board—as the founder of VISA International, Dee Hock, observed in *Birth of the Chaordic Age*.[242] The failure rate is increasing largely because organizations have not adopted a living systems approach and, in particular, have failed to appreciate the importance of cooperative interdependence at the organization ecosystem level. Focus at the organization level was successful in the past, but accelerating change and increasing complexity have exposed its limits.

If we seek to intervene in escalating organization failure in ways that benefit society as a whole, we must understand life-cycle dynamics at the organization ecosystem as well as organization level. As Moore describes it, the business ecosystem life cycle mirrors the technology S-curve that describes the evolution of an innovation. The Pioneering stage reflects the initially flat, then gradually increasing opportunity present in the difficult start-up period. Expansion reflects the accelerating opportunity produced by replication and experimentation in more diverse contexts. The early-to-middle Authority stage reflects the continued opportunity provided by a consolidated value proposition and broader participation in its improvement. The later Authority stage reflects the flattening of opportunity as the value proposition ages and continuous improvement yields diminishing returns.

At this point the only way to capture new opportunity is through genuine innovation. A successful Renewal reflects a lengthened S-curve. If Renewal does not occur, Death occurs at some point after the top of the curve, usually after a period of decline. A return by that organization ecosystem to Pioneering with a new innovation—or a related Pioneering effort by others—reflects a jump to a new S-curve.

What is really evolving in Moore's view of the business or organization ecosystem is the way a given form of value is provided to society by a particular organization and its web. In play, then, alongside a particular web's way of working with an industry or institution paradigm, is that paradigm itself. As Moore suggests in his chapter on Renewal or Death, the life cycle of an industry or institution paradigm also mirrors the S-curve pattern and can be described as passing through the same general stages as the organization ecosystems working with it. Our use of Christensen's version of the technology S-curve to describe the current state of the architecture and engineering firm business model is an example. Some industries—and the organization ecosystems that populate them—evolve much more slowly than others and remain in the Authority phase for long periods. Public or not-for-profit paradigms and their ecosystems may also tend to evolve very slowly, with a flatter S-curve and a long Authority phase.

In these times of rapid social and technological change, not only organizations and their ecosystems but many of the paradigms of our major institutions are under extreme stress. Whether we are looking at business in the wake of disastrous accounting scandals and corporate collapses, at public education that leaves large numbers of students without basic skills, at health care we can no longer afford, or at democracy itself in a country where the majority of citizens do not vote, our institutions seem to be teetering beyond the top of their S-curves. They appear to be suffering a Renewal or Death crisis in which the path to Renewal or back to Pioneering is not clear, or cannot be agreed upon across deeply conflicting interests. Leading primarily at the organization level appears to be depriving our institutional and industrial paradigms of the degree and cooperative scale of innovation required for their renewal or replacement.

Meanwhile, organizations are working with these struggling paradigms, attempting to find a path to well-being without understanding, first, that they must innovate much more fundamentally than they have imagined, and second, that they need to focus on the organization ecosystem level. Though largely unperceived as yet, ecosystem-level, cooperative leadership may be the greatest emerging need in organizations, holding immense potential for value creation. The nature of an organization's influence on its ecosystem varies widely with its position—as a central, commodity, niche, or keystone player—in that ecosystem.[243] Whatever its position, however, having a better understanding of and working with its ecosystem relationships and dependencies can enhance an organization's efforts to adapt to changing conditions and transform itself. All of the leadership capacities for transformative work in client systems that we will explore in this chapter will include critical focus at the ecosystem level.

Because adaptive challenges are emerging across all sectors of our society, any project of significant scope and scale will likely contain at least some Type III opportunities. The large investments involved ensure that such projects are one of the all-too-few arenas of human endeavor where relatively long-term concerns must be considered. Inherent in consideration of a long-term physical solution are the strategic questions of who we must be or want to be as an organization or as a community at that future time. At least the potential for transformation is thus always present.

In our exploration of the Type III work leadership capacities in the context of transforming your firm, the Type III nature of the work was not in question. As you attempt to evolve the core of your professional practice and reinvent your business model, you have a sense of the adaptive challenge, but no clear problem definition. You have some ideas about how to begin to experiment, but no clear path or ultimate solution. The process is one of guiding the members of your firm *and yourself as a leader* through unknown territory. The scope, scale, and difficulty of the work are daunting, while the possibilities for deep learning are inspiring.

As we work with professionals in understanding Type III work in client systems, however, recognizing situations that truly represent Type III work becomes an issue. On the one hand, like all of us, engineers and architects often overlook Type III dimensions, preferring to view challenges through the more manageable lenses of technical or collaborative work. On the other hand, these professionals can also tend to see more transformative work than is actually there in work they are already doing, in part because adaptive challenges become technical challenges once they have been successfully met across a variety of situations. The following example illustrates why it is so easy to be misled in either direction.

An engineering firm was working with clients in a branch office of a corporation that had made a centralized decision to move to design-build for all its facilities projects. The branch office clients were faced with learning how to work in this new way with reduced staff. If the professional firm were to assist the clients in this learning, would this represent Type III work? Other groups have figured out how to work in the design-build model, and the professional firm does know quite a bit about it. If yesterday's adaptive work becomes today's technical work, is this really Type II work, with the professional knowing the general solution but requiring collaboration with the clients to bring it into being?

Yes, but. The clients must also work through their own trade-offs in order to be effective with reduced staff, changing internally to be able to perform all future work in this new way. And the professional must change the way he works with the clients as a result. Though the professional does have more knowledge of the parameters of solution than in many Type III situations, we think that the learning and change required of these clients—and the professional himself—puts this work more in the Type III arena. Because Type II and Type III work exist on a continuum, the definition may squirrel in one direction or the other.

As Type III work goes, this example presents a relatively low level of difficulty—the general direction is clear, the group is bounded and not too diverse, and the issues involve relatively low stakes compared to many other situations. In considering whether to offer Type III work as a distinct and separately valued service, firms should ask themselves what level of scope, scale, and difficulty they are willing to take on. At the same time, they should develop means of being very clear with clients about the unknown dimensions as well as deeper learning and internal change of Type III work that require a more complex form of leadership.

Type III situations will present themselves in different ways. A few clients will know that this difficult kind of work is what they want and need. For example, pursuing the Ford Motor Company's $2 billion makeover of its River Rouge plant, cited in chapter 15, as a project that involved only technological innovation would be almost unimaginable. Indeed, McDonough and Braungart quickly found themselves also involved in adaptive work in Heifetz's sense. The city of San Jose and San Jose State University had already decided to pursue the major social innovation of a joint library when they engaged ABA. Although these clients did not realize the extent of necessary adaptive work that could be done as part of the planning process, Type III dimensions were clear from the outset.

Social and political controversy surrounding a project can also place it in the Type III arena right at the start. When a resolutely "no-growth" city experienced ad hoc growth for which it was not prepared, civic leaders and citizens were faced with basic infrastructure decisions that challenged their fundamental values. Clients dealing with such situations are likely to recognize the need for a different kind of help that has a different kind of value. In a third type of situation, clients will be at a major developmental milestone, such as the decision to expand to other locations or to merge existing organizations and modify facilities accordingly. In situations like this, the need for some form of transformative work is likely, even though it has not been explicitly articulated.

In many cases, however, the need for Type III work will emerge from what was thought to be a Type I or Type II situation. In a recent example, environmental engineers had been hired to solve the problem of a highly contaminated watershed that affected a large city's water supply. When the technical solution was completed, stakeholders realized that it didn't really solve the problem. What did the community want to do with and in addition to the technical solution? This struck us a case where an engineering firm could provide valuable leadership on a societal scale. To do so, the firm would need to be prepared to offer transformative work as a separate and differently valued service. It would also need to be prepared to recognize when the nature of the work that needs to be provided is changing and to take the initiative in proposing and negotiating a new value proposition.

No less than Type I and Type II work, Type III work in the built environment context can vary enormously in scope, scale, and difficulty, as even our few examples attest. Not all situations will require all five of the leadership capacities that we have discussed in the context of transforming your firm. However, we believe that most situations will require most of them, at least to some degree. You might want to refer once again to table 16.3 to review the Type III leadership capacities and the shifts of mind for getting out ahead upon which they depend.

Through our exploration of the social capacities for higher value technical work and the leadership capacities for collaborative work, you now have a fuller understanding of the hierarchy of competencies and skills entailed in the leadership capacities for transformative work. While transforming your firm is essential training ground, learning to offer transformative work to clients will require substantial additional effort. This chapter will highlight further aspects of the advanced leadership capacities (presented in chapter 19) that come into play in providing transformative work to client systems.

Because its demands are great and its stakes high, perhaps only a small percentage of firms will venture into this area. Our aim in this chapter is not to provide detailed instruction but rather to help firms decide whether they want to pursue transformative work. Even if you decide not to offer Type III work, you can leverage the knowledge you gain here and through the experience of transforming your firm. You might form a synergistic joint venture with a consulting firm that does Type III work, or at least advise clients when they need that kind of help and work effectively alongside those who provide it.

Our exploration of transformative work leadership capacities is a work in progress. Though we are learning more all the time, no one fully grasps how to facilitate transformation in organizations, let alone at the more complex level of organization ecosystems. As the living systems sciences have been teaching us, such processes are nonlinear and somewhat mysterious. If it were easy, we would already be doing it! With this chapter, even more than chapter 19, we are venturing to the leading edge of our own Type III effort to transform the built environment professional services industry. We hope our ideas give you a point of departure with which to forward our shared enterprise.

Strategic Conversation

The exhortation to "know your clients' business" has been relentless for many years now, along with the aspiration to become "a strategic partner" in that business, whether health care, education, government, a non-profit, or a corporate enterprise. Few technical professionals have achieved these lofty goals, for both the mental and business model reasons discussed at length throughout this book. While technical and collaborative work benefit enormously from deep knowledge of your clients' businesses, transformative work requires it. Strategic conversation is the primary means by which you acquire the strategic knowledge you need in order to identify and perform transformative work. Leading such conversation is also one of the key processes you can use for immersing clients in transformative experience. At the moment a Type III work assignment crystallizes—either at the beginning or at a later point in a project—you will begin to be able to achieve these twin goals in one motion.

Strategic conversation with clients takes place in the context of their mind-sets, organization culture, prior strategic planning activities, and current challenges. If they have not done any formal strategic or scenario planning, the informal conversations you hold can help them begin to answer the basic questions of those disciplines that we discussed in chapter 19. If instead you discover that clients have done some formal planning, your strategic conversations can integrate project dimensions with what they have done, help them place current conditions in that context, or evolve that context to embrace current conditions.

To ensure an organization-ecosystem level consideration of strategic direction, a professional should pose questions from the value network model—viewing the ecosystem from a value creation perspective—as well as from the organization ecosystem model itself. Is the client organization most fruitfully viewed as the center of its own organization ecosystem or as participating in one that is centered around another organization? If not at the center of its own, is its role that of a commodity, niche, or keystone player? Who are the other principal players? What is the state of health of its ecosystem as measured by the number of organizations participating and their economic health?[244] At what stage of maturity is the value proposition underlying that ecosystem? Its own business, if different from that of its ecosystem as a whole? What are the key value criteria that its customers or clients apply to the products, services, or experiences it offers? How can the value of those products, services, or experiences be enhanced? If those products, services, or experiences are becoming less competitive, what new forms of value could the client organization think about providing in the future?

Through this process, you come to understand the organization and its ecosystem more deeply. Clients further explore and express their identity and aspiration as an organization while better understanding forces in their larger ecosystem contexts that pose adaptive challenges, either now or in the probable future. In being invited to make or return to these fundamental considerations, clients experience—or re-experience—their organization's potential to shape its circumstances, take a leadership role in its ecosystem, and provide value to society. This experience plows optimal ground for transformative work.

We have found that many architects and engineers—especially if they have been working as a subconsultant—feel boxed in by clients who view their role as one limited to technical work. It is hard to know whether this resistance to initiating strategic conversation resides more in themselves or in their clients. In any case, we have observed that when professionals break through this barrier, they have been quite surprised at the delighted response from many clients. Clients are often as mired in operational concerns as professionals and find strategic conversation to be revitalizing. Clients have been particularly delighted to have professionals inquire into the nature and state of their organization ecosystem. This model is new to them and gives them a revealing new way to view their own reality. As with strategic conversation in your firm, the key to success and high impact is the spirit and process of inquiry rather than advocacy.

Another type of challenge for leading strategic conversation arises in such settings as branding discussions, in which clients now involve architects for the purpose of designing a facility that fits their desired image. These discussions are inherently strategic in nature, but often proceed after other key strategic decisions have already been made.

Many questions thus remain off-limits, and architects are further constrained by being participants rather than leaders in this process.

The best time to introduce strategic conversation in a client relationship is undoubtedly right at the outset. This opens the possibility of repositioning the firm in more potential rings of the client's value network, as well as reframing the work the client is requesting in a more strategic way. As in our earlier example of the commuter college in earthquake country, consideration of the organization's current state in the context of its ecosystem can turn a technical work opportunity into a transformative one.

If such a reframing has occurred, or if a project has begun with Type III goals included, a professional can determine whether more work on the strategic planning or on the scenario planning side of strategic conversation would be helpful for beginning to immerse the clients in their transformative work. Strategic conversation can thus become a centerpiece of process design for the planning stages of a Type III project—a process to which the professional brings Type II leadership capacities as well. Real-time inquiry into the implications of emerging trends can continue to facilitate transformative learning over the whole life cycle of a Type III project.

The professional with leadership capacity to lead strategic conversation in client systems completely shifts her thought process into her clients' worlds. As in her own firm, she is as able to contribute provocative content as to facilitate deep inquiry in others—all while remaining in a state of not-knowing herself. She is able to initiate transformative work by helping clients define a direction for change that both expresses their organization's identity and aspiration and effectively responds to conditions emerging in their environment. At the same time, she is able to help clients engage in their own process of discovering strategy on an ongoing basis. Because fundamental strategic questions are inherent in the scope, scale, and term of built environment projects, we believe that architect- and engineer-led strategic conversation can provide a new level of value that clients would be more than delighted to embrace.

Systems Thinking

Strategic conversation and systems thinking go hand-in-hand in transformative work. Developing strategic direction with the organization ecosystem and value network models exemplifies what we call Level 1 systems thinking—the kind that is based on understanding the connections and relationships among elements of a system. In considering project-related strategic decisions, clients can supplement these externally oriented Level 1 systems thinking tools with an internally oriented one like Chuck Schaefer's model for a high-performance work system (Figure 19.1). With the help of these tools, the professional can help clients think systemically about how an individual decision may affect various elements of their internal system as well as other organizations and relationships in its ecosystem.

As adaptive challenges are surfaced through these explorations, a professional may recognize patterns of system behavior that are holding the organization back. These patterns may show up as system archetypes, or chronic problems driven by mental models that the professional suspects are no longer serving the client. If the client organization culture is centered in Alignment or Mutuality, involving members from relevant parts of the organization and its ecosystem in a Level 2 systems thinking exercise may be very helpful in opening up new possibilities. Even if clients and other stakeholders have not been exposed to this methodology, the professional who has gained confidence with it will likely be able to guide them to bound the problem, identify the key variables, arrange them in causal loops, and reflect on the mental models that drive the mutual influence patterns.

We want to introduce here a third level of systems thinking: the discipline called system dynamics. Figure 22.1 captures the way in which each level contains the elements of the prior level(s).

MIT's Jay Forrester founded the Level 3 discipline, and it provided the basis for Peter Senge's popularization of Level 2. In system dynamics, causal loop diagrams are constructed from variables, just as in Level 2. But the system dynamicist goes on to quantify the variables and perform calculations that attempt precisely to simulate the system behavior. Successful quantification requires accurate values extrapolated from historical data or appropriate hypotheses, as well as formulas that correctly reflect the way that variables affect one another. While the difficulty of achieving this kind of accuracy is great, without the ability to quantify, we have little means of grasping either the scale or the time frames involved in the dynamics we are considering in a system.

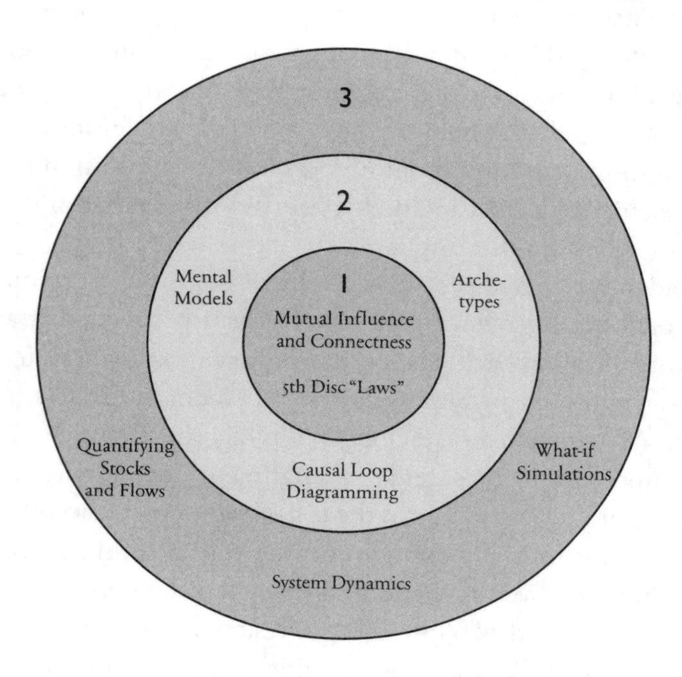

FIGURE 22.1. Three Levels of Systems Thinking

Some of the most famous examples of system dynamics simulations involve ecological issues such as the problem of sustaining fish populations,[245] but quantification is equally important for organizational matters and their implications for the built environment. Imagine being able to simulate the consequences of a potential epidemic for the capacity of a health-care system, or the potential impact on a large business of baby-boom retirements, increasing customer defections, or immigration patterns. Combining system dynamics with scenario planning provides an extremely powerful way to see possible futures for which built environment project clients need to plan.

Just as Senge's version of systems thinking has been underused in the corporate world, the potential of system dynamics to inform policy making has never been realized. Practitioners attribute its underuse to complexity that exceeds the understanding of most policy makers. While the technical aspect of the discipline need not be practiced directly by those seeking to lead decision-making processes, the advanced leadership capacity we are proposing here requires enough understanding of the issues of quantification to guide and make good use of expert practitioners.

We believe that effective use of system dynamics will help guide the most important decisions that are made about our physical environment in the future. We might even hazard the futurist guess that the ability to apply system dynamics to decision making for large human systems will become as essential a leadership capacity for aspiring CEOs as reading a balance sheet has been in the past. We would like to encourage architects and engineers to be on the forefront of this trend.

Organization Ecosystem and Organization Development

In chapter 18, we introduced developmental awareness in part as a basis for achieving a high degree of fit between the products, services, and experiences provided by the consultant and all the dimensions of the client system: individual mind-set, organization culture, and life-cycle stages of the organization and its ecosystem. While fit applies to all three types of work, our purpose here is to explore the unique province of Type III work: helping a client organization and its ecosystem to develop in a desired direction.

Someday soon, in what Moore has called the "Age of Business Ecosystems," we predict that professionals will work for groups of ecosystem-linked organizations as well as for individual ones. As long as clients for Type III work are single organizations, however, much of that work will need to focus at the organization level. Given this requirement, a professional can work at the ecosystem level in two ways. The first is what we have called the second realm of value creation: developing the capacity of a client organization to lead its ecosystem. The second is our third realm of value creation: making direct contributions to the client ecosystem.

With a client organization culture of Alignment or Mutuality, a professional can work at the ecosystem level in both of these ways simultaneously. With a Familial, Transactional, or Self-Expression culture, however, a client organization will be unlikely to be able to lead its system without substantial transformative work. In this case, the professional should pursue the ecosystem level only through his own direct contributions.

In chapter 19, we introduced the leadership capacities for Type III work in the context of transforming your firm. That discussion amounts to an exploration of organization development and provides the basis for working at the organization level in client systems as well. We focus our discussion here primarily at the ecosystem level, with references back to the organization level. Though our primary focus is client systems, you can also apply these principles for pursuing development of your own firm's ecosystem.

A professional is likely to encounter a wide variety of developmental scenarios and forms of nonlinearity in Type III work situations. We present this leadership capacity as a set of disciplines that can be applied across scenarios and at different times in the unfolding of a project. Many of these disciplines will require little elaboration once highlighted, because of the ways in which they build on the social and leadership capacities we have already discussed. Others involve new material that requires explanation.

Help clients and other stakeholders understand and invest in their ecosystem
Since most clients don't yet think about this level of their system, the first step is to help them see it and appreciate its importance to their own organization's well-being. Humans can't begin to self-organize at a level we don't yet recognize. If you have been conducting systems-thinking-informed strategic conversations incorporating the organization ecosystem model, you have already begun this process. Expanding this effort to

other organizations involved in the ecosystem builds and reinforces the process of the ecosystem coming to identify and know itself as such—to see itself as a unique living system with an identity in its own right.

Once clients and other stakeholders have understood their positions and interdependencies in their ecosystem, a professional should help them understand the ecosystem-level challenges that they share. Is their ecosystem under competitive pressure from other ecosystems? Is their ecosystem resource-poor or resource-rich? Is it healthy or in decline? Does it possess the vitality of ongoing growth and development? An understanding of their shared fate is the strongest basis for developing clients' and other stakeholders' willingness to invest in their relationships and in the ecosystem as a whole. As at the relationship and group levels, investment is a powerful reinforcing process that builds health and developmental momentum.

Determine and make highly leveraged direct contributions

Your own direct contributions to the ecosystem provide another way to build those reinforcing processes. To determine what resources would be most valuable, the professional should use what he has learned from strategic conversations and other observations about an organization ecosystem's nature, current state of health, and adaptive challenges. A state-level Department of Transportation ecosystem, for example, may be rigidly structured, politicized, and power-driven. As a result, it may be closed and relatively slow to change, despite being resource-starved. To make the most of its limited resources, the DOT ecosystem might need more powerful information and tools for prioritizing infrastructure maintenance projects. But it might also need an infusion of ideas to open it up a bit. At the other end of the spectrum, an information technology firm's ecosystem may be relatively unorganized and nonpolitical, operating with distributed power. It may thus be open, turbulent, and resource-rich while struggling with the challenges of instability. This firm's ecosystem might need anchoring relationships in its local economy—more connections to create more access to local employees, suppliers, and customers.

Help the client organization align its life-cycle stage with that of its ecosystem
In 1998 Herman Gyr published his model of the organization life cycle again in *The Dynamic Enterprise*.[246] He and his co-author included a brief look at the relationship between the organization life cycle and the business life cycle as represented by the S-curve. Their diagram closely parallels the functional correlations between the organization life-cycle stages and Moore's organization ecosystem stages, noted in chapter 19. In addition, they comment, "an enterprise is most effective when the business and the organization are developmentally aligned, that is, when one doesn't significantly lag behind the other."[247]

Ideally, an organization's life-cycle stage would be closely aligned with or very slightly ahead of its ecosystem's life-cycle stage. This kind of relationship is most likely to evolve with the kind of proactive ecosystem-level leadership that Moore was attempting to develop in *The Death of Competition*. Figure 22.2 shows this relationship schematically.

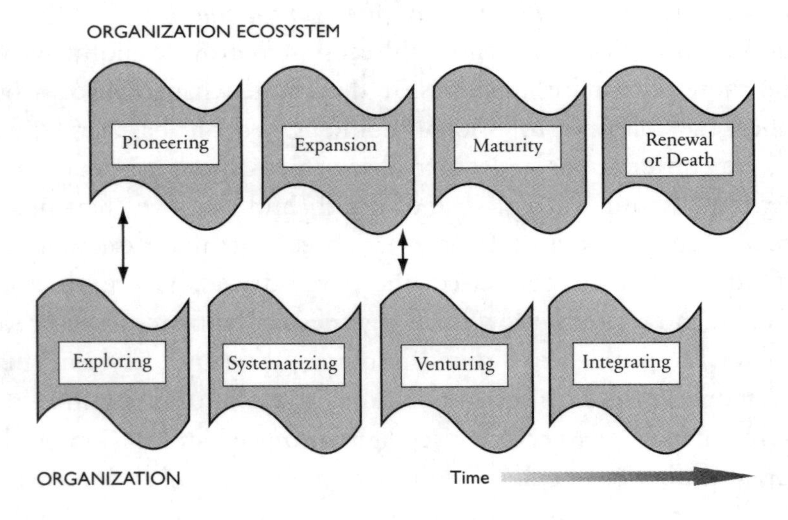

FIGURE 22.2. Proactive Organization Leadership

Being a little ahead of the organization ecosystem life-cycle stage enables the organization to pull the ecosystem, much as an effective leader in an organization is able to pull rather than push its members in a desired direction. However, an organization may make the mistake of getting too far ahead of the larger forces of its ecosystem. Lisa Friedman and Herman Gyr presented the following example in *The Dynamic Enterprise*:

> During the 80's [an oil industry supplier] had grown tremendously and had anticipated continued growth. In the 90's, however, as oil companies streamlined contracts with suppliers [this supplier] realized that it had placed people in roles to handle complexity and growth that had never materialized. They had built the organization ahead of the business and the business never caught up. It was time to retrench . . . [248]

As a niche player, this supplier had gotten out of touch with the state of the oil-company-centered ecosystems in which it participates.

On the other hand, if an organization's leadership is simply reacting to the dynamics unfolding in its ecosystem, the organization will tend to lag its ecosystem's development, as shown in figure 22.3.

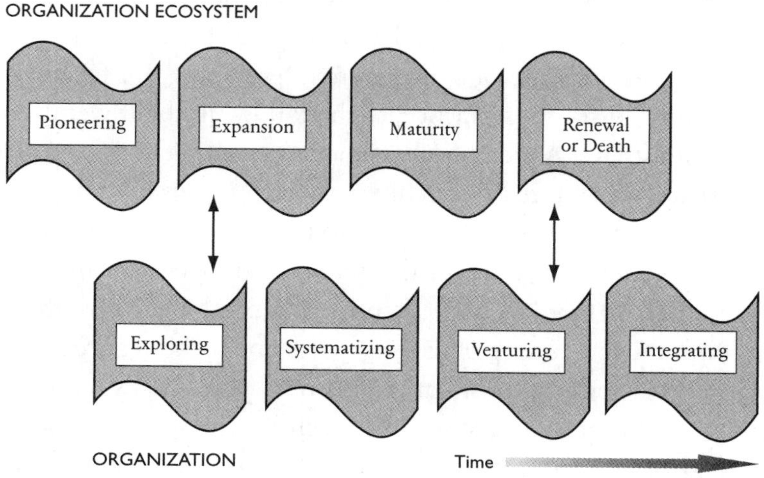

FIGURE 22.3. Reactive Organization Leadership

In another example cited by Friedman and Gyr: "[A high-tech company] had grown to annual revenues of almost $1 billion using systems, policies, and procedures carried over from its start-up days. All individuals . . . made decisions their own way . . . Although they agreed . . . they needed more systems and more uniform work processes . . . the idea of systematizing was equivalent to becoming ossified, rigid . . ."[249] Understanding how its own life-cycle stage aligns or doesn't with the life-cycle stage of its ecosystem enables a client organization to see how it needs to develop. This understanding also enables clients to make project decisions that are strategically aligned with their ecosystem's life-cycle stage. A new facility for this high-tech firm, for example, could balance individual space with group space for beginning to coordinate more. Facilitating such decisions will of course require the professional to work through cultural filters such as the Independent-mind-set-based Self-Expression culture revealed in this high-tech company. As an organization develops more alignment with its ecosystem it becomes more sustainable. It also increases its ability to contribute to and even lead its ecosystem, from any position it might occupy.

To help a client organization make these developmental improvements, the professional must be able to recognize life-cycle stages at both levels and grasp how they are interacting. Developing this facility can be challenging and requires practice, as the following slightly embellished example illustrates. A small high school in a relatively rural American Midwest town has hired an architect to help it develop an expansion plan. This effort has been undertaken to counter the district administration's competing plan to merge this school with two other high schools to form a larger, better equipped entity that can better serve the whole community. A vocal group of parents in the small school are doing everything in their power to obstruct the district's plan, with support from a significant percentage of the school's teachers. The small school appears to be part of its ecosystem's core function as Moore defines it, with the district administration occupying the central position.

While a merger with an entity from another ecosystem might be viewed as a form of Expansion, this ecosystem is attempting to consolidate several of its existing members. Improvements possible with economies of scale often characterize Expansion, but in this case seem rather to express Maturing within existing scale, or moving from Expansion into Maturity with this envisioned project. Fearing its loss of autonomy and control, the small school might be seen in two different ways. On the one hand, it might still be in the Exploring stage, dominated by a Familial culture for which scale itself is suspect. On the other, it might have Systematized itself from the orientation of a Self-Expression culture that now resists Systematizing by anyone else. In either case, it appears to be lagging its ecosystem's early- to mid-Maturity stage.

Helping this organization align its own life-cycle stage with that of its ecosystem will require transformative work. As part of a Type III assignment, the architect might offer to help the dissenting parties determine how best to proceed for the community: whether to keep the center of gravity at the organization level or raise it to the level of the ecosystem as a whole, and if to raise it, how to optimize health and development at that level. If the result is to move ahead with the consolidated larger high school facility, there will be transformative work all along the way to help the small high school adapt. We might even imagine a scenario where, through this work, the resisting school becomes the new school's biggest advocate.

Facilitate developmentally healthy responses to adaptive challenges

In addition to being able to recognize life-cycle stages at the organization and organization ecosystem levels, the professional must grasp the dynamics of the developmental process itself. As Moore explains, it unfolds more as a messy continuum than as discrete stages. If we look closely, we can distinguish early, middle, and late phases within each stage. These internal phases involve challenges more incremental, but no less significant, than those of the major transitions. Major transitions may be more gradual than sudden, and they may involve moving backward as well as forward. A return to Pioneering might occur from an earlier stage than Renewal or Death, as it did for Charles Schwab with the stunning innovation of Schwab.com out of the middle of its

Authority stage. Or an organization in Authority might move back to Expansion in order to pursue a series of growth-oriented acquisitions. Renewal or Death may not be the only stage at which an organization ecosystem confronts mortality. Like an organization, it may face premature death at any point from a variety of possible developmental failures, including getting stuck and beginning to stagnate or being unable to respond adaptively to a major external disruption.

According to the theory of punctuated equilibrium,[250] evolution does not occur in a smooth incremental process. Instead, long periods of relative equilibrium are disturbed by sudden, apparently discontinuous leaps catalyzed by turbulent states. For an organization or its ecosystem, these turbulent states are the developmental turning points that emerge from confronting greater complexity. This new complexity can be caused by internal processes of outgrowing a particular stage, challenges from the external environment, or both. As we have seen in Gyr's organizational paths of impairment, adaptation and transformation, as well as in O'Hara's psychotic, neurotic, and transformational responses at the individual level, three general developmental paths are possible from any point. The leap that IBM's ecosystem made after long foundering in Renewal or Death is an example of a transformation that occurred within a single stage.

Understanding the developmental dynamics unfolding in the client system enables a professional to help clients and other stakeholders understand what they are experiencing as well as the consequences of their various choices. This understanding also provides the basis for leading adaptive work in Heifetz's sense that can enable the system to move toward the edge of chaos where it can be most creative. A combination of developmental understanding and leadership of adaptive work provide the best means we know for facilitating healthy responses to adaptive challenges.

Build the ecosystem into and through the project process

To promote development at the ecosystem level, the professional works with the client organization to identify key ecosystem members for inclusion in the project process. Choosing the right people to participate in the right processes can ensure their meaningful contribution. And developing personal relationships with those individuals and groups can provide additional glue for their involvement and build their stake in the project and their investment in the ecosystem.

Determine how the project can contribute to meeting ecosystem-level challenges
With the help of other members of the ecosystem, a client organization can further clarify challenges being experienced at the ecosystem level and how its project can contribute to meeting those challenges. Some contributions can be made by the physical solution, and some by the human systems outcomes of the transformative work. Conversations among a cross-section of ecosystem members can help to focus the goals of the transformative work.

Shape the project process to support developmental goals
Every project activity provides an opportunity for both task and process outcomes. The art of ongoing transformative work involves taking advantage of literally every project activity to promote not only outcomes related to the built solution but also outcomes related to the human system's developmental goals.

Cultivate a living-systems-friendly philosophy of development
Echoing Maslow's understanding of individuals, Arie de Geus asserts that a living company's purpose is to realize its own unique potential as well as the unique potential of all the individuals in it. As a yet more complex human system, the organization ecosystem likely has this purpose too—to realize its own unique potential as well as the potential of all the organizations and groups within it. The self-actualization level of Maslow's hierarchy is the evolved state of being from which an individual's unique potential can best emerge. The same can be said of the Co-creation stage of a relationship, the Performing stage of a work group or team's development, or the Integrating stage of an organization's life-cycle. A sustained Renewal stage represents the analogous

goal for an organization ecosystem.

A second, and no less important philosophical foundation is fully grasping the essential role and value of each organization ecosystem stage and the way each stage builds on the one before it. As with stages at all levels of living systems, treating an ecosystem currently in Pioneering or Maturity as if it were in—or trying to yank it toward—Expansion or Renewal does not help it develop, and may even impair it. Knowing where the ecosystem is in its life cycle and then helping it consolidate or make its own healthy movement from that state—which may in some cases involve circling back—is the only kind of developmental process that works.

A third important philosophical foundation is to let go of the idea that transformation can be controlled and managed, no matter how much clients may appear to want this reassurance. The turbulent periods out of which developmental leaps are made cannot be predicted but issue from the unique conditions operating between a system and its environment at a given point in time. Cultivating patience, an ability to sense a system viscerally, and a willingness to embrace ambiguity and discomfort are the best means of preparing to support a client system moving through the unnerving and unpredictable transformational process.

A final philosophical foundation is the notion that there may be a time for an organization ecosystem, or a living company, to die. Although members may seek to survive by joining other ecosystems or setting up Pioneering, a return to the beginning is like the natural disaster striking a biological ecosystem. If organization ecosystem death also enables an even more evolved ecosystem eventually to emerge, such death may be Nature's way of ensuring that life on a macro-scale continues to evolve at a significant rate. Some of our ailing institutions may need to die to make way for the new.

Leading Innovation

Understanding the organization ecosystem life cycle illuminates the difficult developmental challenges that confront organizations across all sectors of contemporary society. Those challenges reveal the need for more fundamental innovation than most organizations realize. At stake are not only the organizations themselves but our underlying institutional and industrial paradigms. Though the equilibrium-seeking inertia operating at this society-wide level is enormous, turbulence resulting from paradigm-level stress may be setting up the conditions for developmental leaps. We may take further encouragement from what we know about the diffusion of innovations: transformation at the individual ecosystem level may be the only way to drive change at the larger system level. In any case, vigorous innovation is the best means of driving Renewal at the organization ecosystem level, which in turn is the best way to promote robust longevity at the organization level. Because built environment projects often engage organizations in fundamental ways, the opportunity to lead significant innovation at an organization or organization ecosystem level is often present.

The developmental challenges of organization ecosystems further emphasize that social innovation is at least as important as technological innovation. As we have already mentioned, social innovation is the aspect of Type III work that changes systems by the external means of designing and implementing new processes and behaviors. Though internal adaptive work in Heifetz's sense is one of the key disciplines for organization ecosystem development, those disciplines also include the social innovations of involving more members of an organization's ecosystem in project processes and involving them in new ways.

Some Type III assignments—such as Ford's desire to transform its River Rouge plant through technical innovations for ecological sustainability—will feature innovation from the outset. Others, like ABA's project to merge a city and a university library, begin with the challenge of a social innovation. Still others, like the no-growth town facing unwanted infrastructure expansion, begin with the obvious need for adaptive work, but may present opportunities for social innovation that can promote that internal change. In situations where Type I or Type II assignments morph into Type III opportunities, the professional should always be alert to the role that innovation—either social or technological, or both—can play.

Architecture and engineering firms may be more prepared than they realize to lead social innovation. For example, an architecture firm that has designed many urban community colleges might be in an excellent position to help a particular college begin to achieve Renewal through information-technology-based social innovation in its organization ecosystem. This Type III assignment could be pursued with or without existing forms of Type I or II work, or in combination with Type I innovations.

In chapter 17, we discussed leading innovation in the context of transforming your firm. The first of the disciplines we explored—identify innovation-friendly clients—will take care of itself if you have gotten to the point of a Type III assignment or even a Type I or II assignment that is in the process of morphing. Clients who hire you to lead innovation in their systems will be innovation-friendly—at least to some extent—and may even be some of the same clients with whom you have been able to lead innovation focused on transforming your firm. The other eight disciplines we discussed in chapter 17 can all be applied to the client system context.

Value-rich innovation challenges will likely be inherent in a Type III project definition, but have the potential to further emerge as the project proceeds. Revealing the Type III potential of a Type I or II project may depend on identifying such challenges. Uncovering really powerful ideas for innovation focus may be the most difficult of all the disciplines involved in leading innovation. Practicing this discipline in client systems adds difficulty. The professional must shift her awareness from the worlds of her own clients to the worlds of *their* clients—yet another shift in fundamental frame of reference.

Articulating the innovation-rich challenges in inspiring ways will involve the ability to vividly convey not only why the challenges are important and urgent, but also how the capabilities of the individuals and organizations involved are uniquely qualified to address those challenges. If you are working at the ecosystem level, you will need to learn enough about the key member organizations who will be involved in the project to understand their capabilities. You will also need to organize venues where individuals from the different organizations can hear your message together. This shared experience will lay the groundwork for taking hold of shared purpose and potential synergy.

The degree of isolation necessary for effective incubation will vary depending on the relationship of the specific clients for the Type III work assignment to the rest of their organization and ecosystem. If the effort can be mounted at the ecosystem level, you might want to consider a central project location that is not part of the client organization but that provides easy access to members of the ecosystem who will be participating. If the effort will be centered at the organization level, consider the degree to which the organization has been mobilized behind an innovation effort. If not much, you should consider setting up a skunk works environment for the project, at least initially. If it works anything like the "peace room" created for William McDonough's Type III innovation at River Rouge, as interest builds, the skunk works location may become a magnet for building commitment to the effort.

Gathering the right people also depends on which level is your center of gravity for the project. If you are centering at the ecosystem level, you must hunt out your Innovators and a handful of Early Adopters relative to the project's innovation arena across all the key member organizations.

As you recall from chapter 17, cultivating the conditions for creativity involves optimizing information flow, diversity in ways of thinking, power differential and connectedness among participants, as well as level of anxiety. Initially, you can cultivate these conditions as well as promote state-of-the-art disciplines of individual and group creativity through design of project staffing, structure, and process. As the project unfolds, you can continue these efforts through exercise of ongoing personal influence. These two aspects of leading innovation become more challenging as project scale and number of participants increase. The best way to ensure that sufficient influence of this kind is present is to partner with formal leaders of the client organization and other ecosystem organizations who understand the importance of these and the other disciplines involved in leading innovation. Depending on these leaders' individual mind-sets and organization cultures, you will be more or less able to develop this alignment and the strong partnership that shared leadership will require.

As a result, you may also need to depend on individuals in the organization and its ecosystem who exercise influence without authority. Such individuals are probably the Early Adopters who will help to drive diffusion of the innovations as they begin to gel. Cultivating networks of Early Adopters throughout the organization and its ecosystem—both through your own efforts and through the Early Adopters participating in the core project group—will richly repay any effort you can give it throughout the life of the project. Basing this cultivation on the nature and power of the ideas involved in the innovation challenge is often the best approach.

You will also need strong partnerships with formal leaders in the organization and its ecosystem in order to develop support for prudent risk-taking and resources for addressing any human systems issues that arise in relation to technological innovations. You may be thinking that in order to form such partnerships—and in fact to practice any of the leading innovation disciplines we have discussed—at the ecosystem or even the organization level of client systems, you must have a high degree of formal project authority. Negotiating such authority would in fact be helpful in setting up the terms of the Type III work assignment. At the same time, a great deal can be accomplished from a position of leadership that is established through influence rather than formal authority.[251] The art of leading without authority is central to our fifth leadership capacity for transformative work in client systems, to which we'll now turn.

Leading Whole-System Learning and Adaptive Work

All of the advanced leadership capacities discussed in this chapter involve journeying into the unknown territory of adaptive challenges and guiding others on that path. Some adaptive challenges can be met with external changes alone; for example, those involved in some types of strategic shifts, technological innovations, or even social innovations. Most, however, will involve some degree of internal change in beliefs and perhaps even values underlying behaviors: the double-loop learning that Senge associates with profound change and that Heifetz calls adaptive work.

Taking the whole-system learning and adaptive work leadership capacity from your own firm to client systems is a great challenge. While the key disciplines are essentially the same as the ones discussed in chapter 19, the architect or engineer is now operating in an arena of vastly greater diversity and complexity, with influence that is accordingly diluted. Even when effectively negotiated from the start, project authority does not extend to the whole system involved in achieving the transformative work goals. Leading whole-system learning and adaptive work at the organization level is certainly easier than at the level of its ecosystem.

At the organization level, as in your firm, you will likely be able to use primarily a learning-organization approach if the client culture is Alignment or Mutuality. Even though client organization members in these cultures may not have practiced the personal mastery, mental models, shared vision, and team learning disciplines explicitly, they will be operating from an Interdependent mind-set. In this context, the professional will likely be able to personally model and build enough group practice of the disciplines into project process to drive the learning loop: at least a basic level of seeking and receiving feedback, of self-evaluating and learning from feedback, and then of using that learning to modify thinking and behavior. She can use Heifetz's adaptive work approach to address emotionally charged issues if needed.

If the client organization culture is Familial, Transactional, or Self-Expression, however, Heifetz's adaptive work will be a more effective point of departure. These organizations will be less oriented to learning in an interdependent way. And they will be less likely to recognize that meeting their adaptive challenges will require internal changes in beliefs or values. They will be more tempted to apply a technical fix that can only temporarily moderate the threat to their well-being. Developmentally speaking, a technical fix at best enables an organization or ecosystem to stand still. For these organizations, moving forward within a current stage or to the next stage will likely require adaptive work in Heifetz's sense. A professional can best lead adaptive work in these cultures by learning how to apply Heifetz's principles from a project position that carries no formal authority for the organization.

At the ecosystem level, the professional is unlikely to encounter a critical mass of Interdependent-mind-set based organization cultures. Member organizations and groups tend to exhibit a greater degree of diversity of mind-sets and to have a long distance to go to achieve alignment. That is not to say that introducing a lessons-learned dimension of project process could not be helpful, or that a professional should never attempt to use a dialogue format to surface mental models when working in ecosystem-level settings. Rather, we are suggesting that these approaches are unlikely to be able to effectively address the tough issues that are bound to arise—unless combined with substantial work pursued with Heifetz's approach. Table 22.1 summarizes the principles that Heifetz describes for leading adaptive work when you don't have formal authority in the system you are trying to lead.

The professional who leads without formal authority has the difficult challenge of establishing informal authority solely through influence. This challenge involves many of the same disciplines involved in developing the ecosystem: identifying key stakeholders, building relationships with and among those stakeholders, communicating persuasively in opportunely chosen venues, and walking her talk in all system contexts. On the other hand, she does have the advantage of what Heifetz calls "creative deviance, more freedom from the norms and constraints of formal authority."

The informal leader needs to perform some of the same key adaptive leadership functions that the formal leader performs, from her different vantage point. She must recognize and articulate the adaptive challenge, which requires moving between front-line action and a big picture view available only by stepping back. She must also keep the focus on the issue and not succumb to stakeholders' desire to avoid the problem or have it resolved by someone else. She must ask the tough questions and encourage participation, leadership, and dissent from other people in other parts of the system.

Identify the adaptive challenge
- Access front-line information through your own role and relationships
- See conflicts as clues
- Frame key questions and issues

Identify key stakeholder groups and authority figure(s)
- Get on the balcony
- Look for patterns in the larger view beyond your role and authority
- Understand power relationships
- Grasp the natural boundaries of the holding environment

Operate beyond your authority
- Cultivate relationships and venues
- Leverage existing relationships to develop new ones
- Develop relationship(s) with key authority figures
- Take advantage of latitude for creative deviance

Become a lightning rod of attention
- Identify the impact of the issue on your work and ability to contribute
- Look for moments when the issue surfaces for others or for yourself
- Use yourself as an embodiment of the issue

Mobilize the stakeholders
- Understand the interests of other stakeholders
- Develop an educative strategy that links stakeholder interests to the issue
- Use relationships and venues to educate stakeholders
- Enlist stakeholder participation in actively working the issue

Modulate the provocation
- Expose conflict or let it emerge
- Deepen the debate by asking the tough questions
- Manage conflict if it threatens to shut down collaborative capacity
- Read the authority figure(s) as a barometer
- Partner the authority figure(s) if possible

Take counsel from adversaries
- Delay the impulse to dismiss those who disagree
- Actively inquire into dissenting viewpoints
- Incorporate dissenting interests into adaptive work strategy

TABLE 22.1. Leading Adaptive Work Without Formal Authority

The formal leader can control the pace of work and challenge directly as a means of regulating distress. The informal leader uses provocation—raising and pressing the issue, or backing off of it—by making herself and her own work an example of the issue. Since the informal leader cannot hold the container for the work, she must rely on the formal leader to do that. Hence, she must attune and coordinate her actions with those of the formal leader, in the best-case scenario cultivating a mutually respectful and resilient relationship.

A fascinating, relatively recent example of adaptive work led by an environmental engineer involved the environmental impacts of several hydroelectric dams. The utility, state regulators, and environmental groups were dealing not with issues of a single endangered species, but with a complex range of parameters understood to affect the river ecosystems. Early on, the inability of any one individual or group to figure out the "right" things to do became clear. The engineer, on the staff of the involved utility, held little formal authority relative to the maintenance, operations, and construction decisions that would be made. His authority was not substantially different from that of a consulting engineer. He recognized the need for neutral facilitation of a process that could create a strong working relationship among the stakeholders, enabling them to address these issues in a realm beyond their typically adversarial positions.

Advocating for, gaining agreement to, and standing by this approach, the engineer set the stage for adaptive work. With the help of the facilitators' ability to define and hold the container, over time, he was able to prompt the participants to begin developing enough trust to work effectively together. The planned outcome was to shape a policy that would be administered by an authorized individual. Instead, the group arrived at the creative solution of continuing to meet in order to deal with and make decisions about the emerging complex issues together. As the engineer described the situation, every stakeholder doesn't win every time, yet each does weigh in every time and apparently feels that his views are to some degree incorporated into the outcomes. Each stakeholder has undergone significant internal change to develop new collaborative capacity, and all stakeholders, including the natural environment, are better off as a result.

In many adaptive work situations, as in this example, shared vision has not yet emerged. Articulating an adaptive challenge works as much or more by presenting the current state as no longer acceptable as it does by pulling inspired energy toward a more desired future. Momentum and motivation are generated by the thrust away from a state—in this case, dysfunctional disagreements in the face of urgent problems—that is viewed as increasingly intolerable. Eventually, in the context of this discomfort, a vision can emerge—either from a leader acting without formal authority or from a member of the system. In leading adaptive work in Heifetz's sense, then, a professional would take care not to plant seeds of vision early in the process. Instead, she would concentrate on holding everyone's feet to the fire about what's not working in the current state.

The challenges of leading adaptive work in Heifetz's sense on an organization ecosystem level are substantial. All the Type I social and Type II leadership capacities are needed, along with the more difficult capacities to hold complexity, anxiety, and ambiguity for a system facing significant adaptive challenges. As in your own firm, this level involves working unconditionally and insistently with individuals and small groups. But it adds working with larger, more diverse, and more loosely connected groups. Perhaps most difficult is recognizing when a whole system needs more or less stress in order to move to the edge of chaos where it can be most creative. To achieve this recognition as well as to effectively modulate the stress, a professional needs to be connected to many different people across the system—to be able to hear what they are saying and to sense what they are feeling. Both a strong stomach and a deep appreciation for conflict are essential. The rough waters cannot be navigated without a steadiness grounded in deep faith in the process and in the capacity of the system's members. As Heifetz's second book with co-author Marty Linsky emphasizes, the leader of adaptive work must also have developed a range of ways to protect and care for herself in highly stressful situations.[252]

Learning organization methodology provides a suite of disciplines for achieving conscious learning at the whole-system level. However, few client ecosystems are developmentally ready to pursue transformative work in the Interdependent mind-set these disciplines require. Heifetz's approach to adaptive work provides a bridge from a less conscious mode of learning to a more conscious one. At key moments when this approach has effectively engaged values-based emotional issues, a professional can involve system members in learning organization approaches to reflecting on that experience—to bringing both progress and further challenges to greater awareness.

PART V

Earning a Fair Return on Value Created

23 Pricing and Profit: New Pricing Strategies

Most consultants . . . habitually undercharge for your services and deliver more than you are receiving in remuneration, considering your contribution to success.

—Alan Weiss[253]

A brutal fact of reality for architecture and engineering firms is that prevailing pricing and compensation methods—setting fees on the basis of direct labor cost (whether selling hours on a time-and-materials or lump-sum basis)—provide only minimal profits for most firms. They also reinforce client perceptions that engineering, architecture, and design services are simply commodities to be purchased on the basis of lowest cost. The insufficient returns generated by these pricing methods starve firms of the resources they need to grow and foster a survival mentality in which many professionals are unwilling or unable to apply their unique expertise, dedication, and vision to the complex challenges that confront clients, communities, and society.

Although not the bottom line for a living firm, profit fuels its growth. Profit funds investments in learning, research and development, innovation, and technology. Profit provides a reward for risk taking. To generate a level of profit to sustain the ongoing growth and development of a living firm, the two sides of the new value proposition—creating value and earning a fair return on that value—must evolve together. New sustainable business models require not only the implementation of new value-creating strategies, but also the adoption of new pricing and compensation mechanisms that will enable firms to earn a fair profit for these efforts.

Alternative Pricing Strategies

To earn a fair profit, firms must separate fees from direct labor costs; firms must simply and absolutely move away from selling hours. There are at least five generic types of pricing strategies that can be adopted by firms to make this shift. The first pricing strategy, value-based pricing, establishes price not on the basis of cost, but as a reflection of the benefits or value that will be created or provided for the client. A second related strategy, which we call outcome-based pricing, establishes a contingent fee or profit that depends on the outcome a client actually experiences. A third strategy establishes a retainer fee that reflects the value of an ongoing collaborative relationship between a client and a service provider. The fourth, investment-based pricing, moves firms into a position where they become their own clients, directly leveraging the value they create by taking on additional roles and risks as builders, financiers, and operators of facilities and infrastructure. Finally, firms can use pricing mechanisms that come with new, nontraditional service offerings (tuition for training clients or data maintenance), sales of products (software, data, or furniture), or payment for knowledge and intellectual property transferred or shared with others (copyrights and license agreements).

Value-Based Pricing

Value-based pricing strategies set fees in relationship to the benefits that accrue to clients as a result of the firm's efforts, not on the basis of the costs a firm incurs delivering services. Firms can shift billing rates for high-value services to higher levels, establish lump-sum (fixed fee) arrangements, or adopt innovative pricing methods that are common in other industries.

Firms that are locked into time-and-materials arrangements can recalibrate billing rates to reflect the true worth of high-value services. For example, firms offering strategy services (strategic planning for facilities, real estate, infrastructure, etc.), process facilitation, and other specialized consulting activities are establishing billing rates commensurate with the rates of other management consulting firms retained by those same clients. These rates often have premiums that can range up to double those commonly applied to basic design and analysis services. Some engineering firms are able to add premiums in excess of 20 percent to billing rates for staff providing strategic planning services for GIS deployment versus the rates that firms are able to charge for those same people when they are performing typical engineering services.

Even though many firms have started to provide these high-value services to their clients, it is not uncommon for them to continue to use a single billing rate schedule for all of their work, charging the same rate for an engineer regardless of whether she is sitting in a client's boardroom advising on critical capital spending decisions, or managing a team completing drawings and specifications for a routine design project.

To help clients understand the value of these higher billing rates and avoid confusion about rate schedules, some firms have established separate strategy or consulting groups with distinct identities focused on providing high-value services. Other firms have simply set up multiple billing rate schedules that are applied depending on the type of work engaged by a client.

The predominant means for doing value-based pricing is the adoption of lump-sum (fixed fee) compensation methods that set the fee on the basis of value, not cost. Most past and current lump-sum arrangements are cost-based; fees are negotiated based on a bottom-up analysis of costs (an estimation of direct labor costs, assigned overhead, and desired profit margin). Such contracts at least have the advantage of providing the opportunity to enjoy increased rewards in return for better-than-expected efficiency; that is, if a firm controls its costs it will enjoy increased profits. Even so, this type of lump-sum arrangement seldom reflects either the true value that firms are offering or the risks that it is accepting by agreeing to the contract. In fact, it may, on occasion, actually lead to reduced value (quality) as firms scale back work efforts to the bare minimum required by contract in order to control costs.

In a value-based pricing strategy, lump-sum compensation rests squarely on a determination of the value or benefit that a firm will produce for the client. It also looks beyond the specific project objectives to a consideration of the wider benefits that may accrue to the client and other stakeholders as a result of the achievement of those objectives.

For example, an architect/planner, with extensive experience working in a particular urban redevelopment area, might agree to meet with a real estate developer who is new to this city to brief him on what it takes to successfully develop projects in this locale. Using the prevailing time-and-materials pricing method, the value of the professional's effort would be limited to the relatively small number of hours spent preparing for and meeting with the client, generating, at best, several thousand dollars of compensation. The real estate developer, on the other hand, has benefited considerably. He is able to make a more appropriate go/no-go decision on pursuing projects in this redevelopment. He can better identify significant development risks as well as ways to mitigate them. The developer may also have come away with a list of the key decision makers in the area with whom he will need to build relationships. He may even have lined up potential introductions from the design professional. The developer will also have a better sense of what it takes to expedite and accelerate development schedules, win support and approvals from key regulators and stakeholders, and avoid a myriad of other missteps that could bedevil a developer unfamiliar with the local scene. Who got the good deal here? Was the exchange of value fair?

A value-based pricing strategy would approach this situation in the same manner that surgeons charge for specific medical procedures. The cost of a surgery reflects the value of the procedure to the patient and compensates the surgeon for his knowledge and skill, not just the time he spends in the operating room. In a similar way, the lump sum that Landmark Graphics charges oil and gas companies for its

Decisionarium procedure far exceeds its costs. However, that price is still considered a bargain by those customers that agree to pay it. A value-based lump sum applied to the above example would compensate the architect/planner more fully and fairly for the knowledge, insights, relationships, creativity, and wisdom that he brought to the table and shared with the real estate developer.

More generally, value-priced lump-sum arrangements can be negotiated either for entire projects or portions of those projects. The key to this strategy is an up-front determination of the value or benefits that will be generated for the client as a result of the firm's efforts. Price is negotiated in relationship to those results/benefits, not on the basis of the tasks/activities the firm will perform or its costs.

An interesting variation of this approach combines value-based pricing with traditional time-and-materials compensation, but with an unconventional reversal of conventional practices. It is relatively common for design firms to propose the use of time-and-materials compensation for the early stages of a project. Firms and clients agree that, given that the scope of work is still unclear, both parties would be better off working in this fashion until the design is better defined. Then, a lump-sum agreement is negotiated to cover the remaining construction document and construction administration services based on this design. This works as a risk management strategy, but sells short one of the primary ways in which design firms bring value to the table—their creativity during the conceptualization, problem-solving, and design process—limiting their potential rewards for this high-value phase to the slim profit margins built into billing rates.

According to Phil Edwards, of the Edwards Management Group,[254] increasing numbers of design firms, following the lead of industrial design firms, are reversing this approach and negotiating value-based lump sums for up-front design work, and then using hourly compensation methods for later production efforts. Firms are choosing to negotiate up-front, value-based design fees that are high enough to not only compensate for the scope risk inherent in these early efforts, but to fairly reward them for the value of their applied creativity.

Other firms are willing to accept the higher levels of risk associated with services that have generally been excluded from agreements for design services as long as clients are willing to pay a risk premium that is high enough to make it fair. For example, a structural engineering firm involved in a design-build project for a major new sports facility earned a significant premium on top of its standard design fee by agreeing to include structural steel connection details and specifications as part of its preliminary design drawings, allowing the builder to accelerate the project schedule. The firm also agreed to assist in the development of alternative erection schemes that would make construction easier—work normally restricted from the engineering scope of services. These risks were acceptable to the structural engineer because the firm had engineering and management practices in place that could control and minimize the exposure that would come with them, and they were being fairly compensated for the risks they were willing to assume.

Firms are also adapting compensation arrangements commonly found in other industries or other segments of the design and construction industry, but uncommon for architecture and engineering firms services. The payment of commissions for engineering or design services is one example. Psomas Engineering, a civil engineering firm based in the western United States, is experimenting with the use of commissions in its water resources practice. These compensation arrangements look more like those used by real estate brokers than the traditional fee-for-service model used in engineering contracts. In the past, Psomas has assisted water districts in their search for new water supplies, providing consulting services on a time-and-materials basis. Recently, the firm has struck agreements with selected water districts whereby they are paid a commission (similar to a real estate commission) for the water resources that they are able to bring to the table. Their compensation is based on the number of acre-feet of water they are able to arrange for delivery to the district, not on the number of hours they spend setting up the transfer. Clients are willing to pay these commissions because of the unique nature of this service and Psomas's capabilities. Fees earned through this type of compensation method result in significantly higher profits than can be achieved by using traditional hourly billing rates.

A variation of this commission-based pricing strategy is being pioneered by Barge Waggoner Sumner and Cannon (BWSC), a Nashville-based engineering firm. Its environmental practice group has formed an alliance with a national real estate brokerage firm to assist large industrial companies with the sale of surplus, nonoperating properties. These industrial companies (railroads, utilities, etc.) often have bulging portfolios of excess properties—properties that have been abandoned or are underutilized as a result of past mergers, acquisitions, consolidations, or operational changes—that drain cash instead of contributing to operating profits. In an era when Wall Street places a high value on a company's return on assets, these properties can act as a significant drag on a company's stock price. However, disposing of them can be difficult, particularly if they are brownfield properties with environmental problems.

BWSC's alliance helps these industrial companies move these properties off their books. The alliance helps companies identify properties that are candidates for disposal, assess potential environmental issues and develop solutions to environmental problems that may otherwise loom as deal-breakers, prepare offering memorandums, find potential buyers, and broker sales—turning underperforming assets into cash. The synergy of BWSC's environmental engineering wisdom and regulatory relationships with the real estate company's brokerage and marketing know-how creates the potential of speedier disposal of larger numbers of properties.

This alliance rolls both environmental engineering and real estate brokerage services into a single commission structure to be paid by the industrial company upon sale of a property. Up-front services are paid for as they are delivered, and those payments are subtracted from the commission when a sale is complete.

A major advantage of this compensation arrangement to industrial owners is that they can treat the payments for up-front services (environmental assessments, reports, and regulatory contacts) as commission expenses. In contrast, the fees for those same engineering services, when they are provided directly to the company, are typically viewed as direct costs to its operations and are subtracted from bottom-line profits. As such, managers in those industrial companies generally try to minimize the dollars they spend on environmental engineering services. This includes contracting for these services often only when it is necessary to secure regulatory compliance and the use of procurement procedures that force lower fees by asking engineering firms to compete for these jobs. However, it turns out that those same industrial companies are relatively price insensitive—willing to pay higher fees—when the services are reimbursed as part of a larger commission that is contingent on the sale of the property. The sale generates the cash needed to pay the engineering fees (as part of the commission), thereby eliminating any adverse impact on their operating budgets and the company's bottom line.

For this work, BWSC is able to avoid the stiff price competition in the mainstream environmental engineering market for industrial clients. The firm earns higher fees up-front and a significant kicker at the time of any sale as compensation for both the additional risk the firm accepts and its effectiveness in ensuring that sales are completed. The effective multipliers for BWSC's fees under this program are more than double those typically charged on traditional engineering projects.

To successfully move toward a value-based pricing strategy, the work being performed must be below the line of client performance expectations on Christensen's technology S-curves. If the services have already moved above the line into commodity status, the likelihood of convincing clients that value pricing makes sense is very low. By that time, too many established competitors will be willing to do the work for the low fees that clients demand.

In contrast, firms can find opportunities to successfully launch value-based pricing strategies by identifying emerging project types (or technologies) that are still climbing the performance curve. Segments of the emerging homeland security market and specialized building or infrastructure projects (stadium design or signature bridge design projects) are examples of emerging project types. These high-value potential niches can also include smaller, less glamorous project types. For example, the leak-detection technology for water piping systems that Cavanaugh & Associates, a civil engineering firm based in Winston-Salem, North Carolina, has pioneered for municipal utility use is new enough to be low on the performance curve. Consequently, these services are still immune from the price competition that characterizes much of the municipal engineering market.

Firms can also identify services that are components of larger project delivery processes, but that are still relatively uncommon, such as process facilitation or advanced team-building services. These services may be value-priced directly to owners, though in most cases they will be subcontracted to other prime design professionals or contractors. The Strategies Group at Anderson Brule Architects (ABA) specializes in the facilitation of strategic processes for clients, particularly during the front-end project definition, programming, and conceptual design for municipal, health-care, and library clients. These services have proven equally valuable and price-insensitive, whether delivered directly to owners or nested under a prime architect's contract.

Finally, firms can invent and deploy new disruptive technologies that are attractive to specific types of clients, initiating new technology S-curves and setting new client performance expectations. Frank Gehry discovered a protected niche in the unique computer program his staff developed for his work with high-end museum and institutional clients. The program enables his design process and communicates information describing his complex designs to builders and fabricators. Gehry's firm is now launching a separate company, Gehry Technologies, that will offer this software to other designers. It has the potential of becoming a future disruptive technology as it moves up the performance curve and becomes applicable to more building types.

A word of caution is warranted. In each case, when market opportunities are identified, firms must be careful not to carry over existing self-defeating mental models and contracting processes (time-and-materials or cost-reimbursable contracts) that are neither appropriate to the situation nor advantageous. When clients believe they can have value-creating services for the same prices they pay in commoditized markets, they will lock in contracts that make it difficult for firms to change the basis of compensation in the future. Firms must have the courage to offer alternative pricing strategies from the outset, rather than try to ease clients over to new schemes as they become more familiar with the benefits. Too many firms launch bold new service and design strategies that do, in fact, create significant new forms of value for clients, only to shoot themselves in the foot by pricing them by the hour, thereby ensuring that they will never have the opportunity to capture a fair share of the value they created through their entrepreneurial efforts.

Outcome-Based Pricing

Outcome-based pricing (incentive pricing or contingency pricing) starts with the same premise as value-based pricing, but adds a degree of performance risk to the compensation equation. Performance incentives, structured into the contract between client and professional, are triggered by the extent to which the professional actually meets specified performance requirements.

One form of outcome-based pricing establishes performance metrics for the design and construction process itself. For example, a structural engineer may agree to be paid a share of the savings that she achieves by designing a structural system for a high-rise tower that is less costly than conventional approaches, or an architect could agree to an additional fee that is earned only if the work is completed within a given time frame. These types of incentives are often built into design-build contracts where accountability for cost and time performance can be clearly assigned and rewarded.

The following example shows how this pricing strategy can be successful. A bridge engineer was asked by a general contractor erecting a major new span to provide value-engineering services. Instead of being paid by the hour, the engineering firm agreed to be paid a percentage of the savings it could generate through its analysis and design efforts. Changes to the decking structure incorporating new composite materials generated significant savings for the contractor by reducing the size of the crane that had to be leased for erecting the structure. A percentage of those savings flowed to the engineer, providing a healthy return on the engineer's technical expertise and creativity.

A second form of outcome-based pricing focuses on benefits that accrue to a client organization. In one example of this approach, an environmental engineering firm agreed to a fee that provided only a minimal profit to develop and implement an asset management program for a municipal utility. However, the utility agreed to add a clause to the contract that provided a significant performance incentive if the firm was able to reduce the EPA fines the utility was currently paying. The engineering firm was able to eliminate the EPA fines and earned a substantial bonus that more than tripled the profit it would have normally earned on a project of this type.

A new laboratory project for Ciba-Geigy, the pharmaceutical giant, completed in 1996 on its Tarrytown campus, combined both types of performance incentives. Ciba-Geigy contracted with a design-build team that included the architecture-engineering firm HLW International and the contractor Sardoni Skanska Construction Company. Both agreed to a contract that included a unique set of rewards tied to beating project deadlines, bringing the work in under budget, and meeting quality expectations of building users. The contract put all of the design-build team's profits at risk. However, if the team met the contract's performance goals, HLW would, according to HLW CEO Leevi Kiil, "earn more than what our normal fees would have been."[255]

Schedule incentives tied to milestones for document delivery accounted for one-third of HLW's profits. Meeting a key cost goal at the 50 percent design completion milestone netted HLW the second third of its profits. The final third was staked on a more unusual metric tied directly to quality, measured by the satisfaction of the Ciba-Geigy users who would be moving into the laboratory; the final third of their profit would be awarded if they achieved a 75 percent favorable rating by building users on a 15-question survey. The survey, which asked occupants to rate their satisfaction on such factors as noise, lighting levels, temperature controls, quality of construction, and overall appearance and function, was administered 90 days after the 160 laboratory employees occupied the building. When the results were tabulated, the team scored a satisfaction level of 84 percent, and HLW was rewarded with the maximum incentive possible under the contract.

Based on the success of the Tarrytown project, Ciba-Geigy contracted with the HLW/Sardoni team to perform another project at its High Point, North Carolina, campus with similar incentives. Again, HLW earned all three incentives, earning a score of 93 percent on the user-satisfaction survey. In the end the project director for Ciba-Geigy commented, "Everyone benefited. Sardoni and HLW earned profits and Ciba benefited more so than we would normally because we know our people are satisfied."[256]

Retainer Fees

Retainer fee arrangements can be used when a client wants access to a consultant's talents, knowledge, and wisdom for a specified period. Retainers provide the opportunity of engaging in an ongoing relationship with access to collaboration and support as a part of that relationship. The retainer fee is based on a subjective assessment of the value of the relationship, the degree of access, and the amount of support offered as part of that relationship. The assumption in this arrangement is that access and support will vary within the agreed-upon bounds across the time period, but that the fee will be paid evenly across the time period regardless of the flow of work. There is also the assumption that access and support will be provided on a highly responsive basis. As a result, retainer arrangements often establish a set fee, paid per time period, independent of how much time the consultant actually spends on the project during that period.

Our colleague David Aitken negotiated a retainer agreement for his work with the Greater Toronto Airport Authority (GTAA) on the multibillion dollar expansion of the Toronto International Airport. GTAA originally contracted with David (through Aitken Leadership Group) to provide facilitation services for top-level decision makers on a daily-rate basis. The goal of his work with GTAA was to optimize value for the airport and to improve the efficiency of the project's fast track process by helping key stakeholders make critical decisions and resolve major problems that would inevitably come up as the project proceeded. However, the client quickly discovered that trying to contract for David's assistance on a day-by-day basis was problematic. Because GTAA couldn't predict when the major problems or disagreements would occur, it couldn't schedule David's services in advance. Unfortunately, GTAA's requests for immediate assistance often conflicted with work that David had previously scheduled for other clients, leaving its needs unsatisfied.

To solve this problem and provide the level of access and support that GTAA needed, both parties agreed to switch to a retainer arrangement. The Airport Authority agreed to pay David for a set number of days each month during the upcoming year. In return, David agreed to place GTAA's needs at the top of his priority list and manage his calendar to ensure that he could quickly respond to its requests for help. Regardless of whether the client used the allocated days during a given month, David was still paid the same retainer fee, and the days were not rolled forward into future months. And if the client needed more time, David was compensated for those extra days.

The fairness of this bargain became apparent when several months had passed without the client asking for any of his time. David, a former architect, reflecting widespread attitudes held by many design professionals, began to feel somewhat guilty about being paid without showing up, and even told the client about his feelings. The client pointed out that several months earlier David had responded to a very urgent request by GTAA to help key stakeholders resolve a particularly difficult problem. If left unresolved, that problem could have significantly delayed the project and cost the authority and its stakeholders hundreds of millions of dollars. With David's help, they had been able to quickly reach an agreement about how to proceed, and the client assured David that he had earned his fee many times over in that one meeting.

A civil engineering firm headquartered in a small city paired an innovative new service offering with a retainer pricing arrangement. A major element of the firm's mission, according to firm leaders, is to offer innovative management practices that have worked for their firm as services to its client—to help those clients achieve the same types of benefits the firm receives. After building an internal marketing and communications group, the firm realized that this team was capable of doing much more than simply serving internal needs. They screened their client list and identified the innovators, or clients that might be interested in using the firm's marketing and communications services. The firm's service offering to these clients included a retainer arrangement, requiring payment up front for a set minimum amount of on-demand services. These services could include anything from help in putting together new marketing brochures to consulting work focused on specific communication issues. Services beyond the minimum are paid for separately from the retainer.

A primary selling point for these services is the firm's reputation in the surrounding region as a well-managed business. The firm is viewed as a role model that consistently demonstrates sound management practices that other organizations can learn from and emulate. Clients value its engineering services and also appreciate its leadership in pioneering new business practices. The marketing and communications service is only one of a set of similar best practices that the firm makes available to clients through a preferred-customer program. In the past five years, these offerings have included crisis management planning, mentoring and coaching, and guidance on building a corporate university. For the new marketing and communications service, over 50 percent of the firm's largest clients signed on and made the up-front payment.

Beyond the additional revenue source, these non-engineering services offer other forms of payback to the firm. First, they help build its brand and credibility. For example, at trade shows, experts from the firm teach classes in these best practice areas. Later, back at the firm's booth in the exhibit hall, these staff members are no longer anonymous representatives of an engineering firm. They have become trusted teachers who have a personal connection with many of the people that have just attended their training sessions. Second, the strategy is used to strengthen the firm's relationships with its preferred customers, building bonds that go beyond engineering and construction. Third, the firm is making its organization ecosystem healthier. If the firm's clients are more successful as a result of their adoption of these best practices, then the entire organization ecosystem, including the firm, will be better off.

Investment-Based Approaches

Investment-based approaches to pricing move architecture and engineering firms away from a position of limited risk and reward as the designer or engineer of a project and into a stance where they also bring equity financing to the table. This includes situations where architecture and engineering firms become the sole developers and/or operators of properties or facilities.

Mechanical, electrical, and civil engineering firms have moved into performance management and operations and maintenance contracts with owners (industrial companies, municipal utilities, and health-care organizations, for example) by taking responsibility for design, construction (installation), and the ongoing operation of plant and infrastructure projects. Engineering and architecture firms are becoming equity partners in large real estate development projects, often providing their equity contribution as a share of the value of their services on the project. Large engineering firms are adopting investment positions that take responsibility for the design, construction, financing, and operation of major infrastructure projects, such as municipal water and wastewater systems or toll roads.

At one end of the spectrum of investment-based pricing strategies are organizations like Weston & Sampson Engineers (W&S), a 200-person civil engineering firm based in Peabody, Massachusetts. In 1992 W&S set up a separate company, Weston & Sampson Services, to provide operation, maintenance, trouble-shooting, and start-up services to its water and wastewater clients in Massachusetts and the surrounding New England region.

The Wendel Companies, a 140-person engineering/architecture firm located in upstate New York, is pursuing a similar strategy. The firm recently established Wendel Energy Services to provide a range of energy management and design services, from turnkey design and construction to financing and operation, for energy-related capital projects. Projects range from simple pump or chiller replacements for private companies to larger scale retrofit projects for institutional clients.

Cambridge Seven Associates, a 65-person design firm (urban planning, architecture, and graphic and exhibit design) based in Boston, has pursued investment-based pricing opportunities throughout its history. In the mid-1970s it took ownership of a multimedia show it had produced for Prudential Insurance, called "Where's Boston," which it subsequently managed as a money-making enterprise for 13 years. In the 80s the firm was a co-developer of a $64 million mixed-use project in Cambridge, Massachusetts. More recently it has proposed a turnkey development for a new aquarium in Lisbon, Portugal.[257]

At the other end of the spectrum are large engineering and engineering/contractor companies. CH2M HILL, a large global engineering company, offers its water and wastewater clients a complete package of project conception and development, design, financing, construction, and operations services through its Operations Management International (OMI) division. This 1,400-person group now operates more than 140 facilities worldwide. Washington Group International, a large engineering/construction company, heads the 21st Century Rail Corporation, which in 1996 was awarded a contract to design, build, operate, and maintain the $1.1 billion Hudson-Bergen Light Rail System for the state of New Jersey's NJ TRANSIT. After completion of the design and build phases (the first segments of the system were opened in 2000), the corporation will operate and maintain the system for 15 years.

To be successful in these ventures, firms are adding new people with new capabilities to their staffs. These include construction managers, operating and maintenance personnel, and, of primary importance, financial analysts and investment bankers that can structure these deals and manage their financial risk. Firms are also beginning to recognize that the role of their chief financial officers should shift from internal financial management and control to a new external focus in which they become active participants and leaders in helping the firm discover new ways of creating value for its clients, structure deals, and manage financial risks on projects. If firms can't bring people with these skills in-house, an alternative strategy is to form strategic alliances with organizations that already possess them.

Pricing Strategies for New Products and Services

Some architecture and engineering firms may escape the knot of cost-based pricing by adopting pricing mechanisms that are common to a new type of business the firm has chosen to enter, but are distinctly different from those used within the design professions. Firms which have developed specific software or database applications that solve client problems or support client operations are beginning to understand that this type of knowledge work constitutes a radically different business from engineering or architecture. Instead of selling the hours expended to develop the software or database for a client, firms are realizing that they are better off being paid for their work as intellectual property, like a software company. Microsoft isn't in the business of selling hours.

Increasing numbers of firms are being asked to develop (or are proactively identifying the opportunity to develop) training programs for client organizations, teaching a range of subjects from sustainable design strategies to basic bridge inspection skills and project management roles and responsibilities. Rather than selling the hours that their trainers put in designing and delivering courses, some firms are charging tuition (per seat or per person), like educational institutions. The income potential of this pricing strategy is even greater as firms discover they can reach broader audiences through e-learning courses on the Internet.

For Baker Engineering, headquartered in Pittsburgh, this shift is most apparent in the structural design section of its transportation engineering practice. Its BRADD (Bridge Automated Design and Drafting) software was originally developed for the Pennsylvania Department of Transportation (PennDOT) and paid for on an hourly fee basis. Recently, Baker has implemented a set of pilot programs with PennDOT, which shift the compensation basis for software updates, ongoing software maintenance, and hot-line efforts to a lump-sum reimbursement. Baker is now evaluating opportunities to offer other new software that the firm has developed to both client organizations and other design professionals.

Intellectual Property

Another alternative pricing strategy focuses on reimbursement for intellectual property created and owned by the firm. Architecture and engineering firms are taking a much more proactive stance toward protecting and profiting from their intellectual property. McDonough Braungart Design Chemistry has pursued this type of strategy in the roll-out of a certification program it has developed. The firm not only offers manufacturers its accumulated knowledge and wisdom to rate the eco-effectiveness of fabrics, carpets, and other construction and furniture materials and products, but also uses its brand and industry reputation to represent the significance of that certification to potential buyers.

Meeks + Partners (formerly Kaufman and Meeks), a Houston-based architecture and land planning firm that specializes in housing, has copyrighted specific home designs and patented its designs for affordable housing and communities. The patents cover lot configurations and the placement of certain structures on those lots.[258] One patent covers the design of an apartment complex that reduces construction costs by 20 percent by eliminating breezeways, without violating fire codes or sacrificing green spaces. Another is for a high-density apartment design that incorporates unique construction detailing and parking layout features. Meeks reuses these patented designs in its own work for housing/community developers and also licenses them to builders in other states.

Some design organizations, like the industrial design firm IDEO, actively seek patents on new products designed as in-firm ventures. Gensler, the large architecture and interior design firm, protects its furniture designs as intellectual property and then licenses the designs to manufacturers for production and marketing. MATx, the materials research subsidiary of the Boston-based architecture firm Kennedy & Violich, negotiates royalty payments from manufacturers that contract with it for the development of new products or materials. The royalties provide downstream cash flow to the firm as new product designs move into production.[259] In each case, the value of the intellectual property is commensurate with the potential sales of those products or furniture and is largely decoupled from the labor costs that went into creating them.

A Midwestern engineering firm directly capitalizes on its intellectual property. Landscape architects working for the firm realized that many of its projects included custom-designed landscape elements (trellis, covered walkways, benches, etc.) that had the potential of being sold to a broader customer base. The firm started including a provision retaining intellectual property rights for furniture or any element that could be reproduced and resold as part of its contracts for projects with this potential. The firm then formed a partnership with a local manufacturer to make these units. The engineering firm manages the marketing process, including advertising on the Internet and in such specialty retail catalogs as Brookstone, and the manufacturer makes the units and fulfills the orders. In return, the firm receives an up-front commission on orders, reimbursement for its marketing and sales costs, and a split of the profits. In the case of one trellis design, the partnership offers customers a mass customized modular version that can easily be adapted to specific site conditions and foundation systems. A vice president of the firm reports that professionals in the landscape architecture group, as well as in other practice areas, are now thinking about design differently. They still work to solve the client's problem, but they also look for opportunities to create elements that can be turned into manufactured products.

The same engineering organization has also developed an innovative approach to capitalize on the intellectual property of its Geographic Information Systems (GIS) group. The firm realized that because much of the production work for GIS was being commoditized, it could no longer compete with low-cost providers of those engineering services. To be successful in the GIS market, the firm needed to offer something different to potential clients. The solution to this problem came with the firm's development of a proprietary GIS application that provided dispatch and routing information for emergency vehicles. The software would automatically reroute response teams around traffic accidents or roadwork. Rather than compete for business from large municipalities, the firm decided to focus on small cities that were underserved by other providers of this type of software. The firm sells these clients a 12-month software license, which includes a base number of stations. Yearly renewals and upgrades offer a continuing revenue stream for the firm, similar to the value proposition used by many software companies. The newest version will also include an option for the engineering firm to manage and maintain the client database, for an additional fee.

The firm also realized that the digital databases it assembled for clients, not only through GIS but also through the use of other software in other parts of its engineering practice, offered yet another potential for earning a return on intellectual capital. Certain clients pay the firm to warehouse and maintain their digital data, including services to keep the data current. For many of these clients, the firm has negotiated a right to redistribute this data and information. The engineering firm is now making information and data from these digital warehouses available for resale to other buyers. For example, the firm resells digital aerial photographs of particular sites gathered for one contract to other customers needing this information.

Mix-and-Match Models

It is neither necessary nor desirable to choose only one form of pricing strategy. Firms can tailor the compensation method to specific situations, including mixing and matching alternative pricing arrangements within one contract. For a property development project that had funding in place and was under a tight deadline to get work started, the Chicago-based engineering firm Greeley and Hansen "negotiated five different kinds of pricing into five different elements of the project. Unit prices were set for permit approvals and land acquisitions, lump sums were established for clearly defined work packages, and 3.0 multipliers were set for undefined work."[260]

Maintaining the Status Quo

Some firms will choose to maintain their existing model of practice, continuing to sell hours and use direct labor costs as the basis for their pricing strategies. They will make their peace with the dynamics of competing for work with clients who view their services as commodities. They will also learn to live with the restricted returns inherent with that model of practice. Other firms will reinforce their cost leadership position by growing and capturing an increasing share of the market, driving down their cost structures by sharing administrative overhead costs across higher volumes of work. It is important to note that recent industry history is littered with the carcasses of firms that pursued this strategy only to run out of gas (cash), leaving them prey for the next bigger acquirer.

Moving Toward New Pricing Strategies

The inertia of the existing pricing system poses a major challenge to firms as they shift toward new pricing strategies. The pricing and payment schemes embedded in the current system and the accompanying mental models that guide choices and behaviors of its participants (the professionals, firms, clients, and other industry stakeholders) will weigh down change initiatives. Clients, for their part, have entrenched pricing and payment practices embedded in their own institutionalized procurement processes. This is particularly true for large bureaucratic clients such as transportation departments and federal government agencies. At the same time, the entrenched attitudes of many professionals inhibit their ability to imagine, let alone propose, different pricing arrangements. The financial accounting systems of most architecture and engineering firms, designed around the economic business model of selling hours and monitoring and controlling direct labor costs, further inhibit change efforts. Adjusting those systems to account for incentive bonuses, software sales, or license fees can be a daunting effort.

However entrenched these existing pricing and payment practices may be, firms will be wise to resist the temptation to either co-exist with the status quo or hope that someone else will convince clients to change. As Michael Porter correctly points out, you can't be both a cost leader (offering the lowest cost commodity to customers) and differentiate yourself. If you move forward with new value creation strategies, continuing to compete on the traditional basis of price can only lead to disaster, bankrupting the firm or starving a transformation initiative of the resources essential for success.

To move forward, firms must develop a deep understanding of their clients, their clients' organization ecosystems, and the corresponding value networks; that understanding must be translated into an operational ability to identify, articulate, and provide what clients value. This can include monetary and nonmonetary benefits as well as tangible and intangible outcomes. In most cases, it is not necessary to specify the exact results (or a magic number) a client will receive. It is often sufficient to communicate the future benefit and/or predict the direction of change with respect to key outcomes that support the price a firm merits as a fair return on the value it creates. For Psomas Engineering, the metric was the highly tangible acre-feet of water obtained for a water district. For the HLW/Sardoni design-build team working for Ciba-Geigy, a combination of monetary and nonmonetary metrics addressed cost, schedule, and quality dimensions. For Frank Gehry's work on the Bilbao Museum, it was both the tangible jump in ticket sales and the intangible Bilbao effect that drew visitors to the museum and gave a new identify to members of the surrounding community.

Outcomes can be denominated for project delivery, including construction cost, schedule, and quality of drawings (represented by change orders). More generally, design professionals and consulting engineers can draw from a wide range of other valuation metrics to address value accruing to the client organization itself, including but not limited to the following:

- Higher productivity
- Lower turnover
- Higher morale
- Reduced absenteeism
- Improved image (brand identity)
- Better performance
- Higher sales
- Increased market share
- Greater profits
- Improved financial/operational results
- Decreased costs
- Reduced risk
- Improved employee satisfaction
- Higher customer satisfaction
- Decreased time to market
- Increase in new product ideas generated
- Decreased response time
- Decreased customer complaints
- Improved teamwork
- Less stress

A firm working with an airport client might choose from a wide variety of metrics, including airport revenue, cost reductions, speed of project delivery, energy use, and life-cycle maintenance and operating costs. Terminal designers might assess passenger satisfaction, looking specifically at ease of use, attractiveness, and speed of connections. Firms working with baggage systems might set benchmarks for speed of installation, throughput, and lost bags. Concessions designers can look at sales per square foot and customer satisfaction. Parking garage designers might choose to evaluate efficiency of traffic flow and minimization of bottlenecks.

These metrics of value can be used to support the negotiation of a value-based price, or they can be incorporated directly into outcome-based pricing agreements. In either case, architecture and engineering professionals need to be able to identify appropriate metrics for the situation they are addressing, and to communicate how they will deliver these benefits to their clients. Instead of defaulting to the easy negotiation of selling hours, firms must learn the language of their new value proposition.

Firm leaders must also help their staff overcome deeply held mental models that support the current economic model of practice. Some professionals actively resist new pricing strategies, believing that what looks like an excessive markup on direct labor costs is somehow unethical or unfair to their clients. Leaders must surface these mental models and help their staff understand the imperative of these new business models, as well as the basic principles underlying them. In fact, firms with low levels of profitability may be less able to operate with the highest professional standards. Higher profitability levels are fair when compared with the level of value created. Firms will also need to teach principals and project leaders to be comfortable asserting the firm's interests in receiving these higher returns.

It is important for firms to find the right clients to experiment with new pricing strategies, probably not their mainstream clients. Clayton Christensen's work has shown that many customers (clients) will reject innovative new technologies, preferring small improvements in the cost or customization of existing technologies (services). This short-term myopia often lures service providers along a path toward further commoditization and ultimately sets them up to be overthrown by a new disruptive technology. Consequently, firms experimenting with new pricing strategies need to find niche clients facing unusual situations in which the value of an innovative approach will be both apparent and attractive. After perfecting a new pricing strategy with a few clients, firms can then roll it out to a larger population of clients.

Finally, adopt a prospector's view of the world. Architecture and engineering firms are proactively identifying opportunities for creating value for clients, not waiting for them to ask for help. In many cases, clients are unaware of the opportunity and or need. That is the spirit with which Barge Waggoner Sumner and Cannon, along with its real estate alliance partner, launched its innovative industrial brownfields service. The firm spotted a corporation that had substantial underperforming brownfield properties, figured out a creative way of unlocking value from those assets, and approached it with a new value proposition and pricing strategy. BWSC didn't wait until the industrial company decided to sell a property and then try to convince them to try its new approach. This prospector stance naturally expands the domain in which professional architects and engineers operate. Instead of waiting for projects, firms can be entrepreneurial in areas where they see a need that they, and their professions, can satisfy.

Application to New Ecology of Firms

To complete our overview of the pricing picture, we need to examine the way these new pricing strategies may play out in the new ecology of design firms—the six axes used earlier to describe our strategic response to technology-driven trends.

Look first at the WHO axes. Generally speaking, a service provided to a global client has a higher value than a service provided to a local client. The challenges involved in working at long distance, including overcoming language and cultural differences, are significant. The need for collaboration and facilitation is high, and the potential for learning and knowledge transfer is very high. However, in the global environment, production work may be even more commoditized than in the United States, because cheaper labor may be assumed. Firms may need to separate out work that can be priced according to value (technical work, collaboration, etc.) from services, like documentation, that will be priced as a commodity and/or subcontracted to firms located in developing countries such as India or the Philippines.

In many cases, a service provided to a capital-intensive client should be valued higher than a service provided to a knowledge-intensive client. Processes and procedures for capital-intensive clients are usually more complex and bureaucratic, the requirements for documentation are more demanding, information is harder to get, decision-making is slower, and politics are more entrenched and challenging. While the challenge to work at a higher speed is usually a given in knowledge-intensive organizations where it makes sense to value-price for speed, there are also typically far fewer barriers to achieving that speed. Providing rapid, flexible, information-rich responses to a knowledge-intensive organization may also be a basis for value-based pricing.

Shifting now to the WHAT axes, mass-production and mass-customization provide great potential for providing a fair return to the design professional. While there is some fear of a scenario where designs become downloadable off the Internet—the way Napster made music downloadable for free—the legal trends currently support mechanisms for earning a fair return. In any case, there are ways to make sure a fair price is received before this point. On a value-pricing basis, a mass-producible design has a much higher value than a one-of-a-kind design because of the value inherent in the potential for repetition. As we have learned from the growth of information technologies, markets and profits actually expand as components become cheaper.

A mass-customizable design with designed-in flexibility should have an even higher value. Firms doing this type of work should not only command very high fees for such original designs, but also reap rewards for related value-priced services such as management of prototypes, customization services, and training others to customize using those prototypes. Opportunities for outcome-based pricing are also present in these quadrants, in the form of incentives for speed related to the rollout of mass-customized designs for large retail, consumer service, financial services, and health-care operations across wide geographies (for example, opening 50 new outpatient clinics across the Southwest in 18 months).

Perhaps counterintuitively, there will also be fascinating opportunities for combining mass-production/mass-customization strategies with collaboration and transformational leadership skills to help meet worldwide needs for affordable infrastructure, housing, and sustainable communities. This type of value-creating work could address critical societal problems, while keeping costs far below strategies that rely on one-of-a-kind designs.

On the second WHAT axis, integrated services generally have higher value than segmented services. They typically handle more complexity and risk and, when working across multiple organizations, and contain more potential for the value associated with collaboration. The more integrated the services, the more beneficial outcome-based pricing might be as a component of an overall fee package. If integration extends all the way to financing, operating, and maintaining, then it becomes a scenario for investment-based approaches. The exception to this will be situations in which the service (project delivery technology) has crossed above the line of client performance expectations. As Christensen underscores, pricing at that point will shift to a cost basis and an integration strategy will be problematic. In that circumstance, carving out a niche with a segmented service that is still below the line will provide more opportunities for value-based pricing.

The value complexities associated with the HOW axes are more difficult to generalize. The relative value of a centrally managed vs. a self-organizing/collaborative approach is highly situational. A group that is skilled in the self-organizing/collaborative mode may be able to work much more efficiently than a centrally managed group. Conversely, if the skills are not there, a centrally managed approach may be much more efficient and effective. A self-organizing/collaborative mode may be more effective in providing collaborative or transformational work, or it may not, depending on the complexity of the situation.

Likewise, the relative value of one shop vs. a network (or alliance) is highly situational. While the client for a particular project may value the ease of working with a single organization, no single organization may be able to provide the range of talent required for that project. On the other hand, a client may value the ability of a network/alliance to draw together the best talent for an unusual challenge. One major advantage of a network/alliance approach will be the potential for more effectively managing risk, particularly financial risk, with the selection of partners that bring these skills to the table. On a value-pricing basis, caution should be exercised when factoring in these axes. They might play a significant role, or it might be better to leave them out of the equation.

Business models are holistic in nature. Both sides of a value proposition must evolve together. To convince clients to accept new pricing strategies architects and engineers must be able to deliver higher levels and new forms of value. And, to implement new value creation strategies firms must have the economic resources that flow from these new pricing mechanisms.

Transformation requires experimentation and learning on both sides. As firms are able to deliver genuinely new forms of value through innovation and leadership, new pricing strategies can be put in place. Appropriate remuneration that supports ongoing learning and innovation by professionals will set off a virtuous cycle of value creation that serves firms, clients, and society. Evolving the core of the professions and business model innovation will ultimately produce more non-monetary than monetary value, creating far greater fulfillment for professionals and shaping a brighter, more sustainable future for our world.

Notes

CHAPTER 1: SAILING INTO WHITE WATER

1 The account of the history of Dames & Moore included in this chapter was provided by William (Bill) Moore in an oral history prepared by the Earthquake Engineering Research Institute in 1998. William M. Moore, interview by Stanley Scott, in *Connections: The EERI Oral History Series* (Oakland, Calif., 1998). Quotes by Bill Moore are all drawn from this source.

2 "Dames & Moore Goes Public," *Engineering News Record*, March 16, 1992, 8.

3 "Dames & Moore Tried to Hold On," *Engineering News Record*, May 4, 1999, 16.

4 Nathaniel A. Owings, *The Spaces In Between: An Architect's Journey* (Boston: Houghton Mifflin, 1973).

5 David Maister, "Balancing the Professional Service Firm," *Sloan Management Review* (Fall 1982).

6 "Dames & Moore Goes Public," 16.

7 Ibid.

8 Ibid.

CHAPTER 2: BECOMING A LIVING FIRM

9 Arie de Geus, *The Living Company: Habits for Survival in a Turbulent Business Environment* (Boston: Harvard Business School Press, 1997).

10 Ibid.

11 James C. Collins and Jerry I. Porras, *Built to Last: Successful Habits of Visionary Companies* (New York: HarperBusiness, 1994).

12 De Geus, *Living Company*, 9.

13 Ibid.

14 Howard Gardner, Mihaly Csikszentmihalyi, and William Damon, *Good Work: When Excellence and Ethics Meet* (New York: Basic Books, 2001).

15 David H. Maister, *Practice What You Preach: What Managers Must Do to Create a High Achievement Culture* (New York: Free Press, 2001).

CHAPTER 3: CREATING VALUE, NOT SELLING HOURS

16 Dee Hock, *Birth of the Chaordic Age* (San Francisco: Berrett-Kohler Publishers, 1999), 43.

17 Joan Magretta, "Why Business Models Matter," *Harvard Business Review*, May 2002, 86–92.

18 Ibid.

19 Ibid.

20 Jim Collins, *Good to Great* (New York: HarperCollins, 2001).

21 Clayton M. Christensen, Michael Raynor, and Matt Verlinden, "Skate to Where the Money Will Be," *Harvard Business Review*, November 2001, 73–79.

22 The AIA published new cost-based compensation guidelines in 1975 in the wake of a consent decree to voluntarily restrict itself from imposing standards or policies prohibiting members from submitting price quotations for their services. After a subsequent case against the Virginia Bar Association determined that fee schedules were a form of price fixing, the AIA asked its state affiliates to rescind their published fee schedules and fee curves. Elizabeth Harrison Kubany and Charles D. Linn, "The Fee Dilemma Part I: Why Architects Don't Charge Enough," *Architectural Record*, October 1999, 110–21.

23 Weld Coxe, *Managing Architectural and Engineering Practice* (New York: Van Nostrand Reinhold, 1980).

24 A net multiplier ratio is calculated by dividing net revenue by direct labor cost. It can be used to represent both the productivity of each direct labor hour and the premium clients are willing to pay for those hours. The net covers direct labor cost, overhead expenses, and profit.

25 Clayton M. Christensen, *The Innovator's Dilemma: When New Technologies Cause Great Firms to Fail* (Boston: Harvard Business School Press, 1997).

26 Owings, *Spaces In Between*.

27 James Tobin, *Great Projects: The Epic Story of the Building of America from the Taming of the Mississippi to the Invention of the Internet* (New York: Free Press, 2001).

28 Gardner et al., *Good Work*.

29 Robert Kaplan and David P. Norton, *The Balanced Scorecard* (Boston: Harvard Business School Press, 1996).

30 Jonathan Low and Tony Siesfeld, "Measures That Matter: Wall Street Considers More Than You Think," *Strategy & Leadership*, March–April 1998, 24–28.

31 Andrew Osterland, "Knowledge Capital 2001: Decoding Intangibles," CFO, April 2001, 57–69.

32 Virginia Moreli, "The Variety of Life," *National Geographic*, February 1999, 27–28.

CHAPTER 4: STAKEHOLDER EXPECTATIONS VS. PROFESSIONAL MIND-SET

33 Gardner et al., *Good Work*, 27.

34 Ibid., 30.

35 Ibid., 26.

36 Eliot Freidson, *Professionalism, The Third Logic: On the Practice of Knowledge* (Chicago: University of Chicago Press, 2001).

37 Ibid., 197–222.

CHAPTER 5: UNDERSTANDING NEW SOCIAL COMPLEXITY

38 James F. Moore, *The Death of Competition: Leadership and Strategy in the Age of Business Ecosystems* (New York: Harper Business, 1996), 272.

39 Thomas L. Friedman, *The Lexus and the Olive Tree: Understanding Globalization* (New York: Farrar, Straus and Giroux, 1999).

40 Norman Myers, "The New Consumers," *Population Press* 9, no. 4 (Fall 2003):10.

41 Moore, *Death of Competition*, 272.

42 Ibid., 27.

43 Hock, *Chaordic Age*, 42.

CHAPTER 6: UNDERSTANDING NEW TECHNOLOGICAL COMPLEXITY

44 Clayton M. Christensen and Michael E. Raynor, *The Innovator's Solution: Creating and Sustaining Successful Growth* (Boston: Harvard Business School Press, 2003).

45 Reuters news service, September 3, 2003.

46 This very important point is made by the European scientists who put together the powerful, most widely used framework for sustainability called *The Natural Step*. The earth only became a planet habitable by life as these metals were sequestered beneath its surface.

47 Storm Cunningham, *The Restoration Economy* (San Francisco: Berrett-Koehler, 2002).

48 Marilyn Hampel, From the Editor, *Population Press* 9, no. 4 (Fall 2003): 2.

49 William McDonough and Michael Braungart, *Cradle to Cradle: Remaking the Way We Make Things* (New York: North Point Press, 2002), 90–91.

50 Paul Hawken, Amory Lovins, and L. Hunter Lovins, *Natural Capitalism: Creating the Next Industrial Revolution* (Boston: Little, Brown and Company, 1999), 10–11.

51 www.naturalstep.org

52 Ray C. Anderson, *Mid-Course Correction, Toward a Sustainable Enterprise: The Interface Model* (Atlanta: Peregrinzilla Press, 1998).

53 www.usgbc.org/leed/

54 "Micropower advances," *The Trend Letter*, October 13, 2003, 8.

55 Christensen, *Innovator's Dilemma*, 32.

56 Ibid., 34.

CHAPTER 7: SHIFTS OF MIND FOR CATCHING UP

57 Quoted in John E. Davey, *The Metamorphic Mind: How to Convert Your Thoughts into Physical Reality* (self-published, 2002), 25.

58 Quoted in Mike Jay "Understanding How to Leverage Executive Coaching," *Organization Development Journal* 21, no. 2 (Summer 2003): 9.

59 Ibid.

60 *Webster's Seventh New Collegiate Dictionary*.

61 John C. Gowan's discussion of Graham Wallas's paradigm of creativity, quoted in Willis Harman and Howard Rheingold, *Higher Creativity: Liberating the Unconscious for Breakthrough Insights* (New York: Tarcher-Putman, 1984), 21.

62 Robert Keidel, *Seeing Organization Patterns: A New Theory and Language of Organizational Design* (San Francisco: Berrett-Koehler, 1995).

CHAPTER 8: SHIFTS OF MIND FOR GETTING OUT AHEAD

63 Janine Benyus, *Biomimicry: Innovation Inspired by Nature* (New York: William Morrow, 1997), 247. This now familiar idea of Einstein's has been quoted in many different ways. For example, the well-known systems theorist Russell Ackoff renders it as "without changing our pattern of thought, we will not be able to solve the problems created with our current patterns of thought" in an interview with Robert J. Allio in *Strategy & Leadership* 31, no. 3 (2003): 20.

64 Information on the Myers-Briggs system is available from Consulting Psychologists Press, Inc. (www.cpp-db.com).

65 Peter M. Senge, *The Fifth Discipline: The Art and Practice of the Learning Organization* (New York: Doubleday, Currency, 1990).

66 Michael E. Porter and Mark R. Kramer, "The Competitive Advantage of Corporate Philanthropy," *Harvard Business Review*, December 2002, 56–68.

67 McDonough and Braungart, *Cradle to Cradle*; Hawken et al., *Natural Capitalism*.

68 Benyus, *Biomimicry*, 248–84.

69 Richard T. Pascale, Mark Millemann, and Linda Gioja, *Surfing the Edge of Chaos: The Laws of Nature and the New Laws of Business* (New York: Three Rivers Press, 2000), 7.

70 Benyus, *Biomimicry*, 251.

71 Ibid., 253–54.

72 Ronald A. Heifetz, *Leadership Without Easy Answers* (Cambridge: Harvard University Press, Belknap Press, 1994).

CHAPTER 9: UPDATING THE DOMAINS

73 Gardner et al., *Good Work*, 3.

CHAPTER 10: A NEW ECOLOGY OF FIRMS

74 *Inevitable Surprises: Thinking Ahead in a Time of Turbulence* (New York: Gotham Books, 2003), 166.

75 Benyus, *Biomimicry*, 2.

76 Schwartz, *Inevitable Surprises*, 78.

77 De Geus, *Living Company*, 17.

78 Joseph S. Nye Jr., *The Paradox of American Power* (New York: Oxford University Press, 2002), 85.

79 Friedman, *Lexus and Olive Tree*, 42.

80 B. Joseph Pine II and James H. Gilmore, *The Experience Economy* (Boston: Harvard Business School Press, 1999), 73.

81 Friedman, *Lexus and Olive Tree*, 46.

82 J. William Pfeiffer, PhD, JD, Leonard D. Goodstein, PhD, and Timothy M. Nolan, PhD, *Shaping Strategic Planning: Frogs, Dragons, Bees, and Turkey Tails* (Glenview, Ill.: Scott Foresman & Company, with University Associates, San Diego, 1989).

83 "Architects Go East," *Wall Street Journal*, December 17, 2003, B1.

CHAPTER 11: EXPERIENCE

84 Tom Kelley, with Jonathan Littman, *The Art of Innovation: Lessons in Creativity from IDEO, America's Leading Design Firm* (New York: Doubleday, Currency Books, 2001).

85 Ibid.

86 Robert I. Sutton and Thomas A. Kelley, "Creativity Doesn't Require Isolation: Why Product Designers Bring Visitors Backstage," *California Management Review* (Fall 1997).

87 Pine and Gilmore, *Experience Economy*.

88 Ibid., 8.

89 C.K. Prahalad and Venkat Ramaswamy, *The Future of Competition: Co-Creating Unique Value with Customers* (Boston: Harvard Business School Press, 2004), 137.

90 Based on Chuck Salter, "Digital Decisions," *Net Company* (supplement to *Fast Company*), Winter 2000, 37–43. Also see the Web site of Landmark Graphics, www.lgc.com.

91 Michael Schrage, *Serious Play: How the World's Best Companies Simulate to Innovate* (Boston: Harvard Business School Press, 2000).

92 "Small Worlds to Create Bold, New Ones," *New York Times*, March 1, 2001, D-7.

93 The story of Cambridge Technology Partners' Rapid Solutions Process is adapted from Schrage, *Serious Play*, 126–28.

94 Ibid., 126–28.

95 Prahalad and Ramaswamy, *Future of Competition*, 19–23; visit www.sumerset.com for additional information.

96 Ibid., 21.

97 Ibid., 22.

98 Ibid., 60–61.

CHAPTER 12: TECHNICAL, COLLABORATIVE, AND TRANSFORMATIVE WORK

99 Ronald A. Heifetz and Donald L. Lurie, "The Work of Leadership," *Harvard Business Review* (January–February 1997), 124.

100 "Opening of Historic King Library Attracts Thousands," *SJLibrary.org*, August 4, 2003.

101 The architectural design team for the new San Jose Library included Gunnar Birkerts Architects (design architect), Carrier Johnson (executive architect), and Anderson Brule Architects (associate architect).

102 Jane Light, presentation to the Collaborative Process Institute, San Jose, February 10, 2000.

103 Ibid.

104 Ibid.

105 David Maister, *Managing the Professional Service Firm* (New York: Free Press, 1993).

106 Heifetz, *Leadership Without Easy Answers*.

107 Ibid., 22.

108 Ibid., 14.

109 Ibid., 75.

110 Maister, *Managing the Professional Service Firm*.

CHAPTER 13: ORGANIZATION ECOSYSTEMS

111 Benyus, *Biomimicry*, 2.

112 *1998 Interface Annual Report*, 40.

113 This story is drawn from Anderson, *Mid-Course Correction*.

114 Ibid., 97.

115 *1998 Interface Annual Report*, 36.

116 Anderson, *Mid-Course Correction*, 74.

117 *1998 Interface Annual Report*, 40.

118 *Ibid.*, 31.

119 2002 *Interface Annual Report.*

120 Ibid.

121 Bonnie A. Nardi and Vicki L. O'Day, *Information Ecologies: Using Technology with Heart* (Cambridge: MIT Press, 1999)

122 Anderson, *Mid-Course Correction*, 97.

CHAPTER 14: VALUE NETWORKS AND TECHNOLOGY S-CURVES

123 *Innovator's Dilemma*, xix.

124 D. Sahal, *Patterns of Technological Innovation* (London: Addison Wesley, 1981).

125 Christensen et al., "Skate to Where the Money Will Be."

126 Ibid., 75.

127 Christensen, *Innovator's Dilemma*, 18–22.

128 Mildred Friedman, ed., *Gehry Talks: Architecture + Process* (New York: Universe, 2002), 50.

CHAPTER 15: TECHNICAL INNOVATION AND CREATIVITY

129 McDonough and Braungart, *Cradle to Cradle*.

130 Kieran Long, "It's Official: Star Architects Can Revive Flagging Cities," *World Architecture*, May 2002, 12.

131 Friedman, *Gehry Talks*, 36.

132 Ibid., 141.

133 Ibid., 30.

134 Ibid., 50.
135 Ibid.
136 Suzanne Stephens, "The Bilbao Effect," *Architectural Record*, May 1999, 172.
137 Craig Goehring, "Don't Forget Technology . . . or the Technologist," *Brown and Caldwell Quarterly* (Winter 1999).
138 Ibid.
139 www.mbdc.com/projects/designtextab.html
140 McDonough and Braungart, *Cradle to Cradle*, 108.
141 Kathryn Henderson, "Straw-Bale Building: Using an Old Technology to Preserve the Environment," in *Inventing for the Environment*, ed. Arthur Molella and Joyce Bedi (Cambridge: MIT Press, 2003).
142 Ibid., 191.
143 M.E.A. McNeil, "A Symphony in Straw," *San Francisco Chronicle*, February 3, 2002.
144 Schrage, *Serious Play*, 13.
145 Ibid., 16–17.
146 Stefan H. Thomke, *Experimentation Matters: Unlocking the Potential of New Technologies for Innovation* (Boston: Harvard Business School Press, 2003), 168.
147 Ibid., 168–69.
148 Schrage, *Serious Play*, 17.
149 "BRADD-The Softer Side of Bridge Design," *The Baker Challenge*, January 2002.
150 Recruitment Advisor, *ZweigWhite Publications*, May 20, 2002.
151 John Kao, *Jamming (New York: Harper Business, 1996), 6.*
152 *Fast Company*, December 1999, 244–46.
153 Michael Ray and Rochelle Myers, *Creativity in Business* (New York: Doubleday, 1989), 54.
154 The story of Lockheed's Skunk Works is told from an insider's perspective in Ben R. Rich and Leo Janos, *Skunk Works: A Personal Memoir of My Years at Lockheed* (Boston: Little, Brown, 1994).
155 Andrea Oppenheimer Dean and Timothy Hursley, *Rural Studio: Samuel Mockbee and an Architecture of Decency* (New York: Princeton Architectural Press, 2002).
156 Raul A. Barraneche, "Mind and Matter," *Architecture*, July 2002.
157 Tracy F. Ostroff, "HOK Sustainable Design 'Boot Camp' Enables Firm Designers to Share Green Ideas," *AIArchitect*, February 2002.
158 www.greenmatrix.net
159 Shira P. White, *New Ideas About New Ideas* (Cambridge: Perseus Publishing, 2002).
160 Lisa Heschong, "Skylighting and Productivity" (Pacific Gas and Electric, 1999).
161 www.h-m-g.com
162 "Building on IT," *CIO*, June 15, 2001.
163 www.stanford.edu/group/cife

CHAPTER 16: APPROACHING THE NEW CAPACITIES

164 Warren Bennis, *On Becoming a Leader* (Cambridge: Perseus Books, 1989), 3.

CHAPTER 17: LEADING INNOVATION

165 Donald T. Phillips, *Lincoln on Leadership: Executive Strategies for Tough Times* (New York: Warner Books, 1993).
166 Christensen, *Innovator's Dilemma*.

167 Everett Rogers, *Diffusion of Innovations*, 4th ed. (New York: Free Press, 1995). The curve cited appears as Figure 7-2 on p. 262. While Rogers cautions against assuming that an S-shaped rate of adoption is inevitable and references special conditions that can interfere, he asserts that "in most cases . . . an adopter distribution follows a bell-shaped, normal curve, or is S-shaped on a cumulative basis."

168 Pascale et al., *Surfing the Edge of Chaos*, 70–71.

169 Ibid., 77–92.

170 Ralph D. Stacey, *Complexity and Creativity in Organizations* (San Francisco: Berrett-Koehler Publishers, 1996), 179.

171 Harman and Rheingold, *Higher Creativity*, 21.

172 Kelley, *Art of Innovation*.

173 Pascale et al., *Surfing the Edge of Chaos*, 89.

174 Malcolm Gladwell, *The Tipping Point: How Little Things Can Make a Big Difference* (New York: Little, Brown and Company, 2000).

175 Ibid., 30–59.

176 W. Chan Kim and Renee Mauborgne, "Tipping Point Leadership," *Harvard Business Review*, April 2003, 60–69.

177 Heifetz considers the process used by Rudolph Giuliani and Bratton to bring down crime in New York City to have been overly dependent on "authority" and insufficiently mobilizing of difficult adaptive work by the community as a whole to address trade-offs that would be involved. Bratton's work forms a fascinating case study for comparing both the processes and the results of an approach based on social innovation to one based on adaptive work in Heifetz's sense. Ronald A. Heifetz and Marty Linsky, *Leadership on the Line: Staying Alive through the Dangers of Leading* (Boston: Harvard Business School Press, 2002), 24–25.

CHAPTER 18: DEVELOPMENTAL AWARENESS

178 Robert Kegan, *In Over Our Heads: The Mental Demands of Modern Life* (Cambridge: Harvard University Press, 1994), 5.

179 The terminology of "exterior" vs. "interior" draws on Ken Wilber's Four-Quadrant model, presented in *A Brief History of Everything* (Boston: Shambhala, 1996) and many other of his works.

180 Reported by P.P. Fay and A.G. Doyle in *1982 Annual for Facilitators, Trainers, & Consultants*, University Associates, San Diego.

181 Susan Campbell, *The Couple's Journey: Intimacy as a Path to Wholeness* (San Luis Obispo, Calif.: Impact Publishers, 1980).

182 Herman Gyr, "A Developmental Framework for Guiding Organizational Adaptation and Transformation," *Vision/Action: The Journal of the Bay Area Development Network*, Winter 1992–93, 5–12.

183 Ibid.

184 Stephen R. Covey, *The 7 Habits of Highly Effective People: Powerful Lessons in Personal Change* (New York: Simon & Schuster, Fireside, 1989), 49–51.

185 In the context of modern psychology, this shift was first noted by Erik Erikson in *Childhood and Society* (New York: W.W. Norton, 1950, 1963), 247–74.

186 Roger Harrison, PhD, *Towards the Learning Organization: Promises and Pitfalls*, Harrison Associates, 1992, 20–29.

187 Ibid, 25.

CHAPTER 19: TRANSFORMING YOUR FIRM

188 Harrison Owen, *The Power of Spirit: How Organizations Transform* (San Francisco: Berrett-Koehler, 2000), 60.
189 Noted in Peter Senge et al., *The Dance of Change: The Challenges of Sustaining Momentum in Learning Organizations* (New York: Doubleday, Currency, 1999), 5–10, as well as many other places.
190 For example, "Change Reaction: Understanding Your Response to Organizational Change," developed by John E. Jones and William L. Bearley, available from www.hrdq.com.
191 Cited by John D. Adams at the Organization Development Network Conference, October 19, 1993.
192 C.K. Prahalad and Venkat Ramaswamy have dedicated a whole chapter to this subject in *Future of Competition*, 195–208.
193 Collins and Porras, *Built to Last*.
194 Marco Iansiti and Roy Levien, "Strategy as Ecology," *Harvard Business Review*, March 2004.
195 De Geus, *Living Company*, 22–37.
196 Peter Schwartz, *The Art of the Long View* (New York: Doubleday, Currency, 1991).
197 Useful guidance may be found in Kees van der Heijden, *Scenarios: The Art of Strategic Conversation* (New York: John Wiley & Sons, 1996).
198 Presented at the "Best-in-the-West" conference of the Bay Area Organization Developmental Network, April 20, 1991.
199 *Fifth Discipline*, 57–67.
200 Ibid., appendix 2, 378–90.
201 Gyr, "Developmental Framework," 5–12.
202 Peter Senge et al., *The Fifth Discipline Fieldbook: Strategies and Tools for Building a Learning Organization* (New York: Doubleday, Currency, 1994).
203 Heifetz and Linsky, *Leadership on the Line*, 127.
204 Heifetz, *Leadership Without Easy Answers*, 37.
205 Extracted from Ronald Heifetz and Donald Laurie, "The Work of Leadership," *Harvard Business Review*, December 2001 (Breakthrough Leadership issue), 131–40.
206 For example, Kegan, *In Over Our Heads*.
207 Presentations given at conferences of the World Future Society: "The Psychological Demands of a Global Era" (Washington, D.C., 1999) and "Changing Our Minds: Psychological Visions of the Twenty-First Century" (San Francisco, 2003).

CHAPTER 20: SOCIAL CAPACITIES FOR HIGHER VALUE TECHNICAL WORK

208 Daniel Goleman, Richard Boyatzis, and Annie McKee, *Primal Leadership: Realizing the Power of Emotional Intelligence* (Boston: Harvard Business School Press, 2002), 104–05.
209 *Harvard Business Review*, May-June 1991.
210 Habit 5 in Covey, *7 Habits*, 235.
211 Senge et al., *Fifth Discipline Fieldbook*, 242–46.
212 Senge, *Fifth Discipline*, 195–98.
213 Daniel Goleman, *Emotional Intelligence: Why It Can Matter More Than IQ* (New York: Bantam, 1995).
214 S.E. Taylor et al., "Female Responses to Stress: Tend and Befriend, Not Fight or Flight," *Psychological Review* 107, no. 3 (July 2000): 411–29.
215 Stacey, *Complexity and Creativity*, 180.

216 Roger Fisher and William Ury, *Getting to Yes: Negotiating Agreement Without Giving In* (New York: Penguin, 1981), 17–40.

217 For example, David L. Bradford and Allan R. Cohen, *Power Up: Transforming Organizations Through Shared Leadership* (New York: John Wiley & Sons, 1998). On a more basic level, Sandy Pokras, *Rapid Team Deployment: Building High-Performance Project Teams* (Crisp Publications, 1995), or Peter R. Scholtes, *The Team Handbook: How to Use Teams to Improve Quality* (Joiner Associates, 1988).

218 Douglas Stone, Bruce Patton, and Sheila Heen, *Difficult Conversations: How to Discuss What Matters Most* (New York: Penguin, 1999).

219 Kerry Patterson et al., *Crucial Conversations: Tools for Talking When Stakes Are High* (New York: McGraw-Hill, 2002).

220 Recent sources include Ronald Fernandez, *America's Banquet of Cultures: Harnessing Ethnicity, Race, and Immigration in the Twenty-First Century* (Westport, Conn.: Praeger Publishers, 2000), and Richard D. Lewis, *When Cultures Collide* (London: Nicholas Brealey, 2000). Classics include Charles Hampden-Turner and Alfons Trompenaars, *The Seven Cultures of Capitalism* (New York: Doubleday, Currency, 1993) and two books by Geert Hofstede, *Cultures and Organizations: Software of the Mind* (New York: McGraw-Hill, 1991) and *Culture's Consequences: International Differences in Work Related Values* (Thousand Oaks, Calif.: Sage Publications, 1980). A brief, incisive summary of nine factors of culture appearing in the United States and Canada, East Asia, the Middle East and North Africa, and Latin America can be found in Michael J. Marquardt and Dean W. Engel, *Global Human Resource Development* (Englewood Cliffs, N.J.: Prentice-Hall, 1993), 25–33.

221 The formal version of the Myers-Briggs Type Indicator (MBTI) is available through Consulting Psychologists Press, Inc. at www.cpp-db.com or several other Web sites. A free online version can be found at http://www.humanmetrics.com/cgi-win/Jtypes2.asp. A self-explanatory paper-and-pencil version is the "Personal Style Inventory" available from www.hrdq.com or 800-633-4533.

222 Aubrey Daniels, *Bringing Out the Best in People: How to Apply the Astonishing Power of Positive Reinforcement* (New York: McGraw-Hill, 1994).

223 Senge et al., *Dance of Change*.

224 Covey, *7 Habits*, 188–202.

225 A good resource for this situation is Jeffrey Pfeffer, *Managing With Power* (Boston: Harvard Business School Press, 1992).

226 Ibid., 150–52.

227 Goleman et al., *Primal Leadership* (chapter 7, "The Motivation to Change"), 113–38.

228 Ibid., 110.

229 For example, the Advanced Management Institute for Architecture and Engineering (AMI), at www.ami-institute.com, offers courses and coaching that include 360-degree feedback, as does the Center for Creative Leadership, www.ccl.org.

230 Senge et al., *Dance of Change*, 13–15.

231 For intensive self-development efforts, we recommend Don Richard Riso and Russ Hudson, *The Wisdom of the Enneagram* (New York: Bantam Books, 1999), and their Web site, www.EnneagramInstitute.com for online versions of the typing instrument.

232 Frederic M. Hudson, *The Adult Years: Mastering the Art of Self-Renewal* (San Francisco: Jossey-Bass, 1991).

233 Bennis, *On Becoming a Leader*, 3.

CHAPTER 21: LEADERSHIP CAPACITIES FOR COLLABORATIVE WORK

234 Quoted in Marvin R. Weisbord and Sandra Janoff, *Future Search: An Action Guide to Finding Common Ground in Organizations and Communities* (San Francisco: Berrett-Koehler, 1995).

235 Craig Park, *Design. Market. Grow!* (Alexandria, Va.: Society for Marketing Professional Services, 2002), 162.

236 Wilfred R. Bion, "Selections from 'Experiences in Groups,'" in *Group Relations Reader*, ed. Arthur D. Colman and W. Harold Bexton (GREX, 1975), 11–20.

237 Full academic explication of the Situational Leadership model appears in Paul Hersey and Ken Blanchard, *Management of Organizational Behavior: Utilizing Human Resources* (Englewood Cliffs, NJ: Prentice-Hall, Inc., 1982. A training-friendly version is published in Kenneth Blanchard, *Situational Leadership II* (San Diego: Blanchard Training & Development, 1984).

238 The "Team Learning" chapter in *The Fifth Discipline* provides an excellent introduction to dialogue. The "Team Learning" section in *The Fifth Discipline Fieldbook* provides in-depth guidance for practicing and using dialogue as opposed to "skillful discussion."

CHAPTER 22: TRANSFORMATIVE WORK IN CLIENT SYSTEMS

239 Quoted in Weisbord and Janoff, *Future Search*, 25.

240 Richard Foster and Sarah Kaplan, *Creative Destruction: Why Companies That Are Built to Last Underperform the Market—And How to Successfully Transform Them* (New York: Doubleday, Currency, 2001).

241 Porter and Kramer, "Competitive Advantage of Corporate Philanthropy," 56–68.

242 Hock, *Chaordic Age*.

243 Iansiti and Levien, "Strategy as Ecology."

244 Ibid.

245 A system-dynamics-based game called "Fishbanks" is probably the most widely used simulation in the world. A good source for exploring simulations and other systems thinking resources is Pegasus Communications at www.pegasuscom.com.

246 Lisa Friedman and Herman Gyr, *The Dynamic Enterprise: Tools for Turning Chaos into Strategy and Strategy into Action* (San Francisco: Jossey-Bass, 1998), 165–83.

247 Ibid., 181.

248 Ibid., 181–82.

249 Ibid., 182–83.

250 Originally proposed by Eldredge and Gould in 1972 and fully expounded in Stephen Jay Gould, *The Structure of Evolutionary Theory* (Cambridge: Harvard University Press, Belknap Press, 2002), 725–1024.

251 Two classic and very helpful works in this field are Peter Block, *The Empowered Manager: Positive Political Skills at Work* (San Francisco: Jossey-Bass, 1991), and Allan R. Cohen and David L. Bradford, *Influence Without Authority* (New York: John Wiley & Sons, 1990).

252 Heifetz and Linsky, *Leadership on the Line*.

CHAPTER 23: PRICING AND PROFIT: NEW PRICING STRATEGIES

253 Alan Weiss, *Value-Based Fees: How to Charge and Get What You're Worth* (San Francisco: Jossey-Bass/Pfeiffer, 2002), 9.

254 Edwards Management Group is a financial and operations management consulting and recruiting firm serving design firms based in San Francisco.

255 "Quality Makes the Grade at Ciba-Geigy," *Design-Build Magazine*, January 1998.

256 Ibid.
257 Clifford Pearson, "How to Succeed in Expanded Services," *Architectural Record*, January 1998, 52.
258 "Builders Clone Patented Projects," *Houston Business Journal*, June 14, 1999.
259 Barraneche, "Mind and Matter."
260 David Stone, "The Case for Value Pricing," *Engineering Inc.*, January-February 2003.

Suggested Readings

Anderson, Ray C. *Mid-Course Correction: Toward a Sustainable Enterprise*. Atlanta: Peregrinzilla Press, 1998.

Benyus, Janine M. *Biomimicry: Innovations Inspired by Nature*. New York: William Morrow, 1997.

Capelin, Joan. *Communication by Design: Marketing Professional Services.* Atlanta: Greenway Communications, 2004.

Christensen, Clayton M. *The Innovator's Dilemma: When New Technologies Cause Great Firms to Fail.* Boston: Harvard Business School Press, 1997.

Christensen, Clayton M., and Michael E. Raynor. *The Innovator's Solution: Creating and Sustaining Successful Growth*. Boston: Harvard Business School Press, 2003.

Collins, James C., and Jerry I. Porras. *Built to Last: Successful Habits of Visionary Companies*. New York: HarperBusiness, 1994.

Collins, Jim. *Good to Great*. New York: HarperCollins, 2001.

Covey, Stephen R. *The 7 Habits of Highly Effective People*. New York: Simon & Schuster, Fireside, 1990.

Cramer, James P. *Design Plus Enterprise: Seeking a New Reality in Architecture & Design*, second edition. Atlanta: Greenway Communications, 2002.

Cramer, James P. and Scott Simpson. *How Firms Succeed: A Field Guide to Design Management*, second edition. Atlanta: Greenway Communications, 2004.

de Geus, Arie. *The Living Company*. Boston: Harvard Business School Press, 1997.

Fisher, Roger, and William Ury. *Getting to Yes: Negotiating Agreement Without Giving In*. 2nd ed..New York: Penguin Books, 1991.

Friedman, Lisa, and Herman Gyr. *The Dynamic Enterprise: Tools for Turning Chaos into Strategy and Strategy into Action*. San Francisco: Jossey-Bass, 1998.

Gardner, Howard, Mihaly Csikzentmihalyi, and William Damon. *Good Work: When Excellence and Ethics Meet*. New York: Perseus Books, 2001

Gladwell, Malcolm. *The Tipping Point: How Little Things Can Make a Big Difference*. New York: Little, Brown, 2000.

Hawken, Paul, Amory Lovins, and L. Hunter Lovins. *Natural Capitalism: Creating the Next Industrial Revolution*. Boston: Little, Brown, 1999.

Heifetz, Ronald A. *Leadership Without Easy Answers*. Cambridge: Harvard University Press, 1994.

Hock, Dee. *Birth of the Chaordic Age*. San Francisco: Berrett-Koehler, 1999.

Kegan, Robert. *In Over Our Heads: The Mental Demands of Modern Life*. Cambridge: Harvard University Press, 1994.

Kelley, Tom, with Jonathan Littman. *The Art of Innovation: Lessons in Creativity from IDEO*. New York: Doubleday, Currency, 2001.

Kieran, Stephan, and James Timberlake. *Refabricating Architecture*. New York: McGraw-Hill, 2004.

McDonough, William, and Michael Braungart. *Cradle to Cradle: Remaking the Way We Make Things*. New York: North Point Press, 2002.

Nardi, Bonnie A., and Vicki L. O'Day. *Information Ecologies: Using Technology with Heart*. Boston: MIT Press, 1999.

Pascale, Richard T., Mark Millemann, and Linda Gioja. *Surfing the Edge of Chaos: The Laws of Nature and the New Laws of Business*. New York: Crown Business, 2000.

Pine, B. Joseph, II, and James H. Gilmore. *The Experience Economy*. Boston: Harvard Business School Press, 1999.

Prahalad, C.K., and Venkat Ramaswamy. *The Future of Competition: Co-Creating Unique Value with Customers*. Boston: Harvard Business School Press, 2004.

Rogers, Everett M. *Diffusion of Innovations*. 5th ed. New York: Free Press, 2003.

Schrage, Michael. *Serious Play: How the World's Best Companies Simulate to Innovate*. Boston: Harvard Business School Press, 2000.

Schwartz, Peter. *The Art of the Long View*. New York: Doubleday, 1991.

Schwartz, Peter. *Inevitable Surprises: Thinking Ahead in a Time of Turbulence*. New York: Gotham Books, 2003.

Schwartz, Peter, Peter Leyden, and Joel Hyatt. *The Long Boom: A Vision for the Coming Age of Prosperity*. Reading, Mass.: Perseus Books, 1999.

Senge, Peter M. *The Fifth Discipline: The Art and Practice of the Learning Organization*. New York: Doubleday, Currency, 1990.

———. *The Fifth Discipline Fieldbook*. New York: Doubleday, Currency, 1994.

———. *The Dance of Change*. New York: Doubleday, Currency, 1999.

Stacey, Robert D. *Complexity and Creativity in Organizations*. San Francisco: Berrett-Koehler, 1996.

Swett, Richard N., *Leadership by Design: Creating an Architecture of Trust.* Atlanta: Greenway Communications, 2005.

Tapscott, Don, David Ticoll, and Alex Lowy. *Digital Capital: Harvesting the Power of Business Webs*. Boston: Harvard Business School Press, 2000.

Wheatley, Margaret J. *Leadership and the New Science*. San Francisco: Berrett-Koehler, 1992.

About the Authors

Kyle V. Davy, AIA, helps architecture, engineering, and design firms develop the leadership and management capacities they need to thrive in an increasingly turbulent environment. His innovative learning programs look beyond traditional professional conventions to advance knowledge and wisdom about practice. His clients include many leading firms in the architecture and engineering community. He is a principal faculty member of the Senior Executive Institute offered by the American Counsel of Engineering Companies. He established his consulting practice in 1992 after sixteen years of professional practice as an architect. He holds an MBA from the Stanford Graduate School of Business.

Susan L. Harris, PhD, has been bringing state-of-the-art leadership and organization development to built environment professionals since 1993. Since the early 1980s, she has worked with large and small organizations across many industries as well as in the public and non-profit sectors. Her current practice focuses on leadership and strategy in the context of building learning organizations as living systems. Through her affiliation with the Advanced Management Institute for Architecture and Engineering in San Francisco, she was a principal architect of the Senior Executives Institute created by AMI with the American Council of Engineering Companies. She holds a PhD from the University of California, Berkeley in English literature.

Available from Östberg...

How Firms Succeed: A Field Guide to Design Management
James P. Cramer and Scott Simpson

A hands-on guide to running any design-related business—from a two-person graphics team to middle-management to CEOs of multi-national firms—offering advice on specific problems and situations and providing insight into the art of inspirational management and strategic thinking.

"*How Firms Succeed* is a fountainhead of great ideas for firms looking to not just survive, but thrive in today's challenging marketplace."

—Thompson E. Penney, FAIA
President/CEO, LS3P Architecture, Interior Architecture, Land Planning and President, American Institute of Architects, 2003

Leadership by Design: Creating and Architecture of Trust
Richard N. Swett

Ambassador Richard Swett's groundbreaking new book investigates the unique civic leadership strengths of the architecture profession. Leadership by Design is an eloquent plea to architects, leaders and citizens alike to expand the tool chest as we seek new leadership to design new solutions for the complex challenges facing our nation and the world.

"This book reveals that the 'citizen-architect' has always been in our midst and begins an important dialogue about how that role should be designed for the future."

—Robert A.M. Stern, FAIA
Robert A.M. Stern Architects and Dean, Yale School of Architecture

Almanac of Architecture & Design
James P. Cramer and Jennifer Evans Yankopolus, editors

The only complete annual reference for rankings, records, and facts about architecture, interior design, industrial design, landscape architecture, planning and historic preservation.

"The reader who uses this book well will come away with a richer sense of the texture of the profession and of the architecture it produces."

—Paul Goldberger, The New Yorker

America's Best Architecture & Design Schools

This special 40-page issue of *DesignIntelligence* offers the only ranking of architecture and design schools in the United States. conducted annually since 2000, this study is the only one that polls professional practice leaders—the constituency most qualified to comment on which schools consistency produce the best graduates. National and regional rankings and a commentary and analysis of the current state of design education are just a few of the offerings.

Value Redesigned: New Models for Professional Practice
Kyle V. Davy and Susan L. Harris

In *Value Redesigned*, Davy and Harris reveal a vivid landscape where innovative new models for professional practice are already beginning to flourish, showing firms avenues of escape from the vicious cycle of commoditization and low prestige that is epidemic within the architecture and engineering community. Aligned with the dynamics of the emerging knowledge-based economy, these new models of practice offer bold value propositions, combining new ways of creating value with innovative pricing strategies.

"Value Redesigned is a timely and important book for all professionals and professional firms involved in the built environment projects. This is a must read for any one who cares about the future of their own firm and the future of the industry."

—Al Barkouli, PE
Executive Vice President, David Evans and Associates, Inc.

Design plus Enterprise: Seeking a New Reality in Architecture & Design
James P. Cramer

Using specific examples, *Design plus Enterprise* illustrates how using business principles architects can create better design services—and thereby, a better society. It also demonstrates how smart design can drive economic success.

"This is must reading for every architect...It clearly points out how design and the designer are enriched by recognizing that the profession of architecture is both a business and a way of enhancing the environment"

—M. Arthur Gensler Jr., FAIA
Chairman, Gensler Architecture, Design & Planning Worldwide

Communication by Design: Marketing Professional Services
Joan Capelin

How to communicate—and, especially why—to clients, prospects, staff, and the public is the basis of this powerful book. It is targeted to business principals as well as anyone who aspires to a leadership position in a firm, association, or business joint venture.

" Capelin offers thought-provoking practical lessons in marketing leadership—illustrated by interesting insights and implementable ideas. Read this book, put her advice into action, and your firm will flourish."

—Howard J. Wolff
Senior Vice President/Wimberly Allison Tong & Goo

Design Intelligence

The Design Future Council's monthly "Report on the Future" provides access to key trends and issues on the cutting edge of the design professions. Each month it offers indispensable insight into management practices that will make any firm a better managed and more finically successful business.

"We read every issue with new enthusiasm because the information always proves so timely. No other publication in our industry provides as much useful strategic information."

—David Brody Bond LLP

Order online at www.greenway.us, by phone at 800-726-8603, or by using the order form on the next page.

Order Form

Publication	Price	Qty.	Total
How Firms Succeed: A Field Guide to Design Management	$39.00		
Communication by Design: Marketing Professional Services	$34.95		
Leadership by Design: Creating an Architecture of Trust	$39.50		
Design Plus Enterprise: Seeking a New Reality in Architecture & Design	$29.00		
Almanac of Architecture & Design	$49.50		
Value Redesigned: New Models for Professional Practice	$39.50		
America's Best Architecture & Design Schools†	$34.95†		
DesignIntelligence (1 year subscription)	$365.00		
		Subtotal	
Shipping: $4.95 for 1st item; $1.00 each additional item		**Shipping**	
**GA residents add 6%*		**Tax***	
		Total	

† A PDF version of *America's Best Architecture & Design Schools* may be downloaded immediately from our Web site at www.greenway.us.

Method of Payment

☐ Check ☐ Credit Card

Card Number Exp. Date Signature

Contact/Shipping Information

Name Title

Company

Address

City State Zip

Telephone Fax

Email

Your completed order form may be faxed to us at 770.209.3778 or mailed to:
Greenway Communications, 30 Technology Parkway South, Suite 200, Norcross, GA 30092
All our publications can also be purchased online at www.greenway.us. 800.726.8603

östberg™

Library of Design Management

Every relationship of value requires constant care and commitment. At Östberg, we are relentless in our desire to create and bring forward only the best ideas in design, architecture, interiors, and design management. Using diverse mediums of communications, including books and the Internet, we are constantly searching for thoughtful ideas that are erudite, witty, and of lasting importance to the quality of life. Inspired by the architecture of Ragnar Östberg and the best of Scandinavian design and civility, the Östberg Library of Design Management seeks to restore the passion for creativity that makes better products, spaces, and communities. The essence of Östberg can be summed up in our quality character to you: "Communicating concepts of leadership and design excellence."